A Systems Approach for River and River Basin Restoration

A Systems Approach for River and River Basin Restoration

Editor

Theodore Endreny

MDPI • Basel • Beijing • Wuhan • Barcelona • Belgrade • Manchester • Tokyo • Cluj • Tianjin

Editor
Theodore Endreny
Department of Environmental Resources Engineering
SUNY ESF
USA

Editorial Office
MDPI
St. Alban-Anlage 66
4052 Basel, Switzerland

This is a reprint of articles from the Special Issue published online in the open access journal *Water* (ISSN 2073-4441) (available at: https://www.mdpi.com/journal/water/special_issues/River_Basin_Restoration).

For citation purposes, cite each article independently as indicated on the article page online and as indicated below:

LastName, A.A.; LastName, B.B.; LastName, C.C. Article Title. *Journal Name* **Year**, *Article Number*, Page Range.

ISBN 978-3-03943-631-6 (Hbk)
ISBN 978-3-03943-632-3 (PDF)

Cover image courtesy of Theodore Endreny.

Contents

About the Editor

Theodore Endreny is a Professor in the Department of Environmental Resources Engineering at the State University of New York College of Environmental Science and Forestry (SUNY ESF). His scholarship focuses on developing models to simulate how water, forest, and related natural resources reduce pollution and deliver ecosystem services, with the goal of guiding ecosystem restoration so that it improves human wellbeing and biodiversity. He teaches courses in engineering hydrology and hydraulics, river restoration, and urban restoration and sustainability. He is a member of the National Academies Grand Challenge Scholars Program, Engineers without Borders, provides university service for SUNY ESF, and serves on the editorial board for *nbj Urban Sustainability, Hydrological Processes*, and the *International Journal for River Basin Management*. He was the recipient of a SUNY Chancellor's Internationalization Award, a Fulbright Commission Research and Lecturer Sabbatical award to Cyprus, the SUNY ESF Undergraduate Student Association Distinguished Teacher Award, the US nomination for the IEEE GHTC Global Humanitarian Engineer of the Year, and the Fulbright Commission Distinguished Chair in Environmental Sciences at Parthenope University, Italy. His i-Tree tools research team received the Elsevier Atlas Award for "significantly impacting people's lives around the world" with an article on the value of urban forests in megacities. Endreny earned a BS in natural resources management at Cornell University and then served as a Peace Corps Volunteer in river basin management in Trujillo, Honduras, and as a research associate with the Environmental Law Institute in Washington DC. He earned his MS in soil and water engineering from North Carolina State University as a research scholar with the Environmental Protection Agency, and earned a PhD in water resources engineering from Princeton University as a research scholar with the National Aeronautics Space Administration.

Preface to "A Systems Approach for River and River Basin Restoration"

This book investigates new approaches to river basin restoration as well as the most effective leverage points for achieving meaningful change and expectations in the behavior of those river basins. River and river basin restoration faces significant technical challenges as well as challenges to our conception or paradigm for the purpose of the river basin. This article defines the term restoration as reestablishing the structure and function of an ecosystem, yet invites readers to substitute in the terms rehabilitation, defined as making an ecosystem useful after disturbance, or reclamation, defined as changing the biophysical capacity of an ecosystem. The major technical challenges include the following: (a) the restoration target is often unknown, is not likely an initial or completely natural state, and remains poorly understood; (b) restoration structures should provide multiple functions across seasonal flow regime to benefit humans and biodiversity; (c) restoration spatial scale and complexity should consider local to basin-level issues; (d) restoration resiliency should handle uncertain future drivers related to urbanization and climatic disruption.

Human systems and ecosystems are separated to show their interaction in the United Nations model of the river basin system, recognizing the agency of humans to affect change, i.e., the Anthropocene epoch. In this model, humans depend on services from the same ecosystems we diminish by polluting, harvesting, and other pressures (e.g., climate disruption). Furthermore, humans can assess the actual state of the ecosystem (e.g., water volume or quality) and compare it with our desires for the state of the ecosystem (e.g., below target) For our river basin restoration, which may be active or passive, we used a comparison of the actual and desired state of the system to formulate the paradigm rather than the policy which is a novel addition to this model. This book will demonstrate that the task of formulating our paradigm for the purpose of a river basin is equally if not more important than the list of major technical challenges for river basin restoration.

River basin scientists and engineers have made important progress in recognizing that these challenges are all part of a single complex system, noting that tinkering with elements in one location or time tends to impact the state or function of elements in other parts of the system. Managing the complexity of the river basin system is yet another challenge and progress in this area requires a better understanding of systems and their leverage points.

This book provides new insights on how to strategically use leverage points to restore river basin systems. The most effective leverage point is transcending paradigms and developing new mindsets for working with the complexity of river basin systems. Of course, developing the paradigm is part of the system. So the systems path that scientists and engineers might follow in restoration, as adapted from the UN, is as follows: (1) identifying, understanding, and working with the physical, chemical, and biological processes comprising river basin and river health and delivering ecosystem services; (2) identifying, incorporating, and involving socio-economic values and broader planning and development activities linked to river basin and river health; (3) addressing structure and function relationships at the appropriate scales to address limiting factors to river health; (4) setting clear, achievable, and measurable goals, framed in terms of changes to ecosystem structure and function, the provisioning of ecosystem services, and, where feasible, socioeconomic factors; (5) planning, implementing, and managing to provide resilience to a range of scenarios over time, including changes to climate, land use, hydrology, pollutant loads, and population, so restoration outcomes are sustained over the long term; (6) involving all relevant stakeholders in an integrated approach, addressing land and water issues, and involving interagency and community collaboration, to achieve the greatest benefits; and (7) monitoring, evaluating, adapting, and reporting the actual state of river basin health relative to the desired state, and formulating our paradigm to guide restoration and adaptive management. Together on this journey, we can improve social and ecological systems within our river basins.

Theodore Endreny
Editor

Editorial

Leverage Points Used in a Systems Approach of River and River Basin Restoration

Theodore A. Endreny

Department of Environmental Resources Engineering, SUNY ESF, Syracuse, NY 13210, USA; te@esf.edu

Received: 10 September 2020; Accepted: 11 September 2020; Published: 18 September 2020

Abstract: River basins are complex spatiotemporal systems, and too often, restoration efforts are ineffective due to a lack of understanding of the purpose of the system, defined by the system structure and function. The river basin system structure includes stocks (e.g., water volume or quality), inflows (e.g., precipitation or fertilization), outflows (e.g., evaporation or runoff), and positive and negative feedback loops with delays in responsiveness, that all function to change or stabilize the state of the system (e.g., the stock of interest, such as water level or quality). External drivers on this structure, together with goals and rules, contribute to how a river basin functions. This article reviews several new research projects to identify and rank the twelve most effective leverage points to address discrepancies between the desired and actual state of the river basin system. This article demonstrates river basin restoration is most likely to succeed when we change paradigms rather than trying to change the system elements, as the paradigm will establish the system goals, structure, rules, delays, and parameters.

Keywords: watershed; systems; restoration

---

1. Introduction

River and river basin restoration faces significant technical challenges as well as challenges to our conception or paradigm for the purpose of the river basin. This article defines the term restoration as reestablishing structure and function of an ecosystem, yet invites readers to substitute in the terms rehabilitation, defined as making an ecosystem useful after disturbance, or reclamation, defined as changing the biophysical capacity of an ecosystem. The major technical challenges include: (a) the restoration target is often unknown, is not likely an initial or completely natural state, and remains poorly understood; (b) restoration structures should provide multiple functions across seasonal flow regime to benefit humans and biodiversity; (c) restoration spatial scale and complexity should consider local to basin-level issues; and (d) restoration resiliency should handle uncertain future drivers related to urbanization and climatic disruption [1].

Human systems and ecosystems are separated to show their interaction in the United Nations model of the river basin system (Figure 1), recognizing the agency of humans to affect change, i.e., the Anthropocene epoch [1]. In this model, humans depend on the services from the same ecosystems we diminish by polluting, harvesting, and other pressures (e.g., climate disruption). Furthermore, humans can assess the actual state of the ecosystem (e.g., water volume or quality) and compare it with our desires for the state of the ecosystem (e.g., below target). New to this model is what comes next, that we use this comparison of the actual and desired state of the system to formulate the paradigm rather than the policy for our river basin restoration, which may be active or passive. It is the purpose of this paper to show the task of formulating our paradigm for the purpose of a river basin is equally if not more important than the list of major technical challenges for river basin restoration.

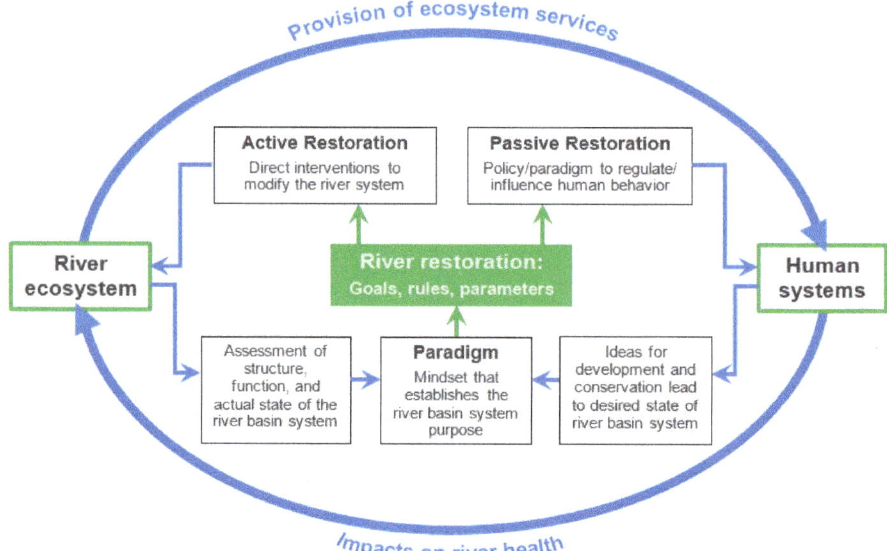

Figure 1. Conceptual model of a river basin system with human systems benefiting from and affecting the ecosystems, and the formulation of a paradigm for river basin restoration based on comparing actual vs. the desired state of the system. Adapted from Speed et al. [1].

2. Evolution of a Systems Approach to River Basins

River basin scientists and engineers have made important progress in recognizing that these challenges are all part of a single complex system [2], noting that tinkering with elements in one location or time tends to impact the state or function of elements in other parts of the system [3,4]. Actually managing the complexity of the river basin system is yet another challenge and progress in this area requires better understanding of systems and their leverage points.

Systems are generally defined as a set of spatially and temporally interconnected parts that respond to internal and external signals, such as those from the surrounding environment. Systems, including those of river basins, typically have a structure that includes stocks (e.g., water levels or quality in basin), inflows (e.g., precipitation or fertilization), outflows (e.g., evaporation or runoff), and positive and negative feedback loops with delays in responsiveness, that all function to change or stabilize the state of the system (e.g., the stock of interest, such as water level or quality). When there are discrepancies between the desired and actual state of the system, managers want to intervene. Too often, a lack of understanding of system structure and functions results in interventions not achieving the desired outcome. Systems theorist and practitioner Meadows [5], who led the Limits to Growth study for the Club of Rome, sought to improve outcomes by identifying twelve strategic leverage points for intervening in any system.

3. Strategic Leverage Points to Advance Restoration of the River Basin System

To arrive at the most strategic leverage points for intervening in systems, Meadows [5] began by observing how we need to look beyond system outcomes (e.g., events such as excessive flooding or pollution) to consider the system structure by which parts are related and the system functions which respond to the rules, opportunities for change, and goals of the system. Together, the structure and function create the state of the system and can be defined as the actual purpose of the system, whether or not if that purpose aligns with our desire. In the end the state of the system, i.e., the outcomes such as water level or quality, tell us the purpose of the system, and if we do not like the outcomes, we either have to change our paradigm or the system's purpose. From this insight emerged the twelve most strategic leverage points for intervening in systems, ranked from least to most effective, and adapted for river basins (see Table 1).

Table 1. Strategic leverage points for intervening in systems, such as river basin restoration, ranked from least to most effective. Adapted from Meadows [5].

Reverse Rank	Leverage Points for Intervening in River Basin Systems
Lever 12	Constants, parameters, numbers (such as subsidies, water rates, standards).
Lever 11	The sizes of buffers and other stabilizing stocks, relative to their flows.
Lever 10	The structure of material stocks and flows (such as transport networks, population age structures).
Lever 9	The lengths of delays, relative to the rate of system change.
Lever 8	The strength of negative feedback loops, relative to the impacts they are trying to correct against.
Lever 7	The gain around driving positive feedback loops.
Lever 6	The structure of information flows (who does and does not have access to information).
Lever 5	The rules of the system (such as incentives, punishments, constraints).
Lever 4	The power to add, change, evolve, or self-organize system structure.
Lever 3	The goals of the system.
Lever 2	The mindset or paradigm that establishes the system goals, structure, rules, delays, parameters.
Lever 1	The power to transcend paradigms.

4. Illustrations of Leverage Points and Their Relative Effectiveness in River Basin Restoration

A systems approach to river basin restoration then involves addressing the grandest and most intractable challenges by transcending paradigms and our mindset, and then, changing the goals of the system. This Special Issue takes us further along the path in understanding the river basin as a complex system, while probing the effectiveness of several important leverage points to affect change.

Rampinelli et al. [6] examined the leverage point #12 of constants, parameters, and numbers related to flooding from the framework of uncertainty analysis, which then provides a framework for updating the leverage point #5 of rules of the system for setting flood policy and managing risk. Rampinelli et al. [6] found that variance in flow data records can be analyzed with a Bayesian approach to separately represent the flood level uncertainty due to uncertainties in flow rate associated with return interval (frequency analysis) and river stage (rating curves). Their method can lead to a more accurate range of expected flood level outcomes to inform restoration and management. Golpira et al. [7] examined leverage points #7 (positive feedback loops) and #11 (sizes of buffers and other stabilizing stocks) as they relate to in-channel boulder placement and subsequent channel bed shear and erosion. Golpira et al., showed that information about this system (leverage point #6) was constrained by the type of instrumentation available to collect data and the subsequent calculation of shear stress (they compared four calculations—reach-average, Reynolds, turbulent kinetic energy (TKE), and modified TKE). Working with different boulder spacing densities and flow levels, which ranged from unsubmerged to fully submerged boulders, it was found that for unsubmerged conditions, the four shear stress equations generated different information, and for submerged conditions, the feedback between boulders and sediment erosion could be controlled by reducing boulder density.

Abebe et al. [8] explored the impact of new rules for the system (leverage point #5) and the length of delays (leverage point #9) to establish a holistic basis for setting flow regimes in Ethiopia's Gumara

River basin, which feeds Lake Tana and is the source of the Blue Nile. The current flow regime, set by an existing set of rules and flow delays that prioritize irrigation, dams, and river regulation, has contributed to dry channels, interrupted fish migration and spawning, and the loss of fishing. The research demonstrates that proposed rules to allocate 10 to 25% of flows to the environment will not achieve ecological health targets without considering the coupling between flow timing (i.e., delays) and flow to achieve naturalization of the flow regime that supports key ecological processes [8]. Liu et al. [9] examined how power to add, change, evolve, or self-organize system structure (leverage point #4) could lead to changes in the rules (leverage point #5) that are fairer and achieve water quality targets. The research used a multi-scale and multi-pollutant waste-load allocation model to explore changes in pollution quotas across 1350 areas within the Xian-jiang river basin of China, finding an allocation that reduced inequality (based on Gini coefficients) yet was more economical and met pollutant thresholds for chemical oxygen demand, ammonia nitrogen, and total phosphorus [9].

Doehring et al. [10] examined the leverage points #11 (the sizes of buffers) and #9 (the length of delays) as they relate to establishing vegetation in riparian buffers across 5 to 34 years, and the emergence of indicators of restoration, such as healthy aquatic microbial communities. The research used paired river reaches in the Waikato region of the central North Island, New Zealand, each pair containing a treatment with exclusion fencing and a control with grazing. Doehring et al. [10] showed that reaches with livestock exclusion led to greater measured riparian shade and greater cotton tensile-strength loss, which is an indicator of better established microbial communities and ecosystem functioning, yet delay in recovery or insufficient buffer size (e.g., only 2% of river basin) led to many missing indicators and makes restoration questionable. Abdi and Endreny [11] created a new river temperature model to identify thermal pollution causes and solutions, which can simulate leverage points #7 to 12 and their impact on outcomes. The i-Tree Cool River model can simulate the shading of riparian vegetation, groundwater and surface water exchange, and the thermal effects of stormwater runoff and green infrastructure treatments, and was written in freely accessible software and with a relatively small number of inputs in order to increase the number of people with access to the information (leverage point #6).

Saulys et al. [12] created an elegant study that monitored the system output signal of nitrate and phosphate concentrations along six rivers of the Nemunas and Venta basins in Lithuania, contrasting reaches straightened for flood drainage with natural unstraightened reaches. By finding the natural unstraightened reaches had statistically higher self-purification rates, defined as reductions in nitrate and phosphate from upstream to downstream, they validated at a national scale the functional importance of curvature in rivers, and to contribute to the change in mindset (leverage point #2) across Europe that self-purification is more important than straight rivers [13]. Zhou and Endreny [14] utilize leverage point #4 (the power to self-organize) to show how restored curvature in bedform topography, without awaiting full channel bank meander restoration, will reorganize the water column flows and restore hydraulic complexity, important for fish, and connectivity with the hyporheic zone, which is beneath the riverbed. Kruegler et al. [15] use leverage points #7 to 12 in a river corridor groundwater model to explore how the hyporheic zone, and its interaction with the river, can be changed and organized (leverage point #4) by riparian drawdown induced by evapotranspiration from riverside plantings. The research provides guidance on how to create opportunities for self-purification of pollutants, using natural resources powered by renewable energies.

5. New Paradigms in River Basin Restoration Include Honoring Water

Incrementally, the research on river basin restoration is bringing about the power to transcend paradigms, which is leverage point #1. This can be seen in the new approach to flood control announced by the US Department of Homeland Security and their Federal Emergency Management Agency, where they use language of "larger-scale migration or relocation" [16] rather than the older paradigm of "rebuild and flood-proof". This new paradigm of yielding to water rather than expecting water to yield to humans approaches the paradigm of honoring water, practiced by the Haudenosaunee

Confederacy of North America, who predate European settlement. The Haudenosaunee make it their duty to protect the water so that water can perform her duties. The Onondaga Nation, a member of the Haudenosaunee Confederacy, lived on the hills surrounding their Sacred Lake rather than settling within the floodplain. By contrast, Europeans who subsequently settled along Onondaga Lake, New York, undertook massive flood control and development projects that have led to the degradation of river basin fisheries and water quality [17]. This Haudenosaunee paradigm of gratitude for water, and its leverage on informing their mindset, is expressed in their Thanksgiving Address, "We give thanks to all the Waters of the world for quenching our thirst and providing us with strength. Water is life. We know its power in many forms, waterfalls and rain, mists and streams, rivers and oceans. With one mind, we send greetings and thanks to the spirit of Water. Now our minds are one." [18].

This Special Issue encourages our readers to utilize these new findings and the leverage points to restore river basin systems. The most effective leverage point is transcending paradigms and developing new mindsets for working with the complexity of river basin systems. Of course, developing the paradigm is part of the system. So the systems path scientists and engineers might follow in restoration (Figure 1), as adapted from the UN [1], is: (1) identifying, understanding, and working with the physical, chemical, and biological processes comprising river basin and river health and delivering ecosystem services; (2) identifying, incorporating, and involving socio-economic values and broader planning and development activities linked to river basin and river health; (3) addressing structure and function relationships at the appropriate scales to address limiting factors to river health; (4) setting clear, achievable, and measurable goals, framed in terms of changes to ecosystem structure and function, the provisioning of ecosystem services, and, where feasible, socioeconomic factors; (5) planning, implementing, and managing to provide resilience to a range of scenarios over time, including changes to climate, land use, hydrology, pollutant loads, and population, so restoration outcomes are sustained over the long term; (6) involving all relevant stakeholders in an integrated approach, addressing land and water issues, and involving interagency and community collaboration, to achieve the greatest benefits; and (7) monitoring, evaluating, adapting, and reporting the actual state of river basin health relative to the desired state, and formulating our paradigm to guide restoration and adaptive management. Together on this journey we can improve social and ecological systems within our river basins.

Funding: This research received no external funding.

Acknowledgments: The author is grateful for the insights provided by the 2016 UN report on river basin restoration [1] and for the development of system leverage points by Meadows [5].

Conflicts of Interest: The author declares no conflict of interest.

References

1. Speed, R.; Li, Y.; Tickner, D.; Huang, H.; Naiman, R.; Cao, J.; Gang, L.; Yu, L.; Sayers, P.; Zhao, Z.; et al. *River Restoration: A Strategic Approach to Planning and Management*; UNESCO: Paris, France, 2016.
2. Cai, X.; Ranjithan, R. Special Issue on the Role of Systems Analysis in Watershed Management. *J. Water Resour. Plan. Manag.* **2013**, *139*, 461–463. [CrossRef]
3. USDA-NRCS. *National Engineering Handbook—Part 654 Stream Restoration Design*; United States Department of Agriculture, Natural Resources Conservation Service: Washington, DC, USA, 2007; pp. 6–36.
4. Mitsch, W.J.; Jorgensen, S.E. *Ecological Engineering and Ecosystem Restoration*; John Wiley & Sons: New York, NY, USA, 2004.
5. Meadows, D. *Thinking in Systems: A Primer*; Chelsea Green Publishing: White River Junction, VT, USA, 2008.
6. Rampinelli, C.G.; Knack, I.; Smith, T. Flood Mapping Uncertainty from a Restoration Perspective: A Practical Case Study. *Water* **2020**, *12*, 1948. [CrossRef]
7. Golpira, A.; Huang, F.; Baki, A.B.M. The Effect of Habitat Structure Boulder Spacing on Near-Bed Shear Stress and Turbulent Events in a Gravel Bed Channel. *Water* **2020**, *12*, 1423. [CrossRef]

8. Abebe, W.B.; Tilahun, S.A.; Moges, M.M.; Wondie, A.; Derseh, M.G.; Nigatu, T.A.; Mhiret, D.A.; Steenhuis, T.S.; Camp, M.V.; Walraevens, K.; et al. Hydrological Foundation as a Basis for a Holistic Environmental Flow Assessment of Tropical Highland Rivers in Ethiopia. *Water* **2020**, *12*, 547. [CrossRef]

9. Liu, Q.; Jiang, J.; Jing, C.; Liu, Z.; Qi, J. A New Water Environmental Load and Allocation Modeling Framework at the Medium–Large Basin Scale. *Water* **2019**, *11*, 2398. [CrossRef]

10. Doehring, K.; Clapcott, J.E.; Young, R.G. Assessing the Functional Response to Streamside Fencing of Pastoral Waikato Streams, New Zealand. *Water* **2019**, *11*, 1347. [CrossRef]

11. Abdi, R.; Endreny, T. A River Temperature Model to Assist Managers in Identifying Thermal Pollution Causes and Solutions. *Water* **2019**, *11*, 1060. [CrossRef]

12. Šaulys, V.; Survilė, O.; Stankevičienė, R. An Assessment of Self-Purification in Streams. *Water* **2020**, *12*, 87. [CrossRef]

13. Anonymous. Re-Engineering Britain's Rivers: Why the Wiggles are Being Put Back into Watercourses, 6 March 2020. Available online: https://www.latestnigeriannews.com/news/8279579/reengineering-britains-rivers.html (accessed on 17 September 2020).

14. Zhou, T.; Endreny, T. The Straightening of a River Meander Leads to Extensive Losses in Flow Complexity and Ecosystem Services. *Water* **2020**, *12*, 1680. [CrossRef]

15. Kruegler, J.; Gomez-Velez, J.; Lautz, L.K.; Endreny, T.A. Dynamic Evapotranspiration Alters Hyporheic Flow and Residence Times in the Intrameander Zone. *Water* **2020**, *12*, 424. [CrossRef]

16. DHS. *Notice of Funding Opportunity Fiscal Year 2020 Building Resilient Infrastructure and Communities;* Department of Homeland Security (DHS): Washington, DC, USA, 2020; p. 37.

17. Effler, S.W. *Limnological and Engineering Analysis of a Polluted Urban Lake: Prelude to Environmental Management of Onondaga Lake;* Springer: New York, NY, USA, 1996; p. 832.

18. King, J.T. The Value of Water and the Meaning of Water Law for the Native Americans Known as the Haudenosaunee Cornell. *J. Law Public Policy* **2007**, *16*, 449–472.

Article

Flood Mapping Uncertainty from a Restoration Perspective: A Practical Case Study

Cássio G. Rampinelli *, Ian Knack * and Tyler Smith

Department of Civil and Environmental Engineering, Clarkson University, 8 Clarkson Avenue, Box 5710, Potsdam, NY 136991, USA; tsmith@clarkson.edu
* Correspondence: cassiorampinelli@gmail.com (C.G.R.); iknack@clarkson.edu (I.K.)

Received: 29 May 2020; Accepted: 7 July 2020; Published: 9 July 2020

Abstract: Many hydrologic studies that are the basis for water resources planning and management rely on streamflow information. Calibration and use of hydrologic models to extend flow series based on rainfall data, perform flood frequency analysis, or develop flood maps for land use planning and design of engineering works, such as channels, dams, bridges, and water intake, are examples of such studies. In most real-world engineering applications, errors in flow data are neglected or not adequately addressed. However, because flows are estimated based on the water level measurements by fitted rating curves, they can be subjected to significant uncertainties. How large these uncertainties are and how they can impact the results of such studies is a topic of interest for researchers, practitioners, and decision-makers of water resources. The quantitative assessment of these uncertainties is important to obtain a more realistic description of many water resources related studies. River restoration in many areas is limited by data availability and funding. A means to assess the uncertainty of flow data to be used in the design and analysis of river restoration projects that is cost effective and has minimal data requirements would greatly improve the reliability of river restoration design. This paper proposes an assessment of how uncertainties related to rating curves and frequency analysis may affect the results of flood mapping in a real-world application to a small watershed with limited data. A Bayesian approach was performed to obtain the posterior distributions for the model parameters and the HEC-RAS (Hydrologic Engineering Center-River Analysis System) hydraulic model was used to propagate the uncertainties in the water surface elevation profiles. The analysis was conducted using freely available data and open source software, greatly reducing traditional analysis costs. The results demonstrate that for the study case the uncertainty related to the frequency analysis study impacted the water profiles more significantly than the uncertainty associated with the rating curve.

Keywords: flood mapping; uncertainty; Bayesian inference; rating curve

---

1. Introduction

River restoration is a prominent area of applied water resources science and involves a variety of modifications in rivers ecosystems and stream riparian zones embracing different purposes to improve hydrologic, geomorphic, and/or ecological processes in degraded watersheds [1,2]. Examples of such overarching purposes include aesthetics, recreation, education, bank stabilization, channel reconfiguration, fish passage, floodplain reconnection, flow modification, land acquisition, instream habitat, and species improvements and management [1]. Regardless of the river restoration goals, the essence behind such initiatives is that restoring rivers to a more natural status is important not only for purely environmental reasons but also to reduce flood and geomorphic risks, besides reducing or avoiding costs of operation, maintenance, and replacement of hard works interventions [3]. Often, these restoration projects are located in areas with limited data resources and have limited financial resources. These limitations greatly reduce the ability to assess the uncertainties in available flow data. A cost-effective means to assess the uncertainty of flow data would help reduce over designing restoration projects to accommodate uncertainties.

In this context, many multi-purpose restoration plans are triggered by flooding and related natural disasters that have a significant impact on socio-economic activities of populations in developed and developing countries around the world [4,5]. Thus, the process of determining inundation extents by the development of flood hazard maps, including the frequency of floods, and how they affect infrastructure and social activities in flood-prone areas is part of any restoration plan. The procedure generally runs through estimating the magnitude of a flood with a particular likelihood, simulating that flood in a hydraulic model, and delineating the resulting flood extents, which are represented as a deterministic boundary [6]. Deterministic flood hazard boundary delineations tend to induce the use of the resulting information for human development patterns closely following them [7,8]. This has revealed shortcomings as a result of neglecting uncertainty when delineating flood boundaries, since flood insurance claims outside regulatory flood hazard boundaries have occurred more frequently [8–11]. Addressing uncertainty can not only support the decision-making process on direct considerations regarding outputs of hydraulic/hydrologic studies but also on data acquisition. Since many projects suffer from a lack of funds, the possibility of including the impact of uncertainty in model outputs allows the use of decision analyses to assess if it is worth investing time and resources to acquire additional data and/or to perform further site investigations to better constrain the uncertainty in model evaluations [12]. From a broader perspective, uncertainty analysis enhances the scientific understanding of hydrodynamic modeling in river, climate, coastal, and environmental systems. Three main steps should be considered to systematically address uncertainty: identification of the sources of uncertainty, quantification of uncertainty from different sources, and proper communication of the uncertainty impact on the study outcomes [7]. Here, our interest focuses on corroborating to the first and second stages of the uncertainty assessment.

Among the different sources of uncertainty, two basic kinds can be typified: natural and epistemic uncertainties [13]. Natural uncertainty is related to the variability of the underlying stochastic process while epistemic uncertainty results from incomplete knowledge about the study processes [13]. Another way of typifying the sources of uncertainty is by focusing on model processes and associating it to the choice of [7]: model structures [14,15], model parameters [14,16], model inputs [14,17,18], validation data [19], change in floodplain landscape over time [8], and change in climate conditions [20,21]. A more thorough definition of the different facets of uncertainty is presented in [22] in terms of aleatory, epistemic, ontological, and linguistic uncertainties.

For this study, we used a broader and more practical perspective in which the uncertainty associated with the development of flood maps is related to the uncertainties that arise from hydrologic, hydraulic, and topographic analyses. The objective of the hydrologic study is to estimate the magnitude of floods for different return periods by applying frequency analysis, regionalization methods, and/or hydrologic modeling when needed. In most real-world engineering applications, errors in flow data are neglected or not adequately addressed, and model outputs are assumed to be deterministic. However, because flows are estimated based on the water level measurements and the rating curve, it is, in reality, affected by uncertainties [23]. The rating curve is a mathematical function that relates stage measurements with discharge values at a given station. Therefore, the rating curve is only an approximation of the real relationship between water levels and discharge values which is reflected in uncertainties in the daily streamflow data. Such uncertainties can be expressed in terms of the rating curve parameters and will be reflected not only on discharge estimated for given return periods but also on the boundary conditions of the hydraulic model. In flood frequency analysis, these uncertainties can be even larger because a relatively significant portion of the data is estimated based on the extrapolation of the rating curve [24], not to mention inherent uncertainty related to the assumptions of the statistical model and hydraulic channel control or flow regime that can be affected by temporary or permanent changes due to seasonal vegetation growth, variation of boundary conditions or hysteresis due to transient flow conditions [25].

As part of the hydraulic study the flow is propagated over the terrain by a hydraulic model. This involves the definition of the boundary conditions for the study reach, which entails setting a rating curve and water surface elevations to the boundaries of the study reach. After defining the hydraulic model to be used, friction parameters to characterize channel and floodplain roughness should also be adjusted, aggregating more uncertainty to the process. Finally, the terrain on which the flow is routed is obtained by a topographic study, which is generally the result of a digital elevation model that merges bed topography (bathymetry) and emerged terrain (topography) based on field survey data and geoprocessing.

How large these uncertainties are and how they can impact the results of the aforementioned studies have been a topic of interest for researchers, practitioners, and decision-makers in the water resources science embracing uncertainties related to boundary conditions [8,26], spatial resolution and model structure [27–31], roughness [32–34], rating curve [25], non-stationarity [35], and general probabilistic approaches to flood mapping uncertainty [36–40]. The quantitative assessment of these uncertainties is important to obtain a more realistic description of many water resources related studies. Bayesian inference is very attractive in these cases because it can easily incorporate the often imprecise knowledge available on the hydraulic behavior of the river into the flood frequency analysis, providing a natural way to not only evaluate the uncertainties in the streamflow sample but also to consider these uncertainties in the estimated flood quantiles [24].

A fully Bayesian model proposed in [24] allows the integrated estimation of the uncertainties from the rating curve and in the flood frequency analysis. Most of the reported studies generally consider such uncertainties separated either for the rating curve parameters [41–43] or for the frequency analysis [44–48]. Although there are a few studies addressing flood mapping uncertainties [6,7,49–51], there is a lack of approaches that investigate how the combined uncertainties related to the parameters from the rating curve and the statistical distributions can affect flood mapping. In this case study, given the limited data available, we focused on the propagation of these two sources of uncertainty in flood mapping, although other sources of uncertainty may also be of relevance.

Additionally, most of the reported studies focus on research applications based on extensive data to support a broad theoretical background to address the different sources of uncertainty. This academic context limits such applications to real-world cases, especially for low-budget engineering projects on small stream reaches. In this paper, we demonstrate, by presenting a practical study case on the East Branch of the Ausable River, Keenee Valley, NY, USA, that even with limited data (poor flow measurements and topography), and a constrained budget, it is possible to obtain reasonable outputs to support decision-making under uncertainty based on flood mapping. The study aimed to quantify to what extent flood maps derived from a hydraulic model can be affected by uncertainties related to the rating curve and frequency analysis as discussed in the Bayesian model proposed in [24] that is capable of combining uncertainties from the rating curve and flood frequency analysis. More specifically, we were interested in generating flood maps for different return periods including the 95% credible intervals for the flood boundaries to the return periods of 3, 100, 1000, and 10,000 years.

2. Materials and Methods

2.1. Site Location

The study area is located in the Noonmark reach of the East Branch Ausable River, at Keene Valley, Essex County, NY. The study reach is approximately 640 m long. The site performed well during hurricane Irene in August 2011, remaining geomorphically stable and recovering quickly. As such, the reach has been used as a model for the restoration of other reaches in the area. It has also been relatively heavily investigated for stream sections in the region. Despite the importance of the reach for river restoration activities in the region, there are relatively limited data, especially flow data, available for the reach. The data collection includes a field campaign conducted by the Ausable River Association, which resulted in 4 surveyed cross-sections along the reach. Figure 1 presents a general overview of the study reach with photos from the 4 cross-sections and a bridge. There are no flow data available in the immediate vicinity of the study reach.

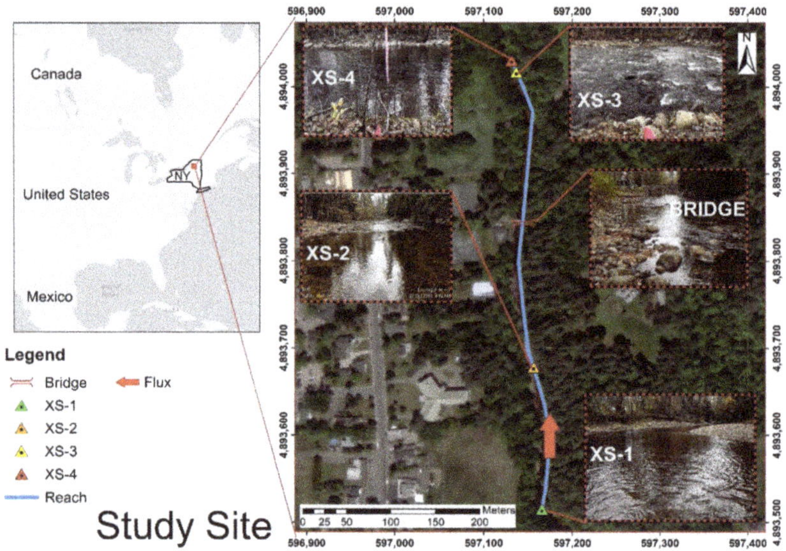

Figure 1. Overview of the study location.

2.2. Digital Elevation Model

A Digital Elevation Model (DEM) for the terrain at the site location was generated based on SRTM (Shuttle Radar Topography Mission) images. The raster images were retrieved from the National Elevation Dataset (NED) provided by the GIS New York State web site [51]. The NED is the primary data product produced and distributed by the United States Geological Survey (USGS). The NED provides seamless raster elevation data of the conterminous United States, Alaska, Hawaii, and the island territories. The NED is derived from diverse source datasets that are processed to a specification with a consistent resolution, coordinate system, elevation units, and horizontal and vertical datums. For this study, the DEM resolution was 1 m. Additional cartographic details can be accessed in [51].

The images were georeferenced and projected to the Datum D/WGS/1984, zone 18N. Some adjustments were necessary to better accommodate the contours to the topographic features. The geoprocessing tools from ArcGIS 10.2 (Esri, Redlands, CA, USA) were set to generate a Triangulated Irregular Network (TIN) for the terrain topography resulting in a Digital Elevation Model (DEM). The extension tool HEC-GeoRAS 10.2 [52] was used to processing geospatial data and to export the cross-sections delineated on the DEM to HEC-RAS 5.0.7 [53]. Extended cross-sections were delineated along the study reach, matching the 4 surveyed cross-sections. Additional cross-sections were also included in between the 4 surveyed ones. Figure 2 shows the resulting DEM, as well as the delineated cross-sections, including the stationing.

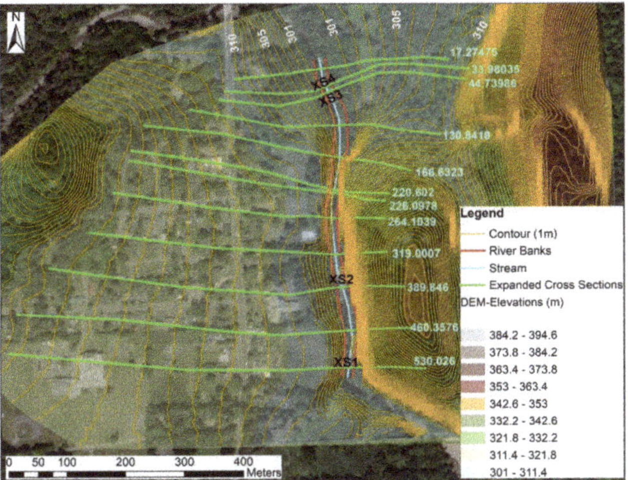

Figure 2. DEM and cross-sections.

2.3. Bathymetric Adjustments

The cross-sections derived from the DEM and those from the field survey were performed independently. Cross-sections based on the DEM were obtained from SRTM images; thus, the bathymetric portion (terrain under the water surface) was not captured by the radar. On the other hand, the cross-sections derived from the field survey, despite representing the riverbed, do not have enough length to cover the flood plain. Despite the impossibility of obtaining a perfect match between them (since there were no topographic benchmarks to do so), a comparison between the surveyed cross-sections and the DEM-retrieved cross-sections indicated the necessity of minor adjustments. These adjustments were performed by inserting the surveyed portion of the cross-section at the center of the riverbed and then merging the stations from the field survey with the cross-sections from the DEM. Additional cross-sections inserted between the surveyed ones were interpolated based on the merged (survey and DEM) cross-sections immediately upstream and downstream. Figure 3 shows an example of a comparison between a surveyed cross-section and DEM-derived cross section, as well as the resulting merged cross-section.

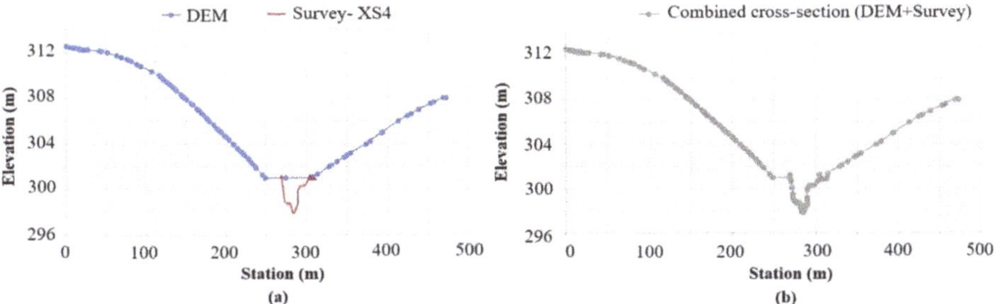

Figure 3. Cross-section XS4 (Station 33.98035): (**a**) comparison between field survey (continuous red line) and DEM (continuous blue line with dots); and (**b**) resulting cross-section after combining field survey and DEM data.

2.4. Rating Curve

2.4.1. Rating Curve Function

The rating curve represents the relationship between water level and flow. In this study, a power equation was used to fit the data. Although the method can be adjusted to represent a more complex situation in which hysteresis due to transient flow regime occurs, here, given the limited data, the flow in the river reach was assumed to be steady, and described by the following equation:

$$Q = a(h - h_0)^c \tag{1}$$

where a depends on the characteristics of the river reach, h is the water surface elevation, h_0 is the elevation associated with zero flow, and c expresses the hydraulic control. The parameters were calibrated by applying the Particle Swarm Optimization algorithm [54] to minimize the Sum of Squared Errors (SSE) between flow data and simulated discharges.

2.4.2. Synthetic Stage–Discharge Data

Since no discharge–stage measurements were provided at the study location, synthetic data for discharge and stage were generated based on a hydraulic simulation performed with HEC-RAS. An interpolated cross-section positioned at station 166.6323 (refer to Figures 2 and 4) was used as a reference to the rating curve. This location was selected to avoid the influence of the downstream boundary condition at cross-section XS4 that was assumed to be the normal depth for the slope of the energy grade line (0.006 was the slope estimated based on the field survey data). For the upstream boundary condition, discharges ranging from 5 to 150 m^3/s were used. Based on test simulations and field data, the Manning coefficient was set to 0.034 for the channel and 0.04 for the banks. Figure 4 presents a profile of the simulated reach with an indication of the location used as a reference to derive the rating curve.

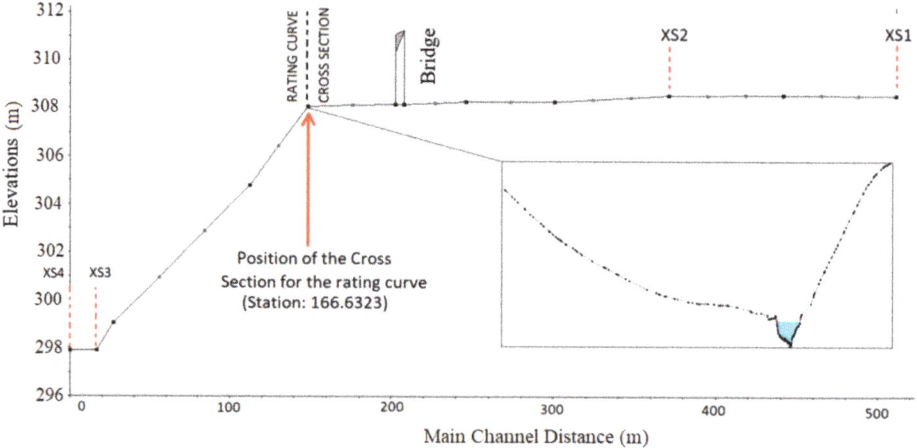

Figure 4. Reach profile with the indication of the surveyed cross-sections and the location of the cross-section used as reference to derive the rating curve.

2.5. Frequency Analysis

2.5.1. Generalized Extreme Values (GEV) Function

The three-parameter Generalized Extreme Value (GEV) distribution [55] was used for modeling the extreme discharges for the return periods 3, 100, 1000, and 10,000 years. This distribution aggregates three asymptotic forms of extreme value distributions (Frechet, Weibull, and Gumbel) in the same expression. The Cumulative Distribution Function (CDF) is given by the following expression:

$$F_X(x) = exp\left\{-\left[1 - k\left(\frac{x-\xi}{\alpha}\right)\right]^{\frac{1}{k}}\right\}; \; k \neq 0 \tag{2}$$

$$F_X(x) = exp\left\{-exp\left[-\frac{(x-\xi)}{\alpha}\right]\right\}; \; k = 0 \tag{3}$$

where κ (kappa), α (alpha), and ξ (xi) represent the shape, scale, and position parameters, respectively. The Frechet and Weibull distributions occur for negative and positive values of κ, respectively (Equations (4) and (5)). When κ = 0, the Gumbel distribution occurs (Equation (6)), in which x can assume any value.

$$Frechet \rightarrow k < 0 \rightarrow \xi + \frac{\alpha}{k} \leq x < +\infty \tag{4}$$

$$Weibull \rightarrow k > 0 \rightarrow -\infty < x \leq \xi + \frac{\alpha}{k} \tag{5}$$

$$Gumbel \rightarrow k = 0 \rightarrow -\infty < x < +\infty \tag{6}$$

The GEV Probability Density Function (PDF) is given by Equations (7) and (8):

$$f_x(x) = \frac{1}{\alpha}\left[1 - k\left(\frac{x-\xi}{\alpha}\right)\right]^{\frac{1}{k}-1} exp\left\{-\left[1 - k\left(\frac{x-\xi}{\alpha}\right)\right]^{\frac{1}{k}}\right\}; \; k \neq 0 \tag{7}$$

$$f_x(x) = \frac{1}{\alpha}exp\left\{\frac{x-\xi}{\alpha} - exp\left(\frac{x-\xi}{\alpha}\right)\right\}; \; k = 0 \tag{8}$$

The parameters from the PDF (α, ξ, κ) can be estimated by the Method of Moments, Method of L-Moments, or the Maximum-Likelihood Method [55]. For this study, the method of L-Moments was used by applying the R package *lmom* [56]. Based on the estimated parameters values, the x_p quantile associated with a discharge and its exceedance probability (p) can be computed by the following expressions.

$$x_p = \xi + \frac{\alpha}{k}\left[1 - (-\ln(p))^k\right]; \; \xi \neq 0 \tag{9}$$

$$x_p = \xi + \alpha[1 - (-\ln(p))]; \; \xi = 0 \tag{10}$$

2.5.2. Extreme Flow Data and Drainage Area Correction Factor

Since data at the study location are unavailable, the USGS flow gauge (04275000), located on the East Branch Ausable River (Latitude 44°26′14.6″, Longitude 73°40′51.5″) downstream of the study reach, was used as a reference for extreme flow data. The daily mean discharge dataset at this gauging station covers the period from 5 September 1924 to 10 September 2019, with a gap between 30 September 1995 and 8 March 2016 (Figure 5). The maximum daily mean discharge (q_{max}) time series at this location was used to compose the extreme value time series. The drainage area at the reference flow gauge is 512.8 km^2, while the drainage area at the study reach is 173.5 km^2. Figure 6 shows both drainage areas.

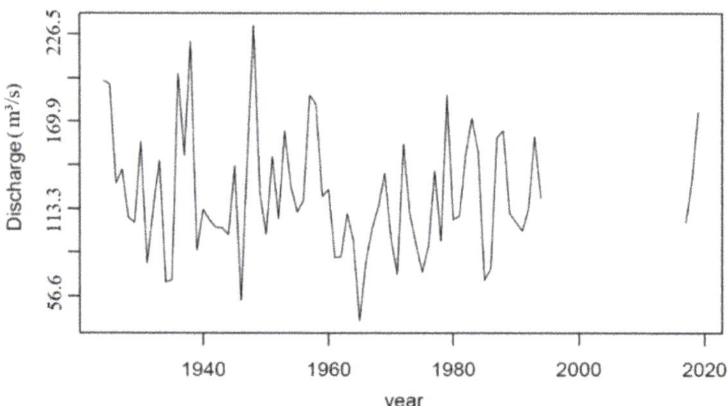

Figure 5. Maximum daily average discharge time series.

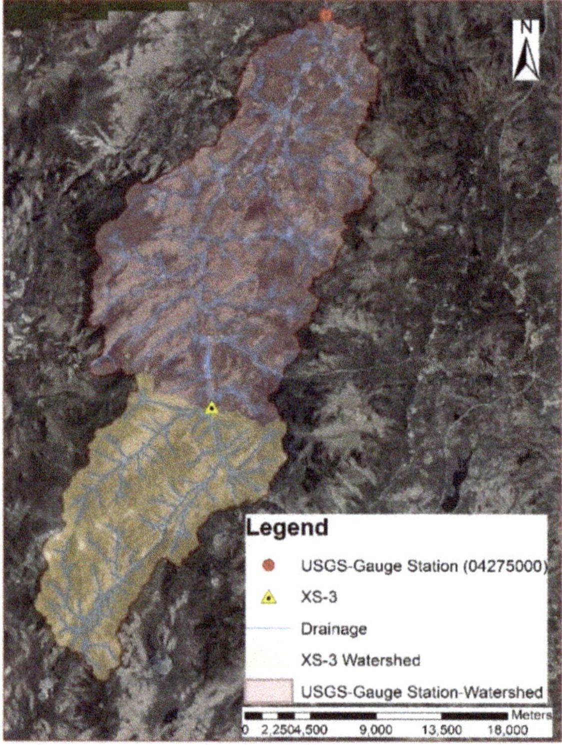

Figure 6. Drainage areas for the reference gauging station and the study site.

The maximum daily mean time series at the USGS gauging station was regionalized to the site location multiplying the discharges by a correction factor given by the ratio (0.338) between the drainage areas. The inverse function of the fitted rating curve was used to derive the associated stages/water elevations time series ($h_{\max}$).

2.6. Bayesian Model

Bayesian inference is a useful approach to estimate the parameters of a given function or distribution conditioned to observed data and prior knowledge related to the model parameters. Based on Bayes theorem, the posterior distributions of the model parameters were computed considering updates on the prior knowledge by applying the likelihood function on the observed data as represented by the following expression.

$$p(\theta|x) = \frac{p(x|\theta)p(\theta)}{\int p(\theta)p(x|\theta)d\theta} \tag{11}$$

where $p(\theta|x)$ represents the posterior distribution of the parameters, $p(x|\theta)$ is the likelihood function, $p(\theta)$ is the prior distribution, and the denominator is a normalizing constant.

2.6.1. Bayesian Rating Curve Model

Based on [25], Bayes theorem for the relation between stage–discharge can be described by:

$$p(\theta,\sigma_f|\widetilde{H},\widetilde{Q}) \propto p(\widetilde{Q}|\theta,\sigma_f,\widetilde{H})p(\theta,\sigma_f) \tag{12}$$

where $p(\theta,\sigma_f|\widetilde{H},\widetilde{Q})$ is the posterior distribution, $p(\widetilde{Q}|\theta,\sigma_f,\widetilde{H})$ is the likelihood, $p(\theta,\sigma_f)$ is the prior distribution, $\widetilde{H}$ and $\widetilde{Q}$ correspond to the measured stage–discharge pairs, θ represents the parameters of the rating curve function (a, h_0, and c), and σ_f represents the standard deviation of the residuals. Since the rating curve function cannot perfectly represent the stage–discharge relation, even if the real values were known, the quality of the fit of the model is related to the magnitude of the residuals that are associated with the standard deviation.

The real discharges $\widetilde{Q}$ for each stage $\widetilde{H}$ can be represented by the sum of the discharge computed by the rating curve based on the stage–discharge function $f(\widetilde{H}_i|\theta)$ and the residuals associated with the discharge measurement and the fit of the rating curve as follows [25]:

$$\widetilde{Q}_i = f(\widetilde{H}_i|\theta) + \varepsilon_i^f + \varepsilon_i^q \tag{13}$$

$$\varepsilon_i^f + \varepsilon_i^q \sim N\left(0, \sqrt{\sigma_f^2 + u_{Q_i}^2}\right) \tag{14}$$

where ε_i^f and ε_i^q correspond to the fit and measurement errors, respectively. Finally, the likelihood function can be represented by the product of density functions over N samples as proposed in [25].

$$p(\widetilde{Q}|\theta,\sigma_f,\widetilde{H}) = \prod_{i=1}^{N} p_N\left(\widetilde{Q}_i\Big|f(\widetilde{H}_i|\theta), \sqrt{\sigma_f^2 + u_{Q_i}^2}\right) \tag{15}$$

2.6.2. Fully Bayesian Model

The fully Bayesian Model presented in [24] combines rating curve and GEV parameters in an integrated posterior distribution that is proportional to the product of the likelihood function based on the observed data by the prior distributions for the parameters. The expression is presented as follows:

$$p\left(a,h_0,c,\sigma_f,\alpha,\xi,k,u_Q|\widetilde{Q},\widetilde{H},H_{max},Q_{max}\right)$$
$$\propto p\left(\widetilde{Q},Q_{max}|a,h_0,c,\sigma_f,\alpha,\xi,k,\widetilde{H},H_{max},u_Q\right)\cdot p\left(a,h_0,c,\sigma_f,\alpha,\xi,k\right) \tag{16}$$

where h_0 and c are the rating curve parameters and σ_f represents the residuals related to the rating curve model, corresponding to the measured stage–discharge pairs $\widetilde{Q}$ and $\widetilde{H}$. H_{max} and Q_{max} are the time series of the maximum annual stages and associated discharges, respectively, in which the latter depends on H_{max}, a, h_0, c, and σ_f for each stage. Since the annual maximum discharge event Q_{max} is independent of the probability of observing $\widetilde{Q}$, one can express the likelihood separated in two terms:

$$p\left(a, h_0, c, \sigma_f, \alpha, \xi, \kappa, u_Q \middle| \widetilde{Q}, \widetilde{H}, H_{max}, Q_{max}\right) \propto p\left(Q_{max} \middle| a, h_0, c, \sigma_f, \alpha, \xi, \kappa, H_{max}\right) \cdot p\left(\widetilde{Q} \middle| u_Q, a, h_0, c, \sigma_f, \widetilde{H}\right) \tag{17}$$

Assuming independent events, one can express:

$$p\left(\widetilde{Q} \middle| u_Q, a, h_0, c, \sigma_f, \widetilde{H}\right) = \prod_{i=1}^{N} p_N\left(\widetilde{Q}_i \middle| f\left(\widetilde{H}_i \middle| \theta\right), \sqrt{\sigma_{fi}^2 + u_{Q_i}^2}\right) \tag{18}$$

$$p\left(Q_{max} \middle| a, h_0, c, \sigma_f, \alpha, \xi, k, H_{max}\right) = \prod_{i=1}^{N} f_{Qmax_i}\left(q_{max_i}\right) \tag{19}$$

where

$$f_{Qmax_i}\left(q_{max_i}\right) = \frac{1}{\alpha}\left[1 - k\left(\frac{q_{max_i} - \xi}{\alpha}\right)\right]^{\frac{1}{k}-1} exp\left\{-\left[1 - k\left(\frac{q_{max_i} - \xi}{\alpha}\right)\right]^{\frac{1}{k}}\right\} \tag{20}$$

$$Qmax_i = f(H_{max}|a, h_0, c) + \varepsilon_i^f \tag{21}$$

$$\varepsilon_i^f \sim N\left(0, \sigma_f^2\right) \tag{22}$$

The standard deviation σ_f is assumed to be heteroscedastic and linearly varying with the discharge, as follows:

$$\sigma_f = \gamma_1 + \gamma_2 Q \tag{23}$$

where γ_1 and γ_2 are parameters to be estimated.

2.6.3. Markov Chain Monte Carlo Simulations

To perform the Bayesian inference, a Markov Chain Monte Carlo (MCMC) algorithm was used to estimate the posterior distributions for the model parameters. The DREAM (Differential Evolution Adaptive Metropolis) algorithm originally presented in [57,58] and available as MATLAB [59] and R [60] packages was employed. Detailed information regarding the DREAM package can be found in [59].

2.7. HEC-RAS Model Set-Up

The HEC-RAS model [53], version 5.0.7, developed by the Hydrologic Engineering Center of the US Army Corps of Engineers was used to perform the hydraulic simulations. The model was chosen since it is freely available and extensively used for hydraulic simulations around the world. For the geometric data, as described in Section 2.2, and Section 2.3, the cross-sections were derived from a hybrid digital elevation model (DEM) resulting from a combination of SRTM images retrieved from the GIS New York State website [51], and 4 surveyed cross-sections. Twelve cross-sections were delimited, as shown in Figure 2, and exported from the DEM to the HEC-RAS with aid of the HEC-GeoRAS 10.2 [53] tool. The interpolation tool of the HEC-RAS Geometric Data module was used to create 12 more intermediate cross-sections to better represent the geometry of the reach channel. Figure 4 shows a profile plot of the study reach indicating by dark black dots the original cross-sections and by light gray dots the interpolated one.

Two sets of simulations were performed. First, simulations considering the entire profile reach were performed to provide synthetic stage–discharge data and allow the derivation of a rating curve, as described in Section 2.4. Second, simulations were performed to generate the flood maps based on four return periods, namely 3, 100, 1000, and 10,000, including the 95% credible intervals for the flood boundaries. For the upstream boundary condition, eight discharges representing the upper and lower boundaries of the 95% credible interval for the four return periods (Table 1) were used.

For the downstream boundary condition, the original profile reach (Figure 4) was trimmed at station 166.6323, where the rating curve (Figure 7) was derived and used as a downstream boundary condition. The Manning values used in the first set of simulations and presented in Section 2.4.2 were the same used for the second set of simulations, and the steady flow regime was considered. The resulting water surface profiles were exported to the ARC-GIS with aid of the HEC-GeoRAS 10.2 [52] tool where the flood maps were prepared.

Table 1. Mean discharges and the thresholds for 2.5% and 97.5% credibility quantiles considering different return periods (RP).

RP	Q (m³/s) 2.5%	Q (m³/s) Med	Q (m³/s) 97.5%
3	41.78	44.97	48.57
100	77	83.61	90.47
1000	97.22	106.76	120.98
10000	105.53	129.67	161.72

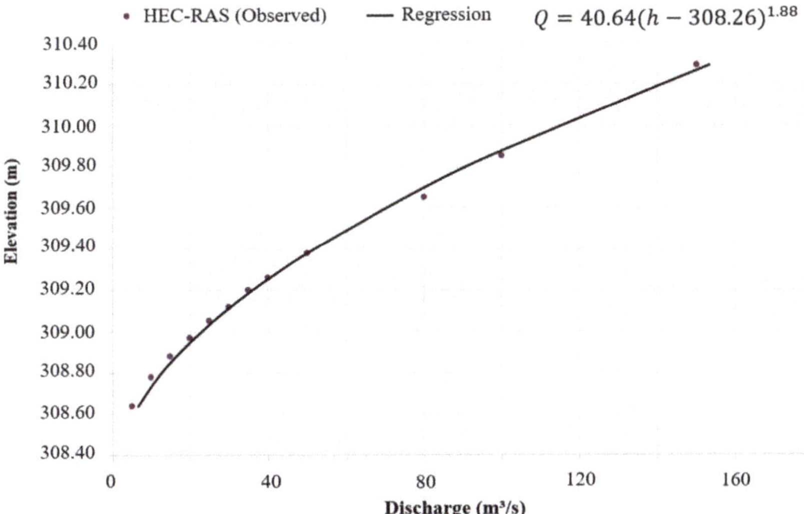

Figure 7. Rating curve for cross-section at station 166.6323.

3. Results

This section presents the results after applying the methodology described in the previous section. First, the outputs obtained for the synthetic rating curve are presented followed by the regionalized extreme data time series. Then, the results for the rating curve and the discharges resulting from the integrated Bayesian approach are shown. Finally, the flood maps are presented including the 95% credible interval.

3.1. Results of the Synthetic Rating Curve

After running the HEC-RAS model as described in Section 2.4.2, the water surface elevations at the rating curve reference cross-section (Figure 4) were obtained and related to the respective discharge, resulting in the synthetic observed flow-stage data used to derive the rating curve function. Based on the methodology described in Section 2.4.1, a potential function was fit to the synthetic observed data resulting in the following optimized parameters: $a = 40.64$, $h_0 = 308.26$, and $c = 1.88$. Figure 7 shows the adjusted curve as well as the synthetic data and the rating curve function, in which Q represents discharge and h represents the water surface elevation.

The synthetic observed flows range from 5 to 150 m³/s while the respective elevations are between 308.64 and 310.29 m, a change of 1.65 m. Based on the field survey and the hydraulic simulations, this is considered a reasonable water level oscillation for the magnitude of the simulated discharges and should be confined to the channel banks. The range of discharges was defined to simulate a possible real-world flow measurement interval since extreme flow measurements are not usually collected. This necessitates the need to extrapolate the rating curve when performing the frequency analysis study.

3.2. Results of the Extreme Data Time Series at the Study Site

Before proceeding with the regionalization of the annual maximum mean daily discharges from the USGS Gauging station referred in Section 2.5.2 to the study site, the consistency of the original daily data flow was evaluated based on the flow-duration curve (Figure 8). The flow-duration curve (FDC) is the relation between the magnitudes of streamflow, q, at a point, and the frequency (probability) with which those magnitudes are exceeded over an extended period.

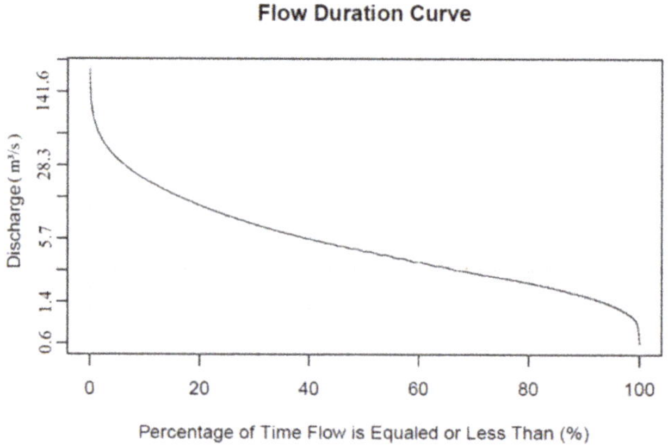

Figure 8. Flow duration curve at USGS flow gauge (04275000).

For the flow duration curve, the low-flow end of the curve is steep and possibly indicative of a small amount of groundwater storage above the channel bed level. Additionally, the general configuration of the obtained flow duration curve is relatively steep, representative of a variable stream. The relatively high variability of the streamflow can also be checked by comparing the discharge that is sustained 50% (4.3 m³/s) of the time with the one that is maintained 99% of the time (1.0 m³/s). The former is more than four times greater than the latter.

The annual maximum mean daily discharges at the USGS flow gauge was regionalized to the study site based on the procedure detailed in Section 2.5.2. Then, the rating curve function presented in Figure 7 was applied on annual maximum mean daily discharges time series. The adopted time interval, considered only the continuous data period from 1924 to 2019, excluding the data gap. The resulting time interval covers 74 years and is deemed reasonable to perform an applied frequency analysis.

The original and resulting extreme data time series regionalized to the study site is presented in Table A1, in Appendix A.

3.3. Results of the Bayesian Approach

By applying the DREAM algorithm to solve the Bayesian model for only the rating curve, as presented in Section 2.6.1, the respective 95% credible interval as well as the posteriors distributions for the rating curve parameters were obtained. The outputs are presented in Figures 9 and 10. As expected, the uncertainty increases with the magnitude of the discharge. The 95% credible interval for parameter a varies between 40 and 58, while, for parameter h_0, it ranges from elevations 308.28 to 308.47 m (approximately 20 cm of variation), and, for parameter, c it varies between 1.57 and 2.

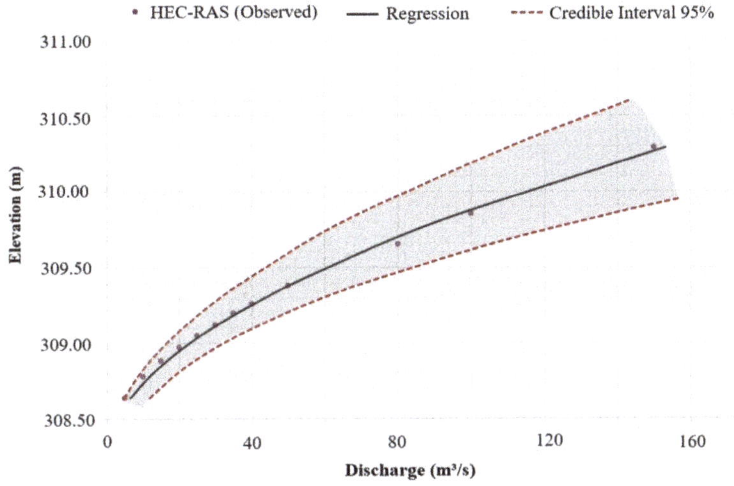

Figure 9. Rating curve for cross-section at station 166.6323 and 95% credible interval.

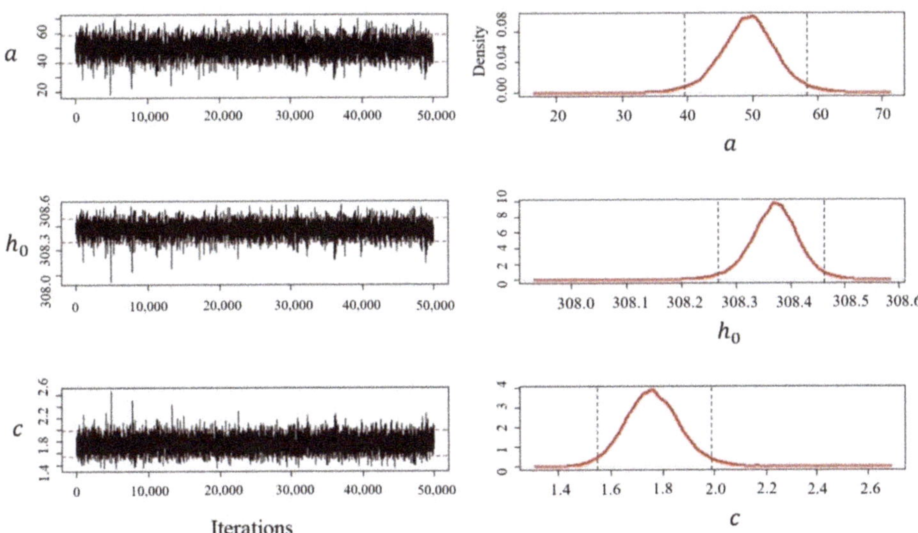

Figure 10. Posteriors distributions for the rating curve parameters and their 95% credible interval.

The fully Bayesian model, as described in Section 2.6.2, was applied resulting in the posterior distributions for the GEV (Figure 11) and for the extreme discharges as a function of their return periods, including the 95% credible interval (Figure 12). The 95% credible interval for the GEV parameter α varies between 3600 and 4700; for parameter ξ, it ranges from 3300 and 3430; and, for parameter κ, it varies between −0.04 and 0.06.

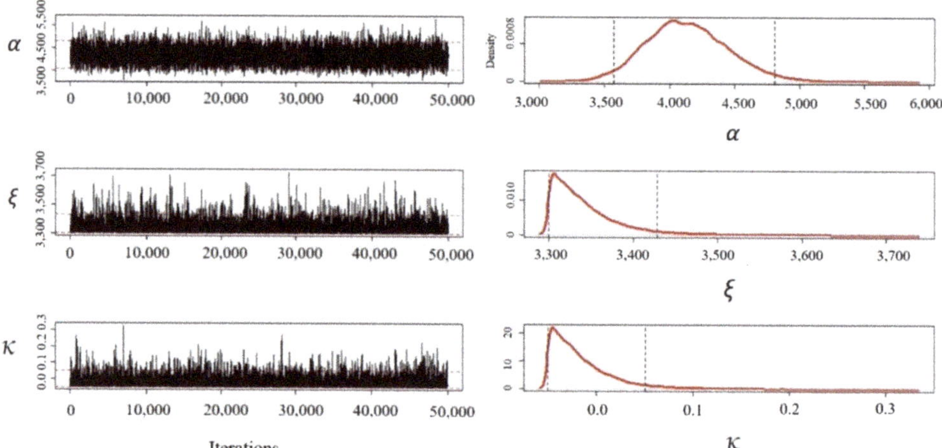

Figure 11. For the GEV parameters and their 95% credible interval.

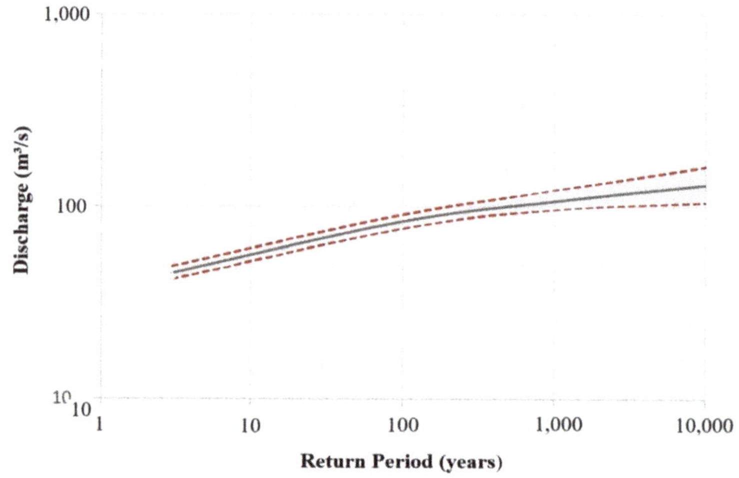

Figure 12. Discharges as a function of the return period and their 95% credible interval.

For the discharge as a function of the return period, as expected, as the magnitude of the flow increases so does the associated uncertainty. Table 1 summarizes the 95% credible limits for the discharge considering the return periods (RP) of 3, 100, 1000, and 10,000 years.

Figure 13 shows the uncertainty reflected on the flood maps for the return periods of 3 (Figure 13a), 100 (Figure 13b), 1000 (Figure 13c), and 10,000 (Figure 13d) years representing the upper and lower boundary of the 95% credible interval. Five buildings are indicated in Figure 13a, two of them are located near the riverbanks (b2 and b3), two of them farther away from the river (b5 and b4), and one of them in an intermediate position (b1). The impact of the simulated flood events and the value of the associated uncertainty is discussed in the next section.

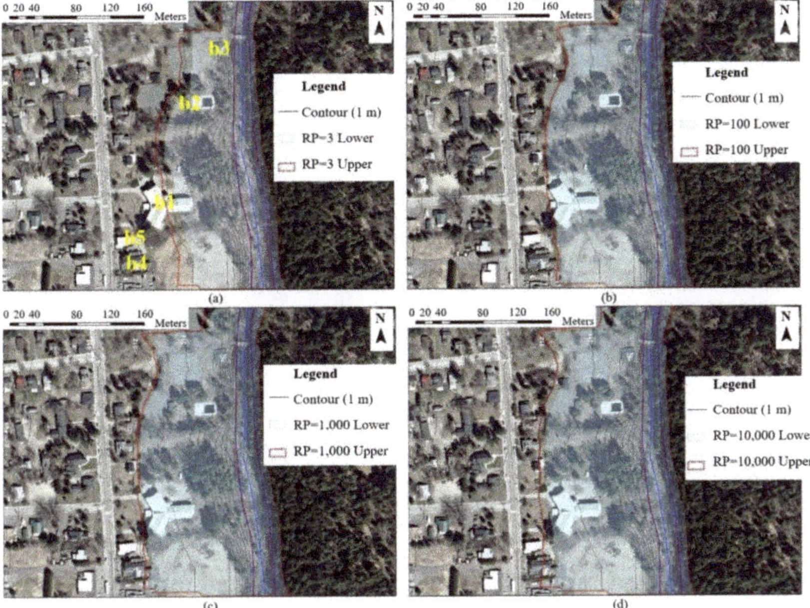

Figure 13. Flood maps and the 95% credible interval for the return periods: 3 years (**a**); 100 years (**b**); 1000 years (**c**); and 10,000 years (**d**).

4. Discussion

Despite some simplifications adopted in the model due to the lack of measured flow, the proposed methodology demonstrates the application of a framework for incorporating the uncertainty related to the rating curve and frequency analysis on the water surface profiles and flood maps to a real-world, data-limited, case study. One of the interesting aspects of the methodology applied in this study case is the fact that all the uncertainty analysis was performed using free open-source data and software. The HEC-RAS hydraulic model was employed to perform the hydraulic simulations, R was used to process the statistical and uncertainty analysis algorithms, and the topographic and hydrometric data were retrieved from free data repositories, provided by governmental agencies. This reveals the importance of open-source initiatives to the dissemination of uncertainty analysis in hydraulic and hydrologic simulations to real-world applications since a majority of studies focus on applications in which data and/or financial resources are not limiting.

The uncertainty related to the rating curve tends to increase as a function of the flow. For this study case, the maximum variation of the water level due to uncertainty was relatively low around 0.6 m (Figure 9). The low magnitude of variability is related to the fact that all data employed to derive the rating curve were artificially generated, which means that, in theory, the true values were known and correspond to the ones produced by the hydraulic model. However, it is expected that, for real measured data, the magnitude of the uncertainty could be significant.

Because the data were synthetically generated, the close fit of the regression model for the rating curve resulted in less uncertainty over the rating curve model parameters, as can be seen in Figure 10. The low degree of uncertainty refers only to the rating curve that is positioned in the control section. However, the fully Bayesian model revealed that there is much more uncertainty related to the parameters of the GEV (Figure 11) than the rating curve model parameters (Figure 10) for this case study. This is reflected in the discharge as a function of the return period (Figure 12 and Table 1) and on the generated flood maps (Figure 13). Since the uncertainty related to the roughness parameter was not considered in this study, it is expected that the consideration of such a source of uncertainty may also affect the uncertainty boundaries and should be evaluated in future studies. Ensuring a reasonable representation of the uncertainty envelope for flooding might be of interest so that river restoration plans adjacent to the stream and riparian zone would not be unexpectedly impacted by flooding in any of the range of inundation values.

To facilitate assessment of the flood map (Figure 13), five buildings were referenced by an alphanumeric code (b1, b2, b3, b4, and b5). For practical purposes, regardless of the return period and uncertainty boundary, buildings b1, b2, and b3 can be expected to be affected. In the case of a three-year return period event (Figure 13a), building b1, near the center of the flood map (Y shaped building), would be partially affected, while a 100-year or greater event (Figure 13b–d) would be completely flood it. For these three buildings, adding uncertainty in the analysis would be indifferent to the set of actions expected from a decision-maker. Additionally, we could infer that it would not be worth investing time and financial resources to acquire more data to constrain the uncertainty limits.

On the other hand, for buildings b4 and b5, depending on the return period the uncertainty boundary starts intersecting building locations. The backyard of building b4 is intercepted by the uncertainty boundary for a 100-year return period event. For a 1000-year event, building b4 is intercepted by the 95% credible interval boundary, while the backyard of building b5 starts being affected. Finally, for the worst-case scenario, a 10,000-year return period event, both buildings are impacted by the flood uncertainty boundary. For these two buildings, in case of a deterministic flood mapping approach, they would be considered out of the flooded area. However, when uncertainty is incorporated, these buildings can be considered affected by the flood events, depending on the return period and the related uncertainty. This may be useful to consider the possibility of evaluation of the costs and value of collecting more information in constraining the uncertainty for enhanced decision making.

5. Conclusions

This work presents a case study to address uncertainty related to flood mapping, due to the rating curve and the frequency analysis. The quantification of the uncertainty on flood mapping for the return periods of 3, 100, 1000, and 10,000 years, including the 95% credible intervals, was performed. The Bayesian approach was applied making it possible to express the rating curve and GEV parameters as statistical distributions and addressing the impact of such uncertainty in flood mapping.

Although the uncertainty related to the rating curve did not significantly propagate through the upstream water elevation profile, in this study case, it does not mean this conclusion could be extended to any case. It is expected that regions with real measured flows should result in higher uncertainty for the rating curve. In addition, in this study case, the control section of the bridge also reduced the influence of the downstream condition on upstream cross-sections. Future studies might address such issues by including a more accurate digital elevation model, implementing in situ measuring discharge campaigns, and quantifying the uncertainty related to the topography, data regionalization, and hydraulic roughness. We highlight that acquiring more data, specifically stage–discharge measurements, is of paramount importance to validate the uncertainty analysis related to the rating curve and flow discharges. Additionally, the digital elevation model and the Manning roughness coefficient may be significant contributors to the total uncertainty.

Despite the short extension of the study reach, the data limitations, and few possible affected infrastructure, when uncertainty was incorporated into the analysis, some buildings that could not be considered affected by the flooded area in a deterministic approach became vulnerable to flood. Depending on the importance and use of the study outputs, it would be of interest to study the value of collecting more information for constraining the uncertainty and making a decision. This study demonstrates that, even in real-world case studies that suffer data limitations (both flow measurements and topography surveys), it is possible to obtain useful outputs to better inform river restoration projects and support decision-making in water resources management.

Author Contributions: Conceptualization, C.G.R. and I.K; methodology, C.G.R.; software, C.G.R.; validation, C.G.R.; formal analysis, C.G.R.; investigation, C.G.R.; resources, C.G.R.; data curation, C.G.R.; writing—original draft preparation, C.G.R.; writing—review and editing, C.G.R., I.K., and T.S.; visualization, C.G.R.; supervision, I.K., and T.S.; project administration, I.K. All authors have read and agreed to the published version of the manuscript.

Funding: This research received no external funding. However, financial support for the first author was provided by Clarkson University.

Acknowledgments: The authors would like to thank the Ausable River Association, Keene Valley, NY, USA for providing bathymetric and sediment data, as well as insight into the characteristics of the river reach. Additionally, the authors express their gratitude to Ana Luisa Nunes de Alencar Osorio Castañon and Prof. Dirceu Silveira Reis Jr. for sharing the R programming code that was used as a reference to support this study.

Conflicts of Interest: The authors declare no conflict of interest.

Appendix A

For this table, Column 1 (Q USGS) shows the annual maximum mean daily discharges time series at the USGS Gauging station; Column 2 (Q Site) consists of Column 1 multiplied by the drainage factor (0.338) to regionalize the discharges to the study site; Column 3 (Hmax) shows Column 2 associated water surface elevation at the rating curve cross-section; and Column 4 (Year) is the corresponding year.

Table A1. Annual maximum mean daily discharges and associated stages at the rating curve cross-section.

Q USGS (m^3/s)	Q Site (m^3/s)	H_{max} (m)	Year	Q USGS (m^3/s)	Q Site (m^3/s)	H_{max} (m)	Year
196.24	66.40	309.56	1924	81.27	27.50	309.08	1961
194.25	65.73	309.56	1925	81.84	27.69	309.08	1962
129.41	43.79	309.30	1926	110.15	37.27	309.22	1963
138.75	46.95	309.34	1927	92.03	31.14	309.13	1964
107.60	36.41	309.21	1928	40.21	13.61	308.82	1965
104.21	35.26	309.19	1929	78.15	26.45	309.06	1966
156.59	52.99	309.42	1930	100.24	33.92	309.17	1967
77.59	26.25	309.06	1931	114.40	38.71	309.24	1968
113.55	38.42	309.23	1932	136.20	46.09	309.33	1969
144.70	48.96	309.37	1933	94.01	31.81	309.14	1970
65.41	22.13	308.99	1934	70.23	23.76	309.02	1971
66.83	22.61	309.00	1935	155.18	52.51	309.41	1972
201.05	68.03	309.58	1936	110.15	37.27	309.22	1973
147.81	50.02	309.38	1937	89.76	30.37	309.12	1974
221.72	75.03	309.65	1938	71.92	24.34	309.02	1975
85.80	29.03	309.10	1939	88.63	29.99	309.11	1976
112.70	38.14	309.23	1940	138.19	46.76	309.34	1977
105.91	35.84	309.20	1941	92.03	31.14	309.13	1978
101.09	34.21	309.18	1942	186.89	63.24	309.53	1979
100.81	34.11	309.17	1943	106.19	35.93	309.20	1980
96.28	32.58	309.15	1944	108.45	36.70	309.21	1981
141.02	47.72	309.35	1945	147.53	49.92	309.38	1982
53.52	18.11	308.91	1946	171.88	58.16	309.47	1983
146.96	49.73	309.38	1947	150.65	50.98	309.39	1984
232.20	78.57	309.68	1948	66.54	22.52	308.99	1985
123.18	41.68	309.28	1949	74.19	25.10	309.04	1986
96.28	32.58	309.15	1950	159.14	53.85	309.43	1987
146.96	49.73	309.38	1951	163.95	55.48	309.44	1988
106.47	36.03	309.20	1952	110.15	37.27	309.22	1989
163.67	55.38	309.44	1953	104.49	35.36	309.19	1990
127.14	43.02	309.29	1954	98.83	33.44	309.17	1991
111.00	37.56	309.22	1955	113.27	38.33	309.23	1992
118.65	40.15	309.26	1956	159.99	54.14	309.43	1993
186.89	63.24	309.53	1957	119.78	40.53	309.26	1994
180.66	61.13	309.51	1958	104.21	35.26	309.19	1995
121.20	41.01	309.27	1959	131.96	44.65	309.31	1996
125.73	42.54	309.29	1960	175.85	59.50	309.49	1997

References

1. Wohl, E.; Lane, S.N.; Wilcox, A.C. The science and practice of river restoration. *Water Resour. Res.* **2015**, *51*, 5974–5997. [CrossRef]
2. Wohl, E. What should these rivers look like? Historical range of variability and human impacts in the Colorado Front Range, USA. *Earth Surf. Process. Landf.* **2011**, *36*, 1378–1390. [CrossRef]
3. Nardini, A.; Pavan, S. River restoration: Not only for the sake of nature but also for saving money while addressing flood risk. A decision-making framework applied to the Chiese River (Po basin, Italy). *J. Flood Risk Manag.* **2012**, *5*, 111–133. [CrossRef]
4. Araújo, P.V.N.; Amaro, V.E.; Silva, R.M.; Lopes, A.B. Delimitation of flood areas based on a calibrated a DEM and geoprocessing: Case study on the Uruguay River, Itaqui, southern Brazil. *Nat. Hazards Earth Syst. Sci.* **2019**, *19*, 237–250. [CrossRef]
5. Mondragón-Monroy, R.; Honey-Rosés, J. Urban River Restoration and Planning in Latin America: A systematic review. *Univ. Br. Columbia* **2016**, 1–29. [CrossRef]
6. Stephens, T.A.; Bledsoe, B.P. Probabilistic mapping of flood hazards: Depicting uncertainty in streamflow, land use, and geomorphic adjustment. *Anthropocene* **2020**, *29*, 100231. [CrossRef]
7. Teng, J.; Jakeman, A.J.; Vaze, J.; Croke, B.F.W.; Dutta, D.; Kim, S. Flood inundation modelling: A review of methods, recent advances and uncertainty analysis. *Environ. Model. Softw.* **2017**, *90*, 201–216. [CrossRef]
8. Domeneghetti, A.; Vorogushyn, S.; Castellarin, A.; Merz, B.; Brath, A. Probabilistic flood hazard mapping: Effects of uncertain boundary conditions. *Hydrol. Earth Syst. Sci.* **2013**, *17*, 3127–3140. [CrossRef]
9. Highfield, W.E.; Norman, S.A.; Brody, S.D. Examining the 100-Year Floodplain as a Metric of Risk, Loss, and Household Adjustment. *Risk Anal.* **2013**, *33*, 186–191. [CrossRef]
10. Brody, S.D.; Sebastian, A.; Blessing, R.; Bedient, P.B. Case study results from southeast Houston, Texas: Identifying the impacts of residential location on flood risk and loss. *J. Flood Risk Manag.* **2018**, *11*, S110–S120. [CrossRef]
11. Tyler, J.; Sadiq, A.A.; Noonan, D.S. A review of the community flood risk management literature in the USA: lessons for improving community resilience to floods. *Nat. Hazards* **2019**, *96*, 1223–1248. [CrossRef]
12. Freeze, R.A.; James, B.; Massmann, J.; Sperling, T.; Smith, L. Hydrogeological Decision Analysis: 4. The Concept of Data Worth and Its Use in the Development of Site Investigation Strategies. *Ground Water* **1992**, *30*, 574–588. [CrossRef]
13. Merz, B.; Thieken, A.H. Separating natural and epistemic uncertainty in flood frequency analysis. *J. Hydrol.* **2005**, *309*, 114–132. [CrossRef]
14. Apel, H.; Aronica, G.T.; Kreibich, H.; Thieken, A.H. Flood risk analyses—How detailed do we need to be? *Nat. Hazards* **2009**, *49*, 79–98. [CrossRef]
15. Liu, Y.; Gupta, H.V. Uncertainty in hydrologic modeling: Toward an integrated data assimilation framework. *Water Resour. Res.* **2007**, *43*. [CrossRef]
16. Bates, P.D.; Horritt, M.S.; Aronica, G.; Beven, K. Bayesian updating of flood inundation likelihoods conditioned on flood extent data. *Hydrol. Process.* **2004**, *18*, 3347–3370. [CrossRef]
17. Abily, M.; Bertrand, N.; Delestre, O.; Gourbesville, P.; Duluc, C.M. Spatial Global Sensitivity Analysis of High Resolution classified topographic data use in 2D urban flood modelling. *Environ. Model. Softw.* **2016**, *77*, 183–195. [CrossRef]
18. Savage, J.T.S.; Bates, P.; Freer, J.; Neal, J.; Aronica, G. When does spatial resolution become spurious in probabilistic flood inundation predictions? *Hydrol. Process.* **2016**, *30*, 2014–2032. [CrossRef]
19. Stephens, E.M.; Bates, P.D.; Freer, J.E.; Mason, D.C. The impact of uncertainty in satellite data on the assessment of flood inundation models. *J. Hydrol.* **2012**, *414–415*, 162–173. [CrossRef]
20. Neal, J.; Keef, C.; Bates, P.; Beven, K.; Leedal, D. Probabilistic flood risk mapping including spatial dependence. *Hydrol. Process.* **2013**, *27*, 1349–1363. [CrossRef]
21. Vaze, J.; Viney, N.; Stenson, M.; Renzullo, L.; van Dijk, A.; Dutta, D.; Crosbie, R.; Lerat, J.; Penton, D.; Vleeshouwer, J.; et al. The Australian water resource assessment modelling system (AWRA). In Proceedings of the 20th International Congress on Modelling and Simulation, MODSIM 2013, Adelaide, Australia, 1–6 December 2013; pp. 3015–3021.
22. Beven, K. Facets of uncertainty: Epistemic uncertainty, non-stationarity, likelihood, hypothesis testing, and communication. *Hydrol. Sci. J.* **2016**, *61*, 1652–1665. [CrossRef]

23. Osorio, A.L.N.A.; Rampinelli, C.G.; Reis, D.S. A Bayesian Approach to Incorporate Imprecise Information on Hydraulic Knowledge in a River Reach and Assess Prediction Uncertainties in Streamflow Data. In Proceedings of the World Environmental and Water Resources Congress, Minneapolis, Minnesota, 3–7 June 2018.

24. Osorio, A.L.N.A. Modelo Bayesiano Completo para análise de frequência de cheias com incorporação do conhecimento hidráulico na modelagem das incertezas na curva- chave. In *Dissertação de Mestrado em Tecnologia Ambiental e Recursos Hídricos*; Publicação PTARH.DM-196/17; Departamento de Engenharia Civil e Ambiental, Universidade de Brasília: Brasília, Brazil, 2017.

25. Le Coz, J.; Renard, B.; Bonnifait, L.; Branger, F.; Le Boursicaud, R. Combining hydraulic knowledge and uncertain gaugings in the estimation of hydrometric rating curves: A Bayesian approach. *J. Hydrol.* **2014**, *509*, 573–587. [CrossRef]

26. Pappenberger, F.; Matgen, P.; Beven, K.J.; Henry, J.B.; Pfister, L.; Fraipont, P. Influence of uncertain boundary conditions and model structure on flood inundation predictions. *Adv. Water Resour.* **2006**, *29*, 1430–1449. [CrossRef]

27. Liu, Z.; Merwade, V.; Jafarzadegan, K. Investigating the role of model structure and surface roughness in generating flood inundation extents using one- and two-dimensional hydraulic models. *J. Flood Risk Manag.* **2019**, *12*. [CrossRef]

28. Dimitriadis, P.; Tegos, A.; Oikonomou, A.; Pagana, V.; Koukouvinos, A.; Mamassis, N.; Koutsoyiannis, D.; Efstratiadis, A. Comparative evaluation of 1D and quasi-2D hydraulic models based on benchmark and real-world applications for uncertainty assessment in flood mapping. *J. Hydrol.* **2016**, *534*, 478–492. [CrossRef]

29. Georgakakos, K.P.; Seo, D.J.; Gupta, H.; Schaake, J.; Butts, M.B. Towards the characterization of streamflow simulation uncertainty through multimodel ensembles. *J. Hydrol.* **2004**, *298*, 222–241. [CrossRef]

30. Zarzar, C.M.; Hosseiny, H.; Siddique, R.; Gomez, M.; Smith, V.; Mejia, A.; Dyer, J. A Hydraulic MultiModel Ensemble Framework for Visualizing Flood Inundation Uncertainty. *J. Am. Water Resour. Assoc.* **2018**, *54*, 807–819. [CrossRef]

31. Horritt, M.S. A methodology for the validation of uncertain flood inundation models. *J. Hydrol.* **2006**, *326*, 153–165. [CrossRef]

32. Papaioannou, G.; Vasiliades, L.; Loukas, A.; Aronica, G.T. Probabilistic flood inundation mapping at ungauged streams due to roughness coefficient uncertainty in hydraulic modelling. *Adv. Geosci.* **2017**, *44*, 23–34. [CrossRef]

33. Pappenberger, F.; Beven, K.; Horritt, M.; Blazkova, S. Uncertainty in the calibration of effective roughness parameters in HEC-RAS using inundation and downstream level observations. *J. Hydrol.* **2005**, *302*, 46–69. [CrossRef]

34. Werner, M.G.F.; Hunter, N.M.; Bates, P.D. Identifiability of distributed floodplain roughness values in flood extent estimation. *J. Hydrol.* **2005**, *314*, 139–157. [CrossRef]

35. Call, B.C.; Belmont, P.; Schmidt, J.C.; Wilcock, P.R. Changes in floodplain inundation under nonstationary hydrology for an adjustable, alluvial river channel. *Water Resour. Res.* **2017**, *53*, 3811–3834. [CrossRef]

36. Aronica, G.T.; Franza, F.; Bates, P.D.; Neal, J.C. Probabilistic evaluation of flood hazard in urban areas using Monte Carlo simulation. *Hydrol. Process.* **2012**, *26*, 3962–3972. [CrossRef]

37. Wu, X.Z. Probabilistic solution of floodplain inundation equation. *Stoch. Environ. Res. Risk Assess.* **2016**, *30*, 47–58. [CrossRef]

38. Pedrozo-Acuña, A.; Rodríguez-Rincón, J.P.; Arganis-Juárez, M.; Domínguez-Mora, R.; González Villareal, F.J. Estimation of probabilistic flood inundation maps for an extreme event: Pánuco River, México. *J. Flood Risk Manag.* **2015**, *8*, 177–192. [CrossRef]

39. Di Baldassarre, G.; Schumann, G.; Bates, P.D.; Freer, J.E.; Beven, K.J. Flood-plain mapping: a critical discussion of deterministic and probabilistic approaches. *Hydrol. Sci. J.* **2010**, *55*, 364–376. [CrossRef]

40. Romanowicz, R.; Beven, K. Estimation of flood inundation probabilities as conditioned on event inundation maps. *Water Resour. Res.* **2003**, *39*, 1–12. [CrossRef]

41. Reitan, T.; Petersen-Øverleir, A. Bayesian power-law regression with a location parameter, with applications for construction of discharge rating curves. *Stoch. Environ. Res. Risk Assess.* **2008**, *22*, 351–365. [CrossRef]

42. Reitan, T.; Petersen-Øverleir, A. Bayesian methods for estimating multi-segment discharge rating curves. *Stoch. Environ. Res. Risk Assess.* **2009**, *23*, 627–642. [CrossRef]

43. Coles, S.G.; Tawn, J.A. Modelling Extremes of the Areal Rainfall Process. *J. Royal Stat. Soc.* **1996**, *58*, 329–347. [CrossRef]

44. Kuczera, G. Comprehensive at-site flood frequency analysis using Monte Carlo Bayesian inference. *Water Resour. Res.* **1999**, *35*, 1551–1557. [CrossRef]

45. Reis, D.S.; Stedinger, J.R. Bayesian MCMC flood frequency analysis with historical information. *J. Hydrol.* **2005**, *313*, 97–116. [CrossRef]

46. O'Connell, D.R.H. Nonparametric Bayesian flood frequency estimation. *J. Hydrol.* **2005**, *313*, 79–96. [CrossRef]

47. Neppel, L.; Renard, B.; Lang, M.; Ayral, P.A.; Coeur, D.; Gaume, E.; Jacob, N.; Payrastre, O.; Pobanz, K.; Vinet, F. Flood frequency analysis using historical data: Accounting for random and systematic errors. *Hydrol. Sci. J.* **2010**, *55*, 192–208. [CrossRef]

48. Paiva, R.C.D.; Collischonn, W.; Tucci, C.E.M. Large scale hydrologic and hydrodynamic modeling using limited data and a GIS based approach. *J. Hydrol.* **2011**, *406*, 170–181. [CrossRef]

49. Winter, B.; Schneeberger, K.; Huttenlau, M.; Stötter, J. Sources of uncertainty in a probabilistic flood risk model. *Nat. Hazards* **2018**, *91*, 431–446. [CrossRef]

50. Koivumäki, L.; Alho, P.; Lotsari, E.; Käyhkö, J.; Saari, A.; Hyyppä, H. Uncertainties in flood risk mapping: A case study on estimating building damages for a river flood in Finland. *J. Flood Risk Manag.* **2010**, *3*, 166–183. [CrossRef]

51. GIS.NY.GOV. New York Government GIS Data Set. Available online: http://gis.ny.gov/gisdata/inventories/details.cfm?DSID=1336 (accessed on 28 March 2020).

52. USACE HEC-GeoRAS 10.2. Available online: https://www.hec.usace.army.mil/software/hec-georas/ (accessed on 20 March 2020).

53. US Army Corps of Engineers Hydrologic Engineering Center—River Analysis System-HEC-RAS 5.0.7. Available online: https://www.hec.usace.army.mil/software/hec-ras/ (accessed on 20 December 2019).

54. Eberhart, R.; Kennedy, J. A new optimizer using particle swarm theory. *Proc. Int. Symp. Micro Mach. Hum. Sci.* **1995**, 39–43. [CrossRef]

55. Griffis, V.W.; Stedinger, J.R. Log-pearson type distribution and its application in flood frequency analysis. I: Distribution characteristics. *J. Hydrol. Eng.* **2007**, *12*, 482–491. [CrossRef]

56. Hosking, J.R.M. L-Moments; R Package, Version 2.8. Available online: https://cran.r-project.org/web/packages/Lmoments (accessed on 22 December 2019).

57. Vrugt, J.A.; ter Braak, C.J.F.; Clark, M.P.; Hyman, J.M.; Robinson, B.A. Treatment of input uncertainty in hydrologic modeling: Doing hydrology backward with Markov chain Monte Carlo simulation. *Water Resour. Res.* **2008**, *44*, 1–15. [CrossRef]

58. Vrugt, J.A.; Ter Braak, C.J.F.; Diks, C.G.H.; Robinson, B.A.; Hyman, J.M.; Higdon, D. Accelerating Markov chain Monte Carlo simulation by differential evolution with self-adaptive randomized subspace sampling. *Int. J. Nonlinear Sci. Numer. Simul.* **2009**, *10*, 273–290. [CrossRef]

59. Vrugt, J.A. Markov chain Monte Carlo simulation using the DREAM software package: Theory, concepts, and MATLAB implementation. *Environ. Model. Softw.* **2016**, *75*, 273–316. [CrossRef]

60. Guillaume, J.; Andrews, F. DiffeRentialEvolution Adaptive Metropolis; R Packge Version 0.4-2. Available online: https://rdrr.io/rforge/dream/man/dream.html (accessed on 15 December 2019).

Article

The Effect of Habitat Structure Boulder Spacing on Near-Bed Shear Stress and Turbulent Events in a Gravel Bed Channel

Amir Golpira, Fengbin Huang and Abul B.M. Baki *

Civil and Environmental Engineering Department, Clarkson University, Potsdam, NY 13699, USA; golpira@clarkson.edu (A.G.); fehuang@clarkson.edu (F.H.)
* Correspondence: abaki@clarkson.edu; Tel.: +1-315-268-4156

Received: 10 April 2020; Accepted: 14 May 2020; Published: 16 May 2020

Abstract: This study experimentally investigated the effect of boulder spacing and boulder submergence ratio on the near-bed shear stress in a single array of boulders in a gravel bed open channel flume. An acoustic Doppler velocimeter (ADV) was used to measure the instantaneous three-dimensional velocity components. Four methods of estimating near-bed shear stress were compared. The results suggested a significant effect of boulder spacing and boulder submergence ratio on the near-bed shear stress estimations and their spatial distributions. It was found that at unsubmerged condition, the turbulent kinetic energy (TKE) and modified TKE methods can be used interchangeably to estimate the near-bed shear stress. At both submerged and unsubmerged conditions, the Reynolds method performed differently from the other point-methods. Moreover, a quadrant analysis was performed to examine the turbulent events and their contribution to the near-bed Reynolds shear stress with the effect of boulder spacing. Generally, the burst events (ejections and sweeps) were reduced in the presence of boulders. This study may improve the understanding of the effect of the boulder spacing and boulder submergence ratio on the near-bed shear stress estimations of stream restoration practices.

Keywords: boulder spacing; submergence ratio; near-bed shear stress; Reynolds shear stress; turbulent events; stream restoration

---

1. Introduction

Bed shear stress plays a determinant role in the incipient motion of sediment. The bed shear stress has been the focus of many studies in both laboratory flumes and rivers. The majority of the existing studies were conducted in the flumes due to the controlled flow and bed conditions. However, the field investigation on the measuring of the bed shear stress is relatively limited. The direct measurements on the bed shear stress were often conducted by using the shear plate [1–5]. The indirect measurement of the bed shear stress depends on the velocity in the inner layer, lower 20% of the flow depth [6]. The common approaches to calculate the bed shear stress include, but are not limited to, the reach-averaged bed shear [7], logarithmic law of the wall [8], drag force [9], Reynolds stress [10], and turbulent kinetic energy (TKE) methods [11]. In a laboratory study, the bed shear stress estimated from several methods were compared for a simple boundary layer and a complex flow [12]. It was found that the Reynolds and TKE methods showed the most appropriate results for a simple boundary layer and a complex flow, respectively [12]. In another study, Reynolds stress was calculated by using the double-averaged methodology based on Reynolds temporal and spatial averaging [13]. The study concluded that the maximum Reynolds stress could better represent the bed shear stress than the linearly extrapolated Reynolds stress profile [13]. In the field measurements, acoustic Doppler current profiler (ADCP) has been commonly used for the measurement of velocity profiles to provide estimates of shear stress [8,14,15]. The bed shear stress was estimated from the logarithmic law of the wall and depth-averaged velocity in the Lower Fraser River, Canada, by using obtained data from an ADCP [15]. The field investigation indicated that using depth-averaged velocity paired with the zero-velocity height based on bed grain size gives more reliable results on the shear stress than using the logarithmic law of the wall [8].

The in-stream habitat structure boulder is one of the most favorable measures for river/stream restoration projects. The Natural Resources Conservation Service (NRCS 1998) demonstrated different utilization of boulders for channel restoration designs [16]. The boulders in the channel enhance the flow heterogeneity and modify the sediment transport process such that fish habitat, channel bed/bank stabilities, and water quality are improved [17]. Five stream restoration projects have been performed by using boulders to shed light on the benefits of adding boulders for local hydraulic conditions, thermal conditions, and total transient storage [18]. Although stream/river restoration has become a worldwide phenomenon and booming enterprise [19], our knowledge of the flow hydrodynamics associated with boulders is very limited [17]. The understanding of the impact that in-stream boulders bring about on the bed shear stress and fish habitat requires systematic investigation. The quantitative description of how the boulder spacing and boulder submergence ratios affect the bed shear stress, which is the key factor for bed sediment transport, would be useful for the channel restoration design.

The few existing studies measuring the bed shear stress with a disruption such as boulders shed light on its impact on flow configurations [20–26] and sediment transport [23,27–29]. Boulders have been documented to create spatial variability in bed shear stress [28,30–32]; and the presence of boulders decreases the sediment transport capacity [33,34]. The flume measurements over the gravel bed indicated that near-bed flow plays a key role in the flow structure fed into the outflow, and the increasing Reynolds number promotes large-scale flow structures [26]. The experimental studies demonstrated that a single boulder or an array of boulders affect the mean and turbulent flow characteristics [22,24,30]. It has been reported that the presence of boulder deaccelerates flow and leads to a significant deviation of streamwise mean velocity from the classic logarithmic law [22]. A boulder array generally leads to a decreased total bedload transport rate due to the reduction of available grain shear stress caused by the boulder form drag [24]. Several bed load equations have been proposed for boulder-bed channels considering the effects of sediment availability, boulder protrusion, bed roughness, the variability of shear stress, and boulder spacing [28,29]. Moreover, the boulder submergence ratio can be considered as a key parameter that affects the mean and turbulent flow characteristics as well as the local sediment transport patterns [23,30,35]. For a partially submerged flow, an array of boulders may reduce 5–20 times the bedload rate compared to conditions without boulders [36]. It has been reported that under the unsubmerged and fully submerged conditions, the deposition of the sediment may significantly vary downstream and upstream of boulders [35].

From the turbulence aspect, many recent studies have primarily focused on the effects of coherent flow structures (i.e., turbulent bursting events) on the flow field and sediment transport with or without the presence of large roughness elements such as boulders [28,37–42]. Turbulent events may affect bedload and suspended load sediment movement, as well as fish behavior and swimming performance, which can be of great importance in the design of natural fish habitat structures [43]. Turbulent events as a cycle of ejections, sweeps, outward, and inward interactions contribute to the Reynolds shear stress. Ejections transfer slow-moving fluid into the outer layer while sweeps bring high-momentum fluid into the bed region [44]. Ejection events have been reported important in sediment entrainment into the water column, while sweeps are associated with bedload transport [40]. Quadrant analysis has been widely used to highlight the significance of turbulent events, specifically around large roughness elements [21,36,38,45]. However, there is still a lack of knowledge on the turbulent events variation and their contribution to the Reynolds shear stress under different boulder spacing and submergence ratios.

This study is motivated by the need to improve our understanding of how the boulder spatial distribution, which is frequently employed in stream restoration practices, affects the near-bed shear stress and its estimations. This study experimentally investigates the effects of boulder spacing and boulder submergence ratio on the near-bed shear stress in a flume with a gravel bed. Four different methods were used to estimate the reach-averaged near-bed shear stress and its spatial distribution. Besides, the performance of methods was compared with each other. Moreover, dominant turbulent events and their contribution to the near-bed Reynolds shear stress was studied through a quadrant analysis to deepen the understanding of near-bed turbulence under varying boulder spacing and submergence ratio.

2. Materials and Methods

2.1. Experimental Setup

Experiments were conducted in the Ecohydraulics Flume, a 13.0 m long, 0.96 m wide, water-recirculating flume with a gentle slope (S_0) of 0.5% located in the Water Resources Engineering Laboratory at Clarkson University. Measurements were taken in a section with a length of 2.40 m located 3.80 m downstream from the flume entrance to ensure that the turbulent flow is hydraulically fully developed. Figure 1 shows a scheme of the flume. Natural semi-spherical shape boulders with an equivalent diameter (D) of 12.50 cm were used as large roughness elements. Experiments were carried out over a gravel bed with $d_{90} = 27.50$ mm, $d_{50} = 17.21$ mm, and $d_{10} = 9.37$ mm, where d_p refers to the grain size for which p% of particle sizes are finer. The critical bed shear stress for the median sediment size (d_{50}) was calculated according to [46] formula, and it was 12.38 N/m^2, at which incipient sediment motion takes place. As projected, no sediment movement was observed at any set of experiments.

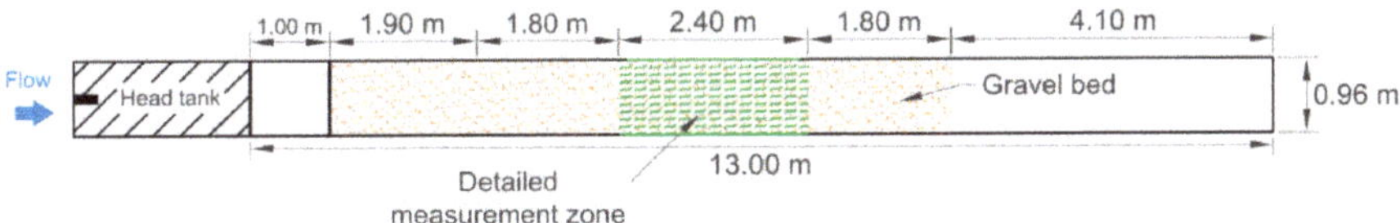

Figure 1. A schematic plan view of the flume. The hatched area shows the detailed measurement zone.

Four series of experiments (hereafter called scenarios) were designed with boulder-to-boulder spacing (from center to center) as a varying parameter to represent different flow regimes. Covered boulder-to-boulder spacing was infinity (no boulder), 10D (large spacing), 5D (medium spacing), and 2D (small spacing). The large and medium boulder spacing referred to an isolated flow regime while the small boulder spacing corresponded to a wake-interference or even a skimming flow regime [38]. Table 1 lists four scenarios and the hydraulic parameters associated with each scenario. Each scenario was conducted at 60 L/s and 100 L/s flow rates. Boulders were unsubmerged at 60 L/s with a submergence ratio (H/D) between 0.73 and 0.78, while at 100 L/s, boulders were fully submerged with a submergence ratio between 1.25 and 1.32 (Table 1). Figure 2 illustrates the measurement points and position of boulders in each scenario in the measurement zone. For the boulder-scenarios, the boulders were placed along the centerline of the channel in a single straight array. First, the central boulder at point X4Y3 was placed; afterward, the other boulders were arranged relative to the central boulder. A total of 35 measurement points were selected covering an area from −4D to 4D (relative to the central boulder in the measurement zone) in the streamwise direction (x) and from −2D to 2D in the spanwise direction (y) of the central boulder. The distance of all the measurement points from the walls was greater than 10 cm to minimize the wall effects [28]. It is worth mentioning that 35 measurement points were measured for the no-boulder scenario; however, for the boulder-scenarios, some of the measurement points in the centerline of the boulders were blocked by the boulders. In addition, Figure 2 shows that the points immediately upstream and downstream of a boulder have an equal distance from the boulder; however, due to the irregularity in the boulder shape, these distances were not perfectly the same.

Table 1. Characteristics of the experiments.

Scenario	Flow Rate (L/s)	Boulder Spacing	Reach-Averaged Flow Depth (m)	Submergence Ratio
S1-60		Infinity	0.082	-
S2-60		10D	0.092	0.73
S3-60	60	5D	0.098	0.78
S4-60		2D	0.093	0.74
S1-100		Infinity	0.151	-
S2-100		10D	0.157	1.25
S3-100	100	5D	0.165	1.32
S4-100		2D	0.161	1.29

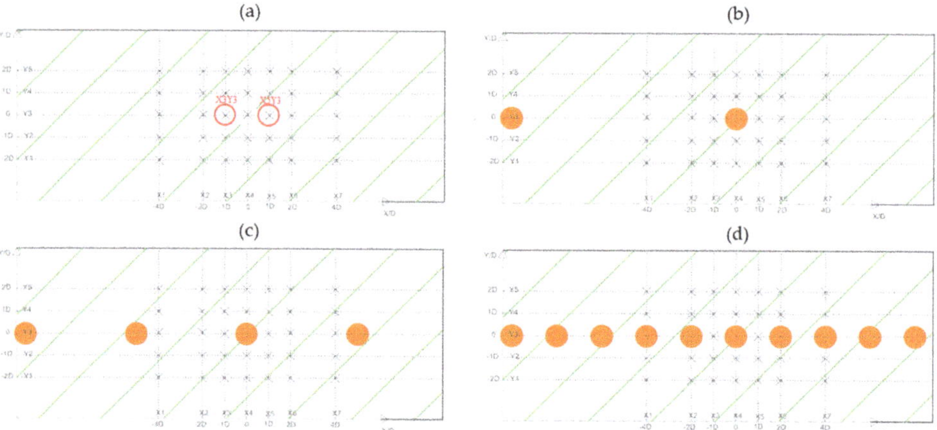

Figure 2. A schematic plan of boulder arrangement in the measurement zone (hatched area) for (**a**) no boulder, (**b**) 10D boulder spacing, (**c**) 5D boulder spacing, and (**d**) 2D boulder spacing. Cross marks show measurement points for each setup, and large circles show the position of idealized (assumed fully spherical) boulders. Two points X3Y3 and X5Y3 have been marked in (**a**) as common measurement points in the centerline of boulders in all scenarios.

The velocity measurements were taken in a relative depth (z/H) around 0.20–0.25, at the beginning of the outer region of flow [47]. Although measurements closer to the bed are preferable, the use of such a relative depth to capture the near-bed flow characteristics has been justified in a similar study [42]. Additionally, acoustic Doppler velocimeter (ADV) measurement about 3 cm from the bed to acquire higher levels of accuracy within the turbulent flows has been recommended [48], a threshold that led to a z/H around 0.20–0.30 in this study.

2.2. Data Filtering

Three-dimensional instantaneous velocities were collected using a Vectrino probe (Nortek ADV) at a sampling rate of 25 Hz for three minutes to ensure statistical significance. Aliased points were despiked using a phase-space threshold filter proposed by [49]. This filter can produce clean signals in highly turbulent flows [50]. The manufacturer (Nortek) recommends COR ≥ 70% and SNR ≥ 15 dB for reliable turbulence measurements. However, in the case of high SNR values—a valid condition in this study with an average SNR value of greater than 50 dB—COR values less than 70% can still provide reliable data [51]. As suggested by [52], COR values less than 70% in the turbulent flows are not necessarily an indicator of the low quality. They applied COR ≥ 50% and SNR ≥ 10 dB to investigate the turbulent characteristics around a cluster microform [52]. It also has been shown that for the flow over a rough bed, if at least 70% of data remain after applying a filtering scheme, the Reynolds stress could be measured with a minimum COR of 40% [53]. Following these suggestions, a filtering scheme with COR ≤ 65% and SNR ≤ 20 dB was applied to filter out poor quality data. For the points in location X5Y3 (marked in Figure 2a) at 100 L/s flow rate, a COR of 55% was applied due to the higher turbulence in this point and a significant loss of data in case of applying the main filtering scheme. Such criteria resulted in the remaining 77.1% of data indicating an acceptable portion of data for further analysis. Another possible source of error in ADV data is the Doppler noise, which may affect the turbulence parameters [54]. When the noise level is low, it is expected that the velocity spectra follow the Kolmogorov's −5/3 law by exhibiting a −5/3 slope in the inertial subrange of the velocity spectra [54]. Therefore, as an additional check (also suggested by [50] for highly turbulent flows), the velocity spectra of the measured time series were visually inspected to remove noisy time series with a flat or negative slope in the inertial subrange. None of the points were rejected based on this criterion.

2.3. Calculation of Bed Shear Stress

The first main method to calculate the bed shear stress is the reach-averaged method as below [12]:

$$\tau_0 = \rho g R S_0,\tag{1}$$

where τ_0 is bed shear stress, ρ is water density, g is gravitational acceleration, and R is channel hydraulic radius. Although this method gives a rough estimation of bed shear stress, its performance is questionable, specifically in the presence of boulders, because it is not able to estimate the local variation of the bed shear stress [12].

Bed shear stress can be estimated directly through the Reynolds shear stress [12]:

$$\tau_0 = -\rho \overline{u'w'},\tag{2}$$

where u' and w' are the instantaneous velocity fluctuations of the streamwise and vertical components, respectively. The bar sign over $u'w'$ indicates a time-averaged value. This method is sensitive to the deviations from a two-dimensional uniform flow [12].

Another method is the turbulence kinetic energy (TKE) method as below [12]:

$$\tau_0 = C_1\left(0.5\rho\left(\overline{u'^2} + \overline{v'^2} + \overline{w'^2}\right)\right),\tag{3}$$

where v' is the instantaneous velocity fluctuation of the spanwise component and C_1 is a constant equal to 0.19 [11].

The modified TKE method only uses the vertical component of the velocity fluctuations due to the smaller noise of the measurement instrument in this direction [11]. Therefore, it can be written as below [12]:

$$\tau_0 = C_2\rho\overline{w'^2},\tag{4}$$

where C_2 is a constant equal to 0.9 [11].

Equations (1)–(4) were used to calculate the near-bed shear stress for all the conducted scenarios in this study. Equation (1) only requires the reach hydraulic data while Equations (2)–(4) use the measured data from a single point and are referred to as point-methods in this study.

2.4. Quadrant Analysis

In a quadrant analysis, the velocity fluctuations are decomposed to four quadrants based on their signs [38]:

- Quadrant 1: outward interactions ($u' > 0$, $w' > 0$)
- Quadrant 2: ejection events ($u' < 0$, $w' > 0$)

- Quadrant 3: inward interactions ($u' < 0$, $w' < 0$)
- Quadrant 4: sweep events ($u' > 0$, $w' < 0$).

To remove insignificant events, a hole region in the (u', w') plane can be defined to exclude points within this region. Hole size, H, is a parameter to define the expansion of the hole region as below [55]:

$$H = |u'w'| / |\overline{u'w'}|,$$ (5)

Clearly, a larger hole size means the remaining of more significant events. Many studies have selected a hole size greater than one to distinguish significant events [36,56–59]. Therefore, in this study, a hole size of 2 was selected to distinguish events with a higher magnitude. Points X3Y3 and X5Y3, the common points in all the scenarios (marked in Figure 2a), were selected for a quadrant analysis. The joint frequency distribution of the normalized velocity fluctuations was found for each point. To find the contribution of each quadrant to the Reynolds shear stress, the stress fraction parameter of each quadrant, $S_{i,H}^f$ was found as below [38]:

$$S_{i,H}^f = \frac{1}{T} \int_0^T u'w' I_{i,H}(u',w')dt$$ (6)

where i shows the quadrant number ($i = 1, 2, 3, 4$), T is the total duration of the measurement and I is a conditional function to detect events in each quadrant for a given hole number as below [38]:

$$I_{i,H}(u',w') = \begin{cases} 1, & \text{if } (u',w') \text{ is in quadrant } i \text{ and } |u'w'| \geq H|\overline{u'w'}| \\ 0, & \text{otherwise} \end{cases}$$ (7)

The total portion of each turbulent event, $P_{i,H}$, as a representative of the frequency of turbulent events can be determined as below [38]:

$$P_{i,H} = \int_0^T \frac{\int_0^T |u'w'| I_{i,H}(u',w')dt}{\int_0^T |u'w'|dt}$$ (8)

3. Results and Discussion

3.1. Near-Bed Shear Stress

Equations (1)–(4) were used to calculate the reach-averaged near-bed shear stress for each scenario. Equation (1) gives a single value as the reach-averaged shear stress, but for the point-methods, the average of shear stresses at all the measurement points was calculated as the reach-averaged near-bed shear stress. Figure 3 shows the reach-averaged near-bed shear stress for each method and scenario. The near-bed shear stress estimations using the reach-averaged method (Equation (1)) before adding boulders were 3.46 and 5.63 N/m^2 for the S1-60 and the S1-100 scenarios, respectively. These estimations were higher than the shear stress estimations using the point-methods for no-boulder scenarios. The maximum estimation of the average near-bed shear stress was significantly lower than the calculated critical bed shear stress (12.38 N/m^2). This was expected because no sediment movement was observed in all scenarios. For the no-boulder scenarios (S1-60 and S1-100), the TKE and modified TKE methods resulted in a close estimation for the average near-bed shear stress (a maximum difference of 10% for the S1-60) while estimations using the Reynolds method were slightly higher than the TKE and modified TKE methods. Close estimations using these point-methods were reported by [12]. For scenarios with the largest boulder spacing at 60 L/s (S2-60), the TKE method resulted in a higher bed shear stress while the Reynolds and modified TKE resulted in close estimations. At 100 L/s scenario (S2-100), the Reynolds, TKE, and modified TKE methods showed very similar performance (with a maximum 20% difference between the modified TKE and TKE methods). At 60 L/s, for both medium

(S3-60) and small boulder spacing (S4-60) scenarios, both TKE and modified TKE methods showed a very similar performance while the Reynolds method showed a significantly lower near-bed shear stress. The average near-bed shear stress using the Reynolds method was about 200% and 70% less than estimations of other point-methods for the S3-60 and S4-60, respectively. At 100 L/s, for both S3-100 and S4-100, the modified TKE method led to the highest estimates, while the Reynolds and TKE methods showed similar results. Generally, it can be said that at submerged condition (100 L/s), the difference between estimations using the point-methods was not significant for each scenario, although the modified TKE led to slightly higher estimations for boulder-scenarios. At unsubmerged condition (60 L/s), for both no-boulder and large spacing scenarios, the difference between the point-methods was not significant; however, for the medium and small spacing scenarios, the near-bed shear estimations using the Reynolds method was significantly smaller.

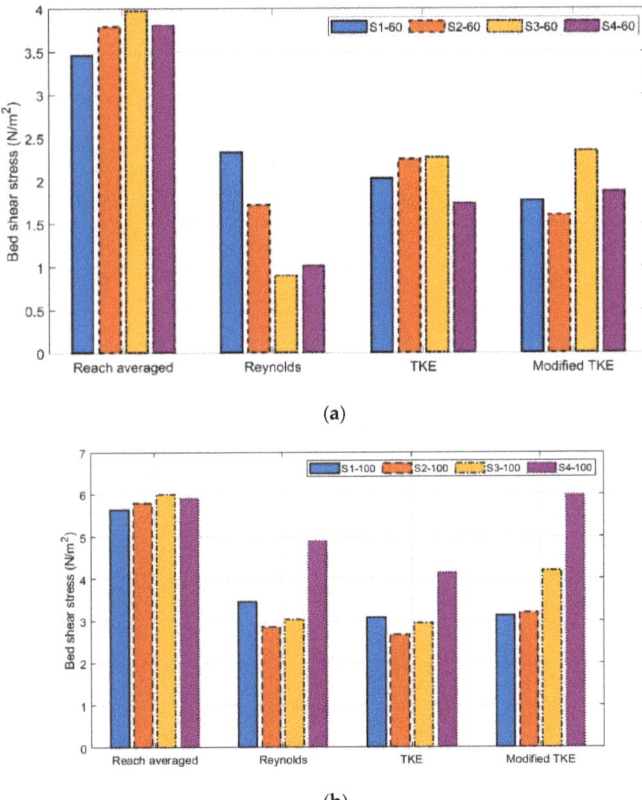

Figure 3. Averaged bed shear stress resulted from different calculation methods for different boulder spacing at (**a**) 60 L/s flow rate, (**b**) 100 L/s flow rate.

The effect of adding boulders and boulder spacing can also be interpreted in Figure 3. The reach-averaged method is directly related to the average flow depth (Equation (1)), and a higher average flow depth resulted in a higher bed shear stress. After adding boulders, the average flow depth slightly increased, and the reach-averaged bed shear stresses slightly increased too. However, for the S4-60 and S4-100 scenarios, the flow depth, and subsequently, the reach-averaged bed shear stress slightly decreased. For the Reynolds method at 60 L/s scenarios, it can be seen that the bed shear stress gradually decreased by decreasing boulder spacing—except for the S4-60 scenario, in which the bed shear stress slightly increased relative to the previous scenario at S3-60. The near-bed shear stress variation for different boulder spacing at 60 L/s using the TKE and modified TKE methods was not similar to the Reynolds method. For the TKE method, the near-bed shear stress slightly increased for the large boulder spacing scenario (S2-60) and remained almost constant for the medium spacing scenario (S3-60); then, it reached the minimum for the small spacing scenario (S4-60). For the modified TKE method, the highest near-bed shear stress belonged to the medium spacing scenario (S3-60), while the other scenarios had a small difference with each other (less than 15%). At 100 L/s, the small spacing scenario (S4-100) experienced the highest near-bed shear stress resulting from all estimation methods. For the Reynolds and TKE methods, the near-bed shear stress decreased for the large spacing scenario (S2-100) and slightly increased for the medium spacing scenario (S3-100); then, it sharply increased for the scenario with the smallest boulder spacing (S4-100). For the modified TKE method, the near-bed shear stress gradually increased after adding boulders and then decreasing boulder spacing; the sharpest increase occurred for the small spacing scenario (S4-100). This trend is in agreement with the findings of [38] in which the total shear stress increased gradually by increasing the number of boulders. In the following sections, a more detailed discussion on the variation of near-bed shear stress for different boulder spacing and estimation methods can be found.

3.2. Relative Performance of the Reynolds, TKE, and Modified TKE Methods

To compare the estimates of Reynolds, TKE, and modified TKE methods in a more compelling way, the near-bed shear stresses from each method for all the points were plotted against each other. Figure 4 shows the near-bed shear stress estimated using the TKE and modified TKE methods for all scenarios. Here, the goal was to understand the similarity between the near-bed shear stress estimated from the point-methods. Therefore, the root mean square error (RMSE) was selected to shed light on the similarity of the results calculated by different methods. As shown in Figure 4a, at 60 L/s scenarios, the RMSE for all the scenarios varied between 0.36 and 0.84 N/m^2 indicating a reasonable similarity between results even after adding boulders and then decreasing the boulder spacing. At 100 L/s (Figure 4b), before adding boulders (S1-100), a low RMSE = 0.18 N/m^2 indicated a good agreement between the TKE and modified TKE methods. After adding boulders, the RMSE significantly increased, and this increase was intensified after decreasing the boulder spacing. Generally, the modified TKE method resulted in higher values of the near-bed shear stress in comparison with the TKE method for the boulder-scenarios (S2-100, S3-100, and S4-100). This difference can be related to the values of C_1 and C_2 constants that still are required to be found for the natural streams [11,12].

Figure 5 shows plots for comparing estimations using the Reynolds shear stress method versus the TKE and modified TKE methods. At 60 L/s and 100 L/s, before adding boulders (S1-60 and S1-100), both TKE and modified TKE methods showed a reasonable agreement with the Reynolds method (At 60 L/s, RMSE = 0.63 and 0.73 N/m^2 for the TKE and modified TKE, respectively. At 100 L/s, RMSE = 0.48 and 0.43 N/m^2 for the TKE and modified TKE, respectively). It was reported that the modified TKE and Reynolds method have a relatively good agreement over a Plexiglas and sand bed [12]. However, estimations using the TKE and modified TKE methods were mostly lower than the Reynolds method, but at a lower range of shear stresses, the values were close to 1:1 line. This behavior in a range of high shear stress is in agreement with [12]; however, for the lower shear stresses, similar studies found systematically higher values from the TKE and modified TKE methods [12,40]. For the large boulder spacing at 60 L/s (S2-60), the agreement between the Reynolds and both TKE and modified TKE methods slightly decreased (RMSE = 0.87 and 0.95 N/m^2 for the TKE and modified TKE, respectively). By decreasing boulder spacing (S3-60 and S4-60), the RMSE increased to higher values indicating a lower similarity between the methods. A reason for the poor performance can be the presence of negative values in the Reynolds method, while values using two other methods are always positive. After adding boulders, it can be found that both TKE and modified TKE methods led to higher estimations than the Reynolds method (except for the modified TKE method at the large boulder spacing). This difference became more significant for the medium and small boulder spacing scenarios.

At 100 L/s for the boulder-scenarios, the RMSE of the Reynolds and TKE methods varied between 1.16 and 1.98 N/m^2, and the RMSE of the Reynolds and modified TKE methods varied between 1.30 to 2.67 N/m^2. The RMSE values for both TKE and modified TKE methods reached their maximum (lower similarity) for the small boulder spacing (S4-100). However, some extreme values resulted from the TKE and modified TKE methods might be the reason behind higher RMSE values. After adding boulders, the TKE method estimates generally were smaller than the Reynolds method, while the modified TKE method usually resulted in higher near-bed shear stress than the Reynolds method.

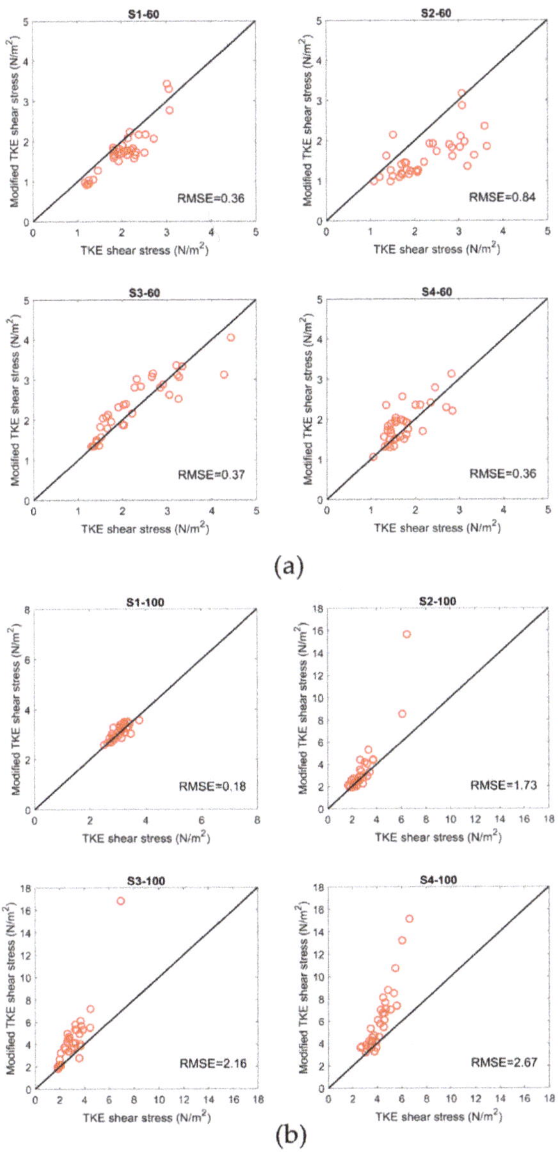

Figure 4. Near-bed shear stress estimated from the turbulent kinetic energy (TKE) method against the modified TKE method for all the measurement points at (**a**) 60 L/s flow rate, (**b**) 100 L/s flow rate. The unit of RMSE is N/m^2.

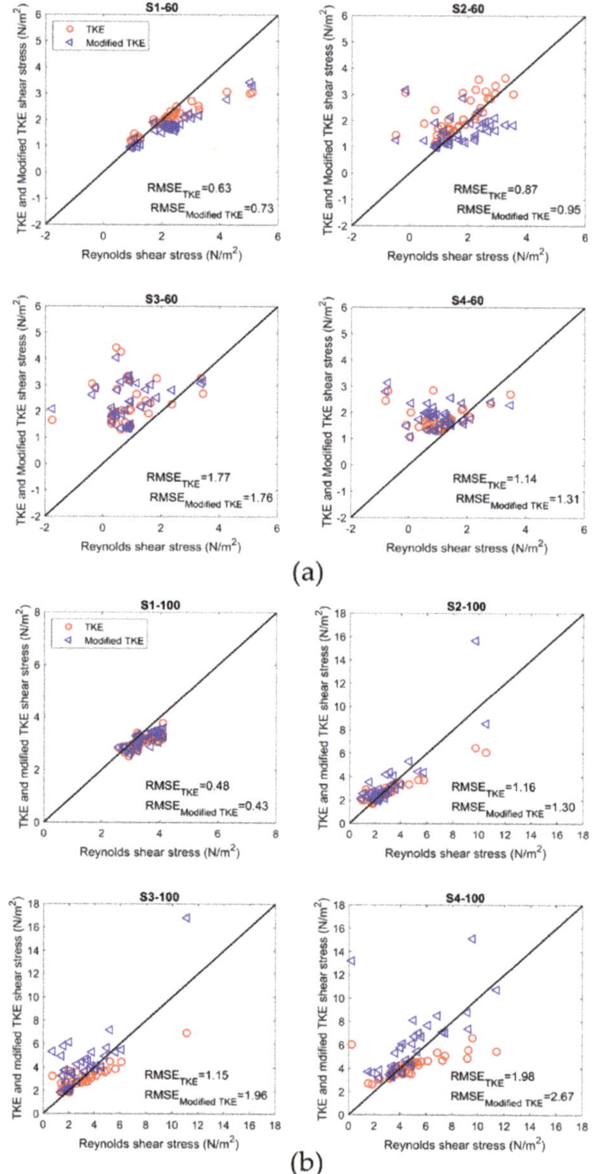

Figure 5. Near-bed shear stress estimated from the Reynolds method against the TKE and modified TKE methods for all the measurement points at (**a**) 60 L/s flow rate, (**b**) 100 L/s flow rate. The unit of RMSE is N/m^2.

3.3. Spatial Distribution of Near-Bed Shear Stress

The spatial distribution of the near-bed shear stress is also of great importance, especially in the presence of boulders that significantly alter local flow fields. On this account, contour maps of estimated local near-bed shear stress using different methods were plotted. Figure 6 shows contour plots of the estimated near-bed shear stress using the Reynolds, TKE, and modified TKE methods at unsubmerged condition (60 L/s). Before adding boulders (S1-60) for all methods, a uniform spatial distribution of the near-bed shear stress was expected. However, as shown in Figure 6, slightly higher near-bed shear stresses can be seen on the left side of the measurement zone before adding boulders. This non-uniform distribution can be attributed to the presence of microforms, placed slightly higher than the mean bed level, on some points of the gravel bed in these series of experiments. For the Reynolds method, after adding boulders, it can be observed that the near-bed shear stress generally decreased in all measurement points. For the large spacing scenario (S2-60), a zone with a highly decreased near-bed shear stress can be observed extended to a distance of 2D downstream of the boulder. For the medium and large boulder spacing scenarios (S3-60 and S4-60), highly decreased shear stresses can be observed along the boulder centerline and extended to the sides of the boulder. It is known that the presence of an unsubmerged boulder within the flow generates intense reverse and irregular flow in the extended wake region [30]. Here, negative shear stresses downstream of the boulder can be attributed to this reverse flow [60]. Additionally, at the smaller boulder spacing with a wake-interference or skimming flow regime, the flow deceleration becomes more pronounced between the boulder [38]. This can be the reason for the low shear stress zone along the boulder centerline for scenarios with a smaller boulder spacing (S3-60 and S4-60) [38]. The spatial distribution of the near-bed shear stress using the TKE and modified TKE methods were similar to each other. The reason for this similarity can be a less significant streamwise velocity due to the reverse flow that led to a more important role of w' in the calculation of shear stress using the TKE and modified TKE shear stress. Therefore, in this condition, the behavior of the TKE method became more similar to the modified TKE method that only depends on w'. However, the TKE and modified TKE spatial distribution deviated from the Reynolds method. In contrast to the Reynolds method, there was a slight increase in the near bed-bed shear stress downstream of the boulders (except downstream of one of the boulders in the S4-60). This increase became more significant for the medium boulder spacing. On the sides of the boulder, the near-bed shear stress did not vary significantly in comparison with the no-boulder scenario.

At 100 L/s (Figure 7), the spatial distribution of the near-bed shear stress using the point-methods was similar to each other, but it was different from 60 L/s. In the large boulder spacing scenario (S2-100), the near-bed shear stress decreased upstream of the boulder, but a significant increase can be seen downstream of the boulder (extended to around 2D) for all the methods. Studies have reported an increased Reynolds shear stress and TKE downstream of a submerged boulder near the bed due to the large-scale vortices in the wake zone and the presence of vertical eddies due to the downwash flow [21,31,61,62]. This increased near-bed shear stress decreased with increasing downstream x/D distance. Similarly, it has been reported this high shear stress zone extended up to the reattachment point, and then the shear stress started to decrease as undisturbed turbulent intensities recovered [31,62]. For the medium boulder spacing scenarios (S3-100), again an increased shear stress zone can be observed downstream of the central boulder for all the methods; however, for the Reynolds method a decrease in the near-bed shear stress observed at X1Y3 at x/D = −4D, downstream of one of the boulders (this boulder has not been shown in the contour plot). Additionally, the length of the high shear stress zone decreased to approximately 1D. Upstream of the boulder, the near-bed shear stress using the Reynolds method decreased, while increased shear stress can be seen upstream of the boulders for the TKE and modified TKE methods. For the small boulder spacing (S4-100), high shear stress zones became more pronounced in the sides of the boulder in this scenario. Increased near-bed shear stresses can also be seen between boulders probably as a result of the stable vortices between boulders in a wake-interference or skimming flow regime [63]. However, for the Reynolds method, a decreased shear stress zone was observed upstream of the central boulder (X3Y3 at x/D = −1D). The reason for this alternate variation of the Reynolds method estimations between boulders for the medium and small boulder spacing remained unclear and required additional experiments. For all scenarios at 100 L/s, it can be seen that the modified TKE method resulted in a higher magnitude of near-bed shear stress, as already shown in Figure 3b. The main reason for this can be attributed to the uncertain values of C_1 and C_2, as mentioned earlier.

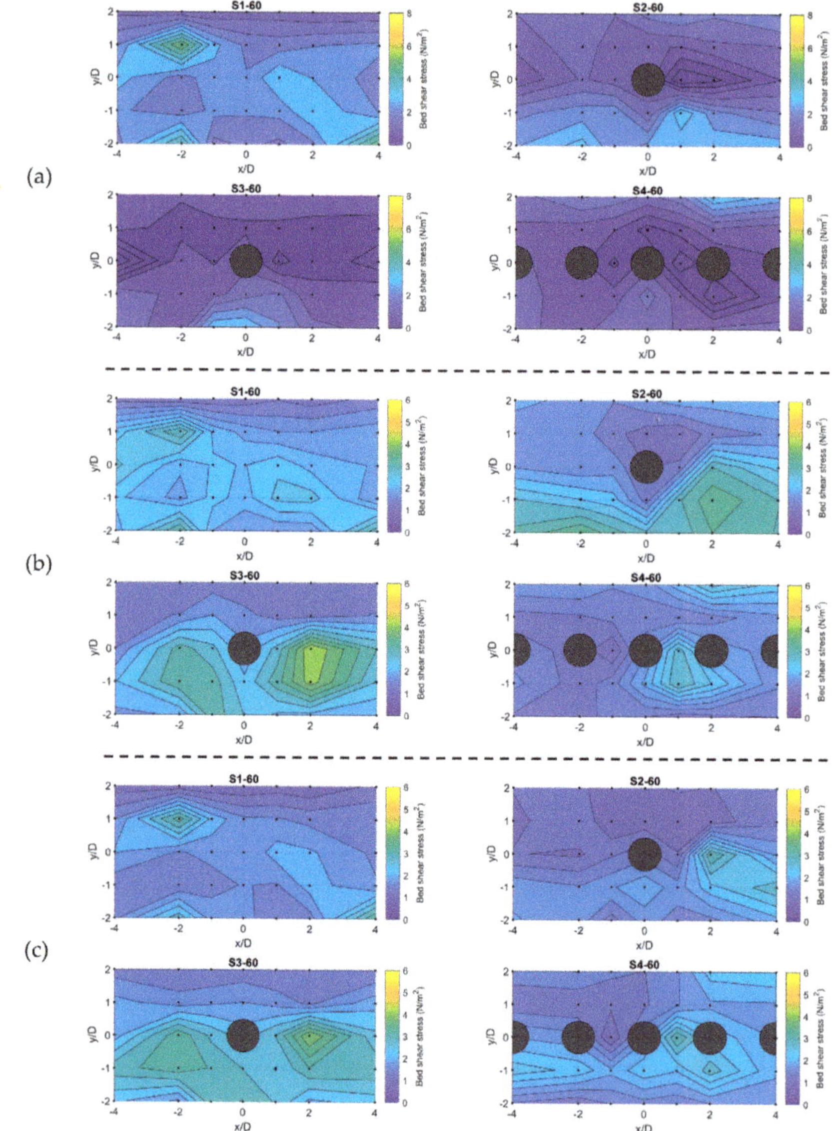

Figure 6. Contour plots of the estimated near-bed shear stress at 60 L/s (unsubmerged condition) for (**a**) the Reynolds, (**b**) TKE, and (**c**) modified TKE methods. Gray circles show the idealized boulders, and small black dots represent the measuring points.

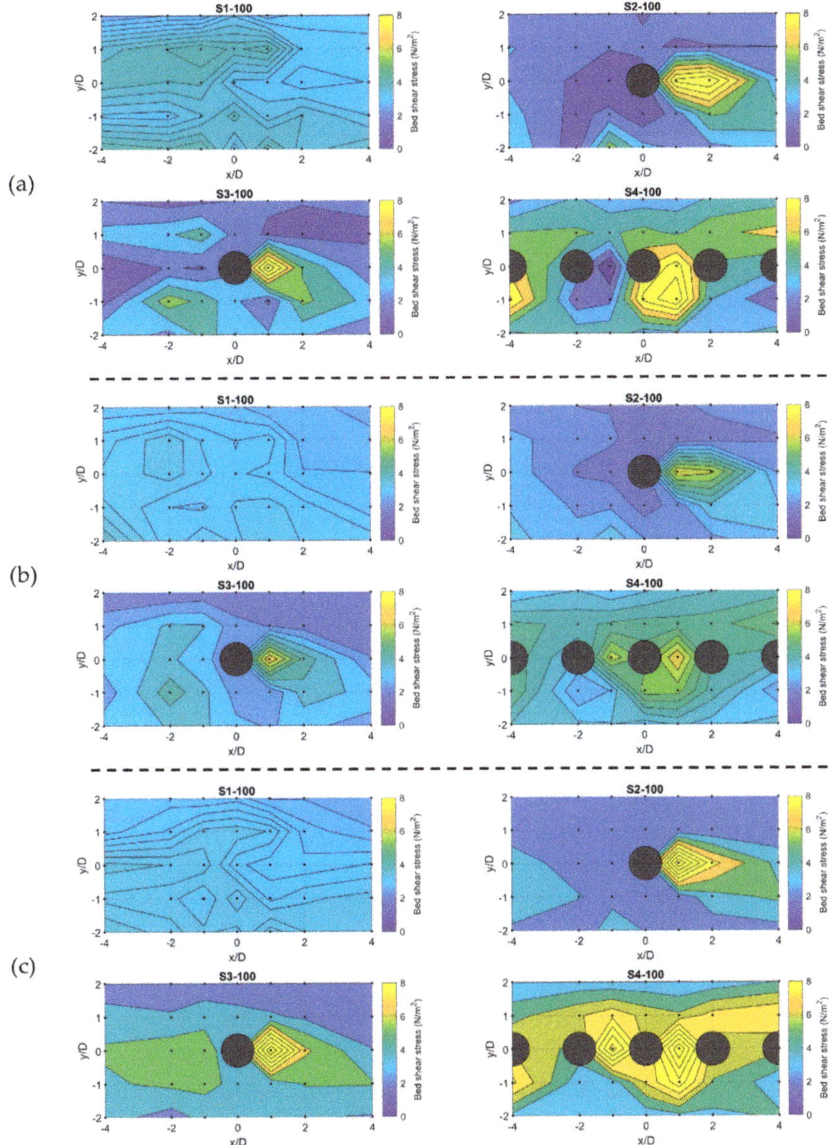

Figure 7. Contour plots of the estimated near-bed shear stress at 100 L/s (submerged condition) for (**a**) the Reynolds, (**b**) TKE, and (**c**) modified TKE methods. Gray circles show the idealized boulders, and small black dots represent the measuring points.

3.4. Near-Bed Turbulent Events

Quadrant analysis was conducted for the common points in all scenarios, X3Y3 and X5Y3. Figures 8 and 9 show the joint frequency distribution of normalized u' and w' at X3Y3 and X5Y3, respectively. u' and w' were normalized by the reach-averaged velocity of each scenario. Tables 2 and 3 complement the results of the quadrant analysis by listing the stress fraction parameter (S_i^f) and the frequency parameter (P_i) for each quadrant at X3Y3 and X5Y3, respectively. Before adding boulders at both points and flow rates, as expected ejection and sweep events were dominant near the bed, and this was in agreement with the previous studies [41,64]. At both flow rates at X3Y3, after adding and then decreasing the boulder spacing, the shear stress fraction of ejections and sweeps increased while the frequency of them gradually decreased. However, sweeps and ejections remained dominant for the large and medium boulder spacing. As can be seen in Figure 8 and Table 2, for the small boulder spacing, all four quadrants showed almost an equal contribution to the Reynolds shear stress and frequency. This is in agreement with the findings of [38] in which they found ejections and sweeps dominant for 10D and 5D spacing, while for 2D spacing each quadrant showed an almost equal frequency upstream of a fully submerged boulder [38]. This can be attributed to the stable vortices and a less pronounced downwash flow at the wake-interference or skimming flow associated with the small boulder spacing.

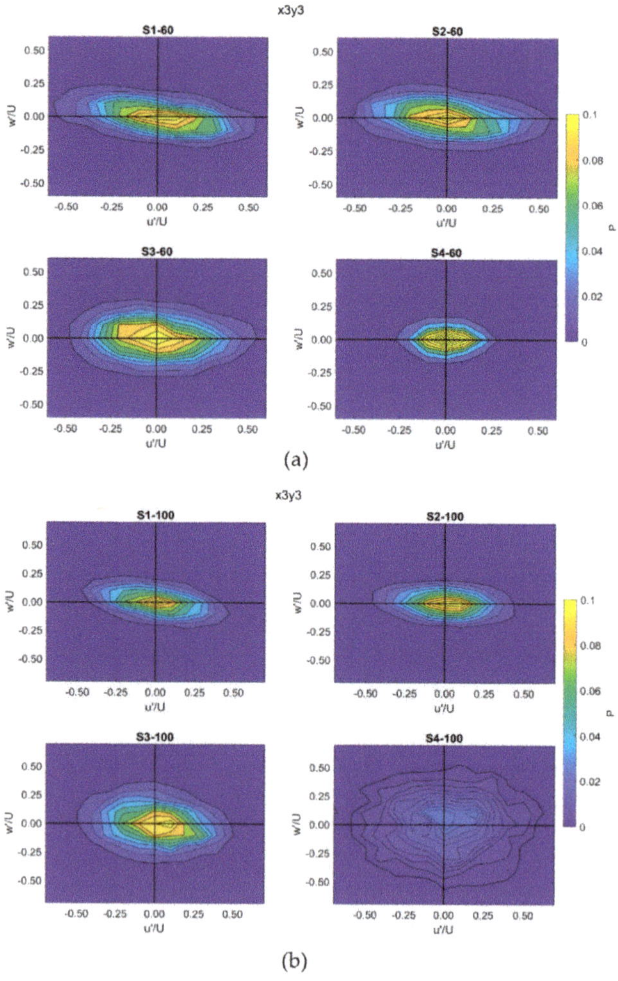

Figure 8. Joint frequency distribution of normalized u' and w' at point X3Y3 at (**a**) 60 L/s flow rate, (**b**) 100 L/s flow rate.

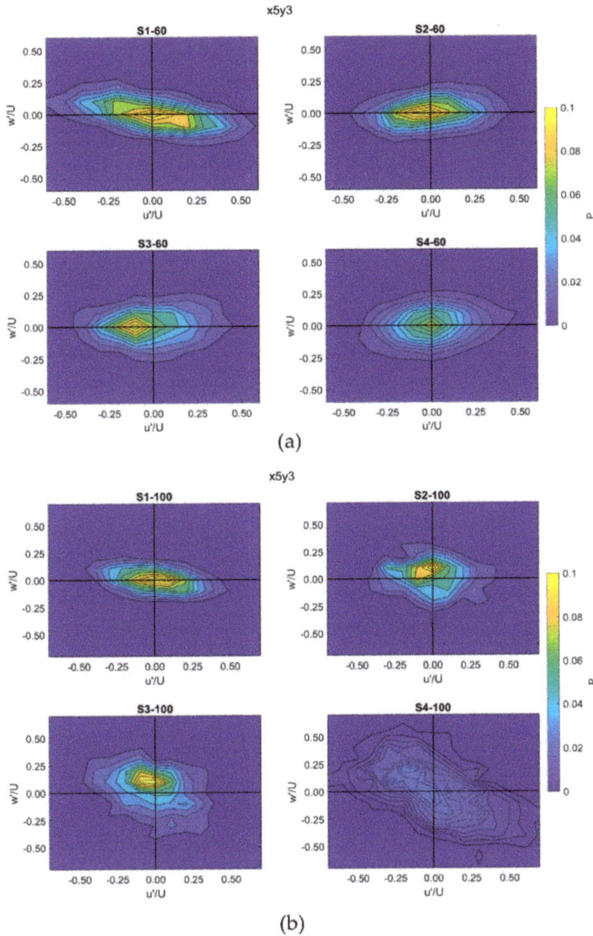

Figure 9. Joint frequency distribution of normalized u' and w' at point X5Y3 at (**a**) 60 L/s flow rate, (**b**) 100 L/s flow rate.

Table 2. Stress fraction and frequency of events for X3Y3 at 60 and 100 L/s flow rates.

Parameter	S^f				P			
Scenario	Outward	Ejection	Inward	Sweep	Outward	Ejection	Inward	Sweep
S1-60	−0.05	0.55	−0.04	0.41	0.03	0.36	0.03	0.27
S2-60	−0.29	0.74	−0.26	0.74	0.12	0.30	0.10	0.30
S3-60	−1.50	2.17	−1.50	1.83	0.21	0.30	0.21	0.25
S4-60	−10.10	12.01	−9.77	8.86	0.25	0.29	0.24	0.22
S1-100	−0.07	0.51	−0.06	0.50	0.04	0.32	0.04	0.31
S2-100	−0.74	1.25	−0.64	1.11	0.18	0.31	0.16	0.27
S3-100	−0.51	1.14	−0.58	0.93	0.14	0.32	0.16	0.26
S4-100	−12.17	12.29	−11.62	12.49	0.25	0.25	0.24	0.26

Table 3. Stress fraction and frequency of events for X5Y3 at 60 and 100 L/s flow rates.

Parameter	S^f				P			
Scenario	Outward	Ejection	Inward	Sweep	Outward	Ejection	Inward	Sweep
S1-60	−0.02	0.45	−0.03	0.38	0.01	0.33	0.02	0.28
S2-60	1.24	−0.57	1.14	−0.84	0.30	0.14	0.27	0.20
S3-60	4.98	−2.92	3.24	−4.30	0.32	0.19	0.21	0.28
S4-60	1.41	−0.90	1.48	−1.00	0.28	0.18	0.29	0.20
S1-100	−0.08	0.59	−0.10	0.49	0.05	0.34	0.06	0.28
S2-100	−0.06	0.22	−0.03	0.72	0.04	0.15	0.02	0.49
S3-100	−0.02	0.23	−0.03	0.62	0.02	0.17	0.02	0.45
S4-100	−0.05	0.39	−0.03	0.49	0.03	0.27	0.02	0.34

At 60 L/s at X5Y3 (Figure 9a), for the large boulder spacing (S2-60), inward and outward interactions were dominant. At the medium boulder spacing (S3-60), outward interactions and sweep events showed higher P_i and S_i^f. Again, at the small boulder spacing (S4-60), outward and inward interactions became dominant. Although 60 L/s indicates an unsubmerged boulder condition, the pattern of changes in the turbulent events was completely in agreement with the results of [38] for the different spacing of a fully submerged boulder. At 100 L/s at X5Y3 (Figure 9b), ejection and sweep events remained dominant in all scenarios while the share of inward and outward interactions was very small. However, as shown in Table 3, between ejection and sweep events, sweeps showed significantly higher P_i and S_i^f. This was not in agreement with the findings of [38] specifically for 5D and 2D boulder spacing [38]. Besides, P_i and S_i^f showed a consistent trend that means P_i kept the same order of dominance of S_i^f (e.g., when S_i^f for ejections and sweeps was the highest, P_i was also the highest for ejections and sweeps). The turbulent events have a considerable effect on sediment transport and resuspension. Predominant burst events (ejections and sweeps) may affect the sediment resuspension and transport around the boulder [39–41]. Specifically, downstream of the boulder, sweeps may remain effective for the bedload transport at the medium and large boulder spacing, while less significant ejections may decrease the sediment entrainment and the fine sediment suspension in these zones. Additionally, adding boulders and reduction of ejection and sweep events may provide resting zones and a path to avoid high ejection zones for the fish [43]. Such areas can be found downstream of the boulders at the medium and large boulder spacing where the burst events reduced in comparison with the no-boulder condition. However, more investigation is necessary to find the role of turbulent events in sediment transport and habitat preference for different boulder spacing.

4. Conclusions

This study investigated the effects of boulder spacing and boulder submergence ratio on the near-bed shear stress estimations and their distributions using laboratory experiments. The following conclusions can be drawn from this study:

1. For all scenarios, the reach-averaged method led to a significantly higher reach-averaged bed shear stress. For the unsubmerged condition, the Reynolds method resulted in a significantly lower near-bed shear stress between the point-methods, while, at submerged condition, all the point-methods showed very similar results.
2. At unsubmerged condition, the effect of the boulder spacing on the variation of near-bed shear stress estimated from the Reynolds method was different from the TKE and modified TKE methods, while, at submerged condition, all of the point-methods showed a similar trend.
3. At submerged condition, the densest boulder spacing led to the highest near-bed shear stress for all point-methods. However, for the unsubmerged condition, maximum near-bed shear stress varied for different methods and boulder spacing.

4. At unsubmerged condition, the TKE and modified TKE methods can be used interchangeably for estimation of the near-bed shear stress in the presence of boulders; however, applying appropriate C_1 and C_2 coefficients is required to obtain more reliable results. For a comprehensive comparison of the Reynolds method with two other point-methods, more measurements are necessary, especially at unsubmerged condition.

5. At unsubmerged condition, denser boulder spacing led to a more uniform contribution of turbulent events to the Reynolds shear stress. At submerged condition, decreased ejection events downstream of boulders in the large and medium boulder spacing may reduce the sediment entrainment and suspension.

This work is based on two boulder submergence ratios and single-point measurements near the bed and placing boulders along the centerline of the channel in a single straight array. To derive general conclusions, further measurements closer to the bed with a denser spatial grid between boulders are necessary. Additionally, a study of the bedload and the suspended load sediment transport within different boulder setups is recommended as the next step.

Author Contributions: Conceptualization, A.B.M.B.; methodology, A.G.; software, A.G.; formal analysis, A.G.; investigation, A.G. and F.H.; resources, A.B.M.B.; data curation, A.G. and F.H.; writing—original draft preparation, A.G. and F.H.; writing—review and editing, A.B.M.B., F.H. and A.G.; visualization, A.G.; supervision, A.B.M.B.; project administration, A.B.M.B.; All authors have read and agreed to the published version of the manuscript.

Funding: This research received no external funding. However, financial support for the first author was provided by Clarkson University.

Acknowledgments: The authors wish to thank Ian Knack (Clarkson University) for his invaluable support and help to build the Ecohydraulics Flume.

Conflicts of Interest: The authors declare no conflict of interest.

References

1. Barnes, M.P.; O'donoghue, T.; Alsina, J.M.; Baldock, T.E. Direct bed shear stress measurements in bore-driven swash. *Coast. Eng.* **2009**, *56*, 853–867. [CrossRef]

2. Howe, D.; Blenkinsopp, C.E.; Turner, I.L.; Baldock, T.E.; Puleo, J.A. Direct measurements of bed shear stress under swash flows on steep laboratory slopes at medium to prototype scales. *J. Mar. Sci. Eng.* **2019**, *7*, 358. [CrossRef]

3. Mirfenderesk, H.; Young, I.R. Direct measurements of the bottom friction factor beneath surface gravity waves. *Appl. Ocean Res.* **2003**, *25*, 269–287. [CrossRef]

4. Park, J.H.; Do Kim, Y.; Park, Y.S.; Jo, J.A.; Kang, K. Direct measurement of bottom shear stress under high-velocity flow conditions. *Flow Meas. Instrum.* **2016**, *50*, 121–127. [CrossRef]

5. Park, J.H.; Do Kim, Y.; Park, Y.S.; Jung, D.G. Direct measurement of bed shear stress using adjustable shear plate over a wide range of Froude numbers. *Flow Meas. Instrum.* **2019**, *65*, 122–127. [CrossRef]

6. Nezu, I.; Nakagawa, H. *Turbulence in Open Channel Flows, IAHR*; Balkema: Rotterdam, The Netherlands, 1993.

7. Babaeyan-Koopaei, K.; Ervine, D.A.; Carling, P.A.; Cao, Z. Velocity and turbulence measurements for two overbank flow events in River Severn. *J. Hydraul. Eng.* **2002**, *128*, 891–900. [CrossRef]

8. Wilcock, P.R. Estimating local bed shear stress from velocity observations. *Water Resour. Res.* **1996**, *32*, 3361–3366. [CrossRef]

9. Williams, J.J. Drag and sediment dispersion over sand waves. *Estuar. Coast. Shelf Sci.* **1995**, *41*, 659–687. [CrossRef]

10. Nikora, V.; Goring, D.; McEwan, I.; Griffiths, G. Spatially averaged open-channel flow over rough bed. *J. Hydraul. Eng.* **2001**, *127*, 123–133. [CrossRef]

11. Kim, S.-C.; Friedrichs, C.T.; Maa, J.-Y.; Wright, L.D. Estimating bottom stress in tidal boundary layer from acoustic Doppler velocimeter data. *J. Hydraul. Eng.* **2000**, *126*, 399–406. [CrossRef]

12. Biron, P.M.; Robson, C.; Lapointe, M.F.; Gaskin, S.J. Comparing different methods of bed shear stress estimates in simple and complex flow fields. *Earth Surf. Process. Landf. J. Br. Geomorphol. Res. Group* **2004**, *29*, 1403–1415. [CrossRef]

13. Lichtneger, P.; Sindelar, C.; Schobesberger, J.; Hauer, C.; Habersack, H. Experimental investigation on local shear stress and turbulence intensities over a rough non-uniform bed with and without sediment using 2D particle image velocimetry. *Int. J. Sediment Res.* **2020**, *35*, 193–202. [CrossRef]

14. Kostaschuk, R.; Villard, P.; Best, J. Measuring velocity and shear stress over dunes with acoustic Doppler profiler. *J. Hydraul. Eng.* **2004**, *130*, 932–936. [CrossRef]

15. Sime, L.C.; Ferguson, R.I.; Church, M. Estimating shear stress from moving boat acoustic Doppler velocity measurements in a large gravel bed river. *Water Resour. Res.* **2007**, *43*. [CrossRef]

16. Federal Interagency Stream Restoration Working Group (15 US Federal Agencies). *Stream Corridor Restoration: Principles, Processes, and Practices*; GPO Item No. 0120-A; FISRWG: Washington, DC, USA, 1998.

17. Kang, S.; Hill, C.; Sotiropoulos, F. On the turbulent flow structure around an instream structure with realistic geometry. *Water Resour. Res.* **2016**, *52*, 7869–7891. [CrossRef]

18. Marttila, H.; Turunen, J.; Aroviita, J.; Tammela, S.; Luhta, P.-L.; Muotka, T.; Kløve, B. Restoration increases transient storages in boreal headwater streams. *River Res. Appl.* **2018**, *34*, 1278–1285. [CrossRef]

19. Henry, C.P.; Amoros, C.; Roset, N. Restoration ecology of riverine wetlands: A 5-year post-operation survey on the Rhone River, France. *Ecol. Eng.* **2002**, *18*, 543–554. [CrossRef]

20. Baki, A.B.M.; Zhu, D.Z.; Rajaratnam, N. Mean flow characteristics in a rock-ramp-type fish pass. *J. Hydraul. Eng.* **2014**, *140*, 156–168. [CrossRef]

21. Baki, A.B.M.; Zhu, D.Z.; Rajaratnam, N. Turbulence characteristics in a rock-ramp-type fish pass. *J. Hydraul. Eng.* **2015**, *141*, 04014075. [CrossRef]

22. Baki, A.B.M.; Zhang, W.; Zhu, D.Z.; Rajaratnam, N. Flow structures in the vicinity of a submerged boulder within a boulder array. *J. Hydraul. Eng.* **2016**, *143*, 04016104. [CrossRef]

23. Papanicolaou, A.N.; Tsakiris, A.G. Boulder effects on turbulence and bedload transport. In *Gravel-Bed Rivers*; John Wiley & Sons, Ltd: Hoboken, NJ, USA, 2017; pp. 33–72.

24. Tsakiris, A.G.; Papanicolaou, A.N.T.; Hajimirzaie, S.M.; Buchholz, J.H.J. Influence of collective boulder array on the surrounding time-averaged and turbulent flow fields. *J. Mt. Sci.* **2014**, *11*, 1420–1428. [CrossRef]

25. Papanicolaou, A.N.; Dermisis, D.C.; Elhakeem, M. Investigating the role of clasts on the movement of sand in gravel bed rivers. *J. Hydraul. Eng.* **2011**, *137*, 871–883. [CrossRef]

26. Hardy, R.J.; Best, J.L.; Lane, S.N.; Carbonneau, P.E. Coherent flow structures in a depth-limited flow over a gravel surface: The role of near-bed turbulence and influence of Reynolds number. *J. Geophys. Res. Earth Surf.* **2009**, *114*. [CrossRef]

27. Van Rijn, L.C. Critical movement of large rocks in currents and waves. *Int. J. Sediment Res.* **2019**, *34*, 387–398. [CrossRef]

28. Papanicolaou, A.N.; Kramer, C.M.; Tsakiris, A.G.; Stoesser, T.; Bomminayuni, S.; Chen, Z. Effects of a fully submerged boulder within a boulder array on the mean and turbulent flow fields: Implications to bedload transport. *Acta Geophys.* **2012**, *60*, 1502–1546. [CrossRef]

29. Yager, E.M.; Kirchner, J.W.; Dietrich, W.E. Calculating bed load transport in steep boulder bed channels. *Water Resour. Res.* **2007**, *43*. [CrossRef]

30. Shamloo, H.; Rajaratnam, N.; Katopodis, C. Hydraulics of simple habitat structures. *J. Hydraul. Res.* **2001**, *39*, 351–366. [CrossRef]

31. Dey, S.; Sarkar, S.; Bose, S.K.; Tait, S.; Castro-Orgaz, O. Wall-wake flows downstream of a sphere placed on a plane rough wall. *J. Hydraul. Eng.* **2011**, *137*, 1173–1189. [CrossRef]

32. Monsalve, A.; Yager, E.M. Bed surface adjustments to spatially variable flow in low relative submergence regimes. *Water Resour. Res.* **2017**, *53*, 9350–9367. [CrossRef]

33. Ghilardi, T.; Schleiss, A. Influence of immobile boulders on bedload transport in a steep flume. In Proceedings of the 34th IAHR World Congress, Brisbane, Australia, 26 June–1 July 2011; pp. 3473–3480.

34. Ghilardi, T.; Schleiss, A. Steep flume experiments with large immobile boulders and wide grain size distribution as encountered in alpine torrents. In Proceedings of the International conference on fluvial Hydraulics, San José, Costa Rica, 5–7 September 2012; Taylor & Francis Group, CRC Press: London, UK, 2012; pp. 407–414.

35. Papanicolaou, A.N.; Tsakiris, A.G.; Wyssmann, M.A.; Kramer, C.M. Boulder Array Effects on Bedload Pulses and Depositional Patches. *J. Geophys. Res. Earth Surf.* **2018**, *123*, 2925–2953. [CrossRef]

36. Papanicolaou, A.N.; Diplas, P.; Dancey, C.L.; Balakrishnan, M. Surface Roughness Effects in Near-Bed Turbulence: Implications to Sediment Entrainment. *J. Eng. Mech.* **2001**, *127*, 211–218. [CrossRef]

37. Shih, W.; Diplas, P.; Celik, A.O.; Dancey, C. Accounting for the role of turbulent flow on particle dislodgement via a coupled quadrant analysis of velocity and pressure sequences. *Adv. Water Resour.* **2017**, *101*, 37–48. [CrossRef]

38. Fang, H.W.; Liu, Y.; Stoesser, T. Influence of Boulder Concentration on Turbulence and Sediment Transport in Open-Channel Flow Over Submerged Boulders. *J. Geophys. Res. Earth Surf.* **2017**, *122*, 2392–2410. [CrossRef]

39. Tang, C.; Li, Y.; Acharya, K.; Du, W.; Gao, X.; Luo, L.; Yu, Z. Impact of intermittent turbulent bursts on sediment resuspension and internal nutrient release in Lake Taihu, China. *Environ. Sci. Pollut. Res.* **2019**, *26*, 16519–16528. [CrossRef] [PubMed]

40. Salim, S.; Pattiaratchi, C.; Tinoco, R.; Coco, G.; Hetzel, Y.; Wijeratne, S.; Jayaratne, R. The influence of turbulent bursting on sediment resuspension under fluvial unidirectional currents. *Earth Surf. Dyn.* **2016**, *5*, 399–415. [CrossRef]

41. Mohajeri, S.H.; Righetti, M.; Wharton, G.; Romano, G.P. On the structure of turbulent gravel bed flow: Implications for sediment transport. *Adv. Water Resour.* **2016**, *92*, 90–104. [CrossRef]

42. Lacey, R.J.; Rennie, C.D. Laboratory investigation of turbulent flow structure around a bed-mounted cube at multiple flow stages. *J. Hydraul. Eng.* **2012**, *138*, 71–84. [CrossRef]

43. Bretón, F.; Baki, A.B.M.; Link, O.; Zhu, D.Z.; Rajaratnam, N. Flow in nature-like fishway and its relation to fish behaviour. *Can. J. Civ. Eng.* **2013**, *40*, 567–573. [CrossRef]

44. Polatel, C. Large-Scale Roughness Effect on Free-Surface and Bulk Flow Characteristics in Open-Channel Flows. Ph.D. Thesis, University of Iowa, Iowa City, IA, USA, 2006.

45. Tan, L.; Curran, J.C. Comparison of Turbulent Flows over Clusters of Varying Density. *J. Hydraul. Eng.* **2012**, *138*, 1031–1044. [CrossRef]

46. Paphitis, D. Sediment movement under unidirectional flows: An assessment of empirical threshold curves. *Coast. Eng.* **2001**, *43*, 227–245. [CrossRef]

47. Nezu, I.; Rodi, W. Open-channel Flow Measurements with a Laser Doppler Anemometer. *J. Hydraul. Eng.* **1986**, *112*, 335–355. [CrossRef]

48. Dombroski, D.E.; Crimaldi, J.P. The accuracy of acoustic Doppler velocimetry measurements in turbulent boundary layer flows over a smooth bed. *Limnol. Oceanogr. Methods* **2007**, *5*, 23–33. [CrossRef]

49. Goring, D.G.; Nikora, V.I. Despiking acoustic Doppler velocimeter data. *J. Hydraul. Eng.* **2002**, *128*, 117–126. [CrossRef]

50. Cea, L.; Puertas, J.; Pena, L. Velocity measurements on highly turbulent free surface flow using ADV. *Exp. Fluids* **2007**, *42*, 333–348. [CrossRef]

51. Wahl, T.L. Analyzing ADV data using WinADV. In Proceedings of the Joint Conference on Water Resources Engineering and Water Resources Planning and Managmenet, Minneapolis, MI, USA, 30 July-2 August 2000.

52. Strom, K.B.; Papanicolaou, A.N. ADV measurements around a cluster microform in a shallow mountain stream. *J. Hydraul. Eng.* **2007**, *133*, 1379–1389. [CrossRef]

53. Martin, V.; Fisher, T.S.R.; Millar, R.G.; Quick, M.C. ADV data analysis for turbulent flows: Low correlation problem. In Proceedings of the Hydraulic Measurements and Experimental Methods 2002, Estes Park, CO, USA, 28 July–1 August 2002; pp. 1–10.

54. Nikora, V.I.; Goring, D.G. ADV measurements of turbulence: Can we improve their interpretation? *J. Hydraul. Eng.* **1998**, *124*, 630–634. [CrossRef]

55. Lacey, R.W.J.; Roy, A.G. Fine-scale characterization of the turbulent shear layer of an instream pebble cluster. *J. Hydraul. Eng.* **2008**, *134*, 925–936. [CrossRef]

56. Buffin-Bélanger, T.; Roy, A.G. Effects of a pebble cluster on the turbulent structure of a depth-limited flow in a gravel-bed river. *Geomorphology* **1998**, *25*, 249–267. [CrossRef]

57. Curran, J.C.; Tan, L. The effect of cluster morphology on the turbulent flows over an armored gravel bed surface. *J. Hydro-Environ. Res.* **2014**, *8*, 129–142. [CrossRef]

58. Liu, D.; Liu, X.; Fu, X.; Wang, G. Quantification of the bed load effects on turbulent open-channel flows: Les-Dem Simulation of Sediment in Turbulent Flow. *J. Geophys. Res. Earth Surf.* **2016**, *121*, 767–789. [CrossRef]

59. Guan, D.; Agarwal, P.; Chiew, Y.-M. Quadrant Analysis of Turbulence in a Rectangular Cavity with Large Aspect Ratios. *J. Hydraul. Eng.* **2018**, *144*, 04018035. [CrossRef]

60. Sadeque, M.A.; Rajaratnam, N.; Loewen, M.R. Flow around Cylinders in Open Channels. *J. Eng. Mech.* **2008**, *134*, 60–71. [CrossRef]

61. Hajimirzaie, S.M.; Tsakiris, A.G.; Buchholz, J.H.J.; Papanicolaou, A.N. Flow characteristics around a wall-mounted spherical obstacle in a thin boundary layer. *Exp. Fluids* **2014**, *55*, 1762. [CrossRef]
62. Zexing, X.; Chenling, Z.; Qiang, Y.; Xiekang, W.; Xufeng, Y. Hydrodynamics and bed morphological characteristics around a boulder in a gravel stream. *Water Supply* **2020**, *20*, 395–407. [CrossRef]
63. Morris, H.M. Flow in rough conduits. *Trans. ASME* **1955**, *120*, 373–398.
64. Nakagawa, H.; Nezu, I. Prediction of the contributions to the Reynolds stress from bursting events in open-channel flows. *J. Fluid Mech.* **1977**, *80*, 99–128. [CrossRef]

 *water*

Article

Hydrological Foundation as a Basis for a Holistic Environmental Flow Assessment of Tropical Highland Rivers in Ethiopia

Wubneh B. Abebe [1,2,*], Seifu A. Tilahun [2], Michael M. Moges [2], Ayalew Wondie [3], Minychl G. Derseh [2], Teshager A. Nigatu [4], Demesew A. Mhiret [2], Tammo S. Steenhuis [2,5], Marc Van Camp [6], Kristine Walraevens [6] and Michael E. McClain [7,8]

[1] Amhara Design and Supervision Works Enterprise, Bahir Dar, P.O. Box 1921, Ethiopia
[2] Faculty of Civil and Water Resources Engineering, Bahir Dar Institute of Technology, Bahir Dar University, Bahir Dar, P.O. Box 26, Ethiopia; satadm86@gmail.com (S.A.T.); micky_mehari@yahoo.com (M.M.M.); minychl2009@gmail.com (M.G.D.); demisalmaw@gmail.com (D.A.M.); tammo@cornell.edu (T.S.S.)
[3] School of Fisheries and Wildlife, Department of Aquatic and Wetland Management, Bahir Dar University, Bahir Dar, P.O. Box 26, Ethiopia; ayaleww2001@yahoo.com
[4] Basin Information System Directorate, Abbay River Basin Authority, Bahir Dar, P.O. Box 1376, Ethiopia; omegadad40@gmail.com
[5] Department of Biological and Environmental Engineering, Cornell University, Ithaca, NY 14850, USA
[6] Laboratory for Applied Geology and Hydrogeology, Ghent University, Pietersnieuwstraat 33, 9000 Gent, Belgium; marc.vancamp@ugent.be (M.V.C.); kristine.walraevens@ugent.be (K.W.)
[7] Department of Water Resources and Ecosystems, IHE Delft Institute for Water Education, Westvest 7, 2611 AX Delft, Países Baixos, The Netherlands; m.mcclain@un-ihe.org
[8] Faculty of Civil Engineering and Geosciences, Delft University of Technology, Mekelweg 5, 2628 CD Delft, Países Baixos, The Netherlands
[*] Correspondence: wubnehb@yahoo.com; Tel.: +25-193-598-2616

Received: 16 December 2019; Accepted: 13 February 2020; Published: 15 February 2020

Abstract: The sustainable development of water resources includes retaining some amount of the natural flow regime in water bodies to protect and maintain aquatic ecosystem health and the human livelihoods and wellbeing dependent upon them. Although assessment of environmental flows is now occurring globally, limited studies have been carried out in the Ethiopian highlands, especially studies to understand flow-ecological response relationships. This paper establishes a hydrological foundation of Gumara River from an ecological perspective. The data analysis followed three steps: first, determination of the current flow regime—flow indices and ecologically relevant flow regime; second, naturalization of the current flow regime—looking at how flow regime is changing; and, finally, an initial exploration of flow linkages with ecological processes. Flow data of Gumara River from 1973 to 2018 are used for the analysis. Monthly low flow occurred from December to June; the lowest being in March, with a median flow of 4.0 m^3 s^{-1}. Monthly high flow occurred from July to November; the highest being in August, with a median flow of 236 m^3 s^{-1}. 1-Day low flows decreased from 1.55 m^3 s^{-1} in 1973 to 0.16 m^3 s^{-1} in 2018, and 90-Day (seasonal) low flow decreased from 4.9 m^3 s^{-1} in 1973 to 2.04 m^3 s^{-1} in 2018. The Mann–Kendall trend test indicated that the decrease in low flow was significant for both durations at $\alpha = 0.05$. A similar trend is indicated for both durations of high flow. The decrease in both low flows and high flows is attributed to the expansion of pump irrigation by 29 km^2 and expansion of plantations, which resulted in an increase of NDVI from 0.25 in 2000 to 0.29 in 2019. In addition, an analysis of environmental flow components revealed that only four "large floods" appeared in the last 46 years; no "large flood" occurred after 1988. Lacking "large floods" which inundate floodplain wetlands has resulted in early disconnection of floodplain wetlands from the river and the lake; which has impacts on breeding and nursery habitat shrinkage for migratory fish species in Lake Tana. On the other hand, the extreme decrease in "low flow" components has impacts on predators, reducing their mobility and ability to access prey concentrated in smaller pools. These results serve as the hydrological foundation for continued studies in the Gumara catchment, with the eventual goal of quantifying environmental flow requirements.

Keywords: environmental flow component; Ethiopia; holistic environmental flow assessment; hydrological foundation; indicators of hydrologic alteration software; Lake Tana

1. Introduction

Hydrologic regimes play an important role in determining the biodiversity of aquatic ecosystems but unwise uses are critically changing them globally [1]. Previous studies confirmed that there are advancements globally in the maintenance of flows in rivers that make water resource uses sustainable [2–10]. Developing countries like Ethiopia are increasingly emphasizing environmental flows and the allocation of water for ecological conservation [8,11].

In using models that are capable to relate flow and ecology at a wider scale, a framework called Ecological Limits of Hydrologic Alteration (ELOHA) was developed [12]. Flow–ecology relationships can apply to rivers of a particular hydrological type with naturally distinctive flow regimes [12,13]. The ELOHA framework involves the establishment of flow-ecology relationships based on hydrological characters and ecological conditions of aquatic ecosystems or watersheds [12]. ELOHA includes four major steps to come up with flow–ecology relationships and quantify environmental flow requirements of water bodies. It starts with hydrological characterization, identification of river types, determination of changes in flow and lastly, establishes relationships of flow changes vis-à-vis ecological processes in each river type using available information [12].

The Nile Basin Initiative (NBI) has developed an environmental flows management framework building on the elements of the ELOHA framework and global best practices [14]. Ethiopia has approved and adopted this framework through its membership in NBI. The NBI environmental flow management framework (NBI-EFMF) includes seven procedural steps in quantifying environmental flow requirements and one of the steps is the establishment of the hydrological foundation. This phase includes the baseline evaluation/modelling of hydrology data for the site/regional environmental flow assessments. Precipitation, flow, evaporation, water abstraction, land use data and other information that may affect flows are used in this phase to characterize baseline flows and potentially describe any differences between these baseline flows and current flows [14].

Water is abstracted in the Blue Nile basin at many locations and more abstractions are planned, impacting the environment [15,16]. For example, in Lake Tana sub-basin, two large dams for irrigation have been completed and two are under construction. Studies have revealed that climate change will affect the water balance of Lake Tana and pose environmental risks unless proper water resource measures are implemented [17,18]. Projected changes in monthly precipitation and temperature in the Tana sub-basin from 15 GCMs (Global Climate Models-General Circulation Models) were analyzed and it was found that four of the nine GCMs indicated a significant decrease in annual stream flow for the 2080–2100 period [17]. In addition, similarly to other studies on impacts of human interventions on sustainable water use, climate change, unmanaged water abstractions and land use change in the Lake Tana Sub-basin threaten the riverine and lake ecosystems [19,20].

The Abbay Basin Authority in its sub-basin master plan preparation has suggested that 10%–25% of river flow be allocated for the environment. However, the suggestion did not consider the flow variabilities and downstream uses of rivers and water bodies (personal communication, Mr. Habtamu Tamir). In addition, a review of the planned environmental flow release of Koga dam (one of the completed irrigation dam projects in Lake Tana sub-basin) showed that the environmental flow release plan is merely the 95 percent exceedance value (Q_{95}) of the Koga River flow record [21]. This method does not consider the dynamic and variable nature of rivers, nor the ecosystem services and social impacts at the watershed scale [21].

Similarly, Gumara river of Lake Tana Sub-basin has pressures which need due attention to sustain the ecosystem. Unmanaged pump irrigation practices, the expansion of eucalyptus trees and sand mining upstream have caused the river to stop flowing in the dry season (Figure 1). Studies found a decline in catch of fish because human interventions on the rivers flowing into Lake Tana affect migration and spawning grounds [22,23]. In addition, studies showed the need for establishing methods for pollution control [24]. Moreover, a study in the sub-humid highlands of Ethiopia indicated that peak sediment influx occurs during the high flows, highlighting the need for land degradation management to protect the health of the aquatic ecosystems [25]. Therefore, the hydrology of the Gumara river and associated floodplain wetlands of Shesher and Welala should be studied to understand the environmental water requirement to restore important aquatic and wetland biodiversity.

Figure 1. Gumara river at the bridge in the Fogera plain where the road from Bahir Dar to Gondar crosses; (**a**) August 2017 at flood stage during the rain phase and (**b**) February 2015 during the dry season when the flow has ceased (courtesy: M. M. Moges).

The objective of this study was to establish hydrological foundation of Gumara river as an initial step in the application of the NBI Environmental Flows framework. Using the Indicators of Hydrologic Alteration (IHA) software, our analysis follows three steps: first, determination of the current flow regime—flow indices and ecologically relevant flow regime; second, naturalization of the current flow regime—looking at how flow regime is changing and finally, an initial exploration of flow linkages with ecological processes.

2. Materials and Methods

2.1. Study Area Description

The Gumara River originates in the afro-alpine vegetation of the Guna mountains above 4000 m.a.s.l. and flows to Lake Tana at 1784 m.a.s.l. (Figure 2). Gumara River catchment is 1376 km^2 and is part of the larger Lake Tana basin. The climate of the area is largely controlled by the movement of the inter-tropical convergence zone (ITCZ), which results in a single rainy season between June and September [26]. The mean annual rainfall over the catchment is 1326 mm year^{-1}. The rivers, before draining to the lake, feed the Welala and Shesher wetlands, which together, cover an area of approximately 8.0 km^2 [27,28]. The flood regime of the wetlands is affected by the abstraction and diking of the Gumara River. Moreover, the wetlands are being encroached by cultivation.

Gumara River is ecologically important as it is the migration habitat of fish of the genus Labeobarbus of the cyprinid family [29–32]. Fifteen unique species of Labeobarbus inhabit the lake [29]. In addition, twelve globally threatened bird species have been identified in Lake Tana and its associated wetlands [33]. Most of the species are recorded in the Shesher and Welala wetlands (Figure 2), which are part of the UNESCO Biosphere reserve areas [34].

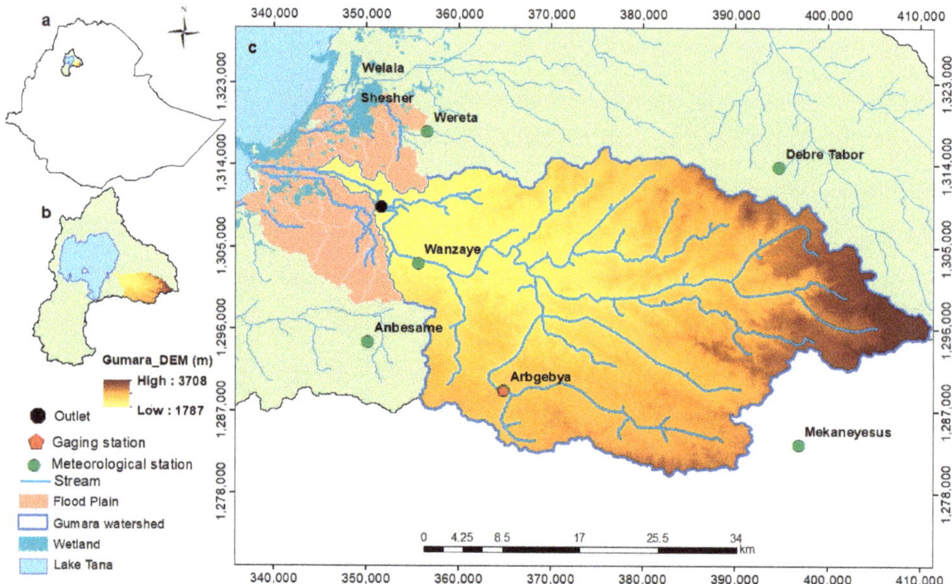

Figure 2. Location map of the study area; (**a**) Ethiopia, (**b**) Lake Tana Sub-basin and (**c**) elevation map of the Gumara watershed; the Gumara floodplain (orange) and Welala and Shesher wetlands (blue) are shown. The boundaries of the Gumara floodplain are approximate due to small elevation differences and depend on the height of the flood.

2.2. Data Collection

2.2.1. Precipitation

Daily precipitation of the Debretabor, Arb-Gebeya, Mekaneyesus, Wanzaye and Anbessame stations was obtained from the Ethiopian Meteorological Agency (EMA). The dataset was not up to date and complete for all stations. Hence, remote sensing precipitation data of the "Climate Hazards Group InfraRed Precipitation with Station Data" (CHIRPS) with 0.05 arc degree resolution were downloaded for the period from 1 January 1981 to 30 September 2019 from Google Earth Engine; cloud computing platform [35]. CHIRPS was chosen because it has daily data for a long record with the best resolution and performance for this location [35].

2.2.2. Stream Flow

The hydrology was characterized for the entire Gumara catchment using the lower gaging station—No. 111006 (see 'outlet' in Figure 2). Flow data for the station were obtained from the Ministry of Water, Irrigation and Electric (MoWIE) from 1960 to 2018. There are large gaps in data for the first 13 years (full years in 1963, and 1967 to 1972), for 5 months in 2015 and for 6 months in 2018. From the total 21,535 days, 3132 (14.5%) were missing. As the annual gaps are too large to fill, we used data from 1973 to 2018.

2.3. Literature Review, Field Observation and Discussions

Literature on fishery and related activities were reviewed. A reconnaissance survey of Gumara catchment and associated wetlands was undertaken and farming communities were questioned on their understanding of flow characteristics and how their livelihoods were connected with the river ecosystem, including pump irrigation, sand mining, vegetation expansion and others.

2.4. Data Analysis

2.4.1. Precipitation and Evapotranspiration Analysis

Areal rainfall of Gumara was estimated from Satellite data of CHIRPS. The performance of CHIRPS was checked by a Pearson correlation test with the available observed data of individual stations around Gumara. Annual, decadal and cumulative rainfall of different durations were calculated using MS EXCEL. The Pearson correlation test was used to visualize the relation of flow and rainfall and Mann–Kendall trend tests were used for trends in rainfall and were performed in SPSS 20, EXCEL and XLstat [36,37]. Cumulative rainfall for generating 20 mm of cumulative direct runoff after the end of the dry season was calculated for trend analysis. In addition, the evapotranspiration over the Gumara catchment was calculated. A synoptic station close to the catchment did not exist to calculate the evapotranspiration from the meteorological data; hence, satellite data of "MOD16A2.006: Terra Net Evapotranspiration 8-Day Global 500 m" was used for the estimate [38].

2.4.2. Flow Analysis Using IHA Software

River flow statistics, components and indices were analyzed using IHA software version 7.1 [39] (The software is developed by the Nature conservacy, Virginia, VA, USA). Setting up and completing an analysis in the IHA involved the use of hydrologic data as input, deciding analysis years and environmental flow component (EFC) thresholds, and water year starting Julian date [39]. Hydrological data from 1973 to 2018 were imported in CSV file format and saved as internal hydrologic file. A project was then created, linked to a single hydrologic data file and used to create and run multiple analysis. The Gumara flow data were not normally distributed (Figure 3) and hence, the non-parametric analysis like medians and coefficient of dispersion were used. The water year was set to start on January 1 and to end on December 31, which is suitable for Gumara River condition.

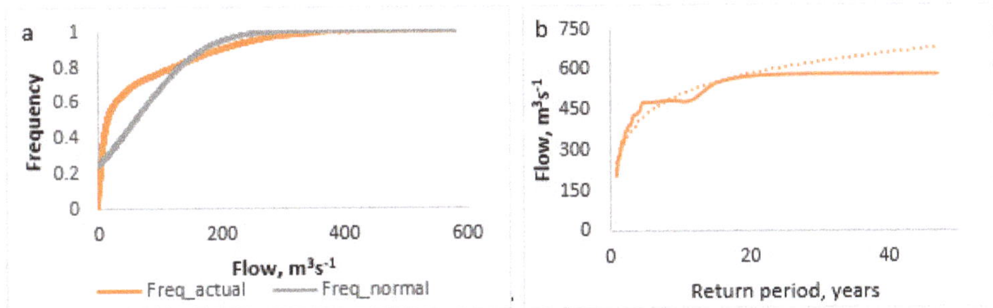

Figure 3. Cumulative Distribution Function (**a**) and Flood Frequency (**b**) Curves of Gumara River at the outlet. The orange line is the actual frequency and the grey line is the normal frequency. The actual frequency is the frequency of a flow value, indicating the number of data at or below it. The normal frequency is the predicted frequency for the normal distribution of the data. The solid line in the Flood Frequency curve is the actual flood flow versus return period and the dotted line is the logarithmic curve fit to estimate flood magnitude at a given return period.

The IHA calculated parameters for five different types of EFCs which are ecologically relevant: extreme low flows, low flows, high flow pulses, small floods, and large floods. This delineation of EFCs is based on the definition given in the software [1,39]. Low flows are calculated from minimum flows within a year [1,39]. Extreme low flows are taken below the 10th percentile flow.

High-flow pulse is calculated as flow between base flow and bankfull discharge, i.e., including any water rises that do not overtop the channel banks. Small floods include all river rises that overtop the main channel but do not include more extreme and less frequent floods (i.e., below 2-year return period). Large floods are flows calculated as above the 10-year return period, which can inundate distant places from riverbanks, such as lagoons and floodplain wetlands.

2.4.3. Environmental Flow Components Threshold Values Setting

A cumulative distribution function (CDF) curve, flood frequency (FF) curves and stage-discharge (rating) curve were used in setting the EFC threshold values to be input in the IHA software (Figure 3). The CDF curve ensured the non-normality of the data indicating that non-parametric analyses need to be conducted. First, the large flood was found from the 10-year return period of the flood frequency curve; it is 483 m^3 s^{-1} which is the 99.93th percentile in the CDF curve.

The bankfull discharge was estimated using the stage-discharge method [40]. The Bankfull stage was measured and found to be 5.6 m in March 2019.

The stage–discharge curve was developed recently as the bed level increase at the gaging station annually and the offset, h_0 changed in time. To do this, h_0 was calculated first from the rating curve (Equation (1)) with measured discharge Q and stage height h for streamflow record from February 1990 to March 2018 containing 52 readings. Secondly, the best fitting curve was found through the 52 h_0 values [41]. By trial and error, the parameters in Equation (1) were changed until a best fit was obtained. The offset is given in Equation (2):

$$Q = 11.5(h - h_0)^{1.9} \tag{1}$$

The best fit ($R^2 = 0.93$) for the offset, h_0 in m was found as

$$h_0 = -0.0002(t - 1990)^3 + 0.0049(t - 1990)^2 + 0.0748(t - 1990) - 0.0819 \tag{2}$$

where t is the year.

The offset, h_0 for years before 1990 was estimated using a linear interpolation applied assuming zero in the beginning (1973) and the h_0 value calculated for 1990 in Equation (2). Hence, the flow data from the ministry was recalculated for the new h_0. Then, the calculated bankfull discharge for 5.6 m stage was found to be 294 m^3 s^{-1} or 97.5% from CDF, which is the maximum for high flow pulse and the starting threshold for small floods. The low flow was found to be 4.8 m^3 s^{-1} (28%) and the maximum extreme low flow was taken as the 10% flow, 0.17 m^3 s^{-1}.

2.4.4. Flow Components and Needs

As a complimentary analysis to IHA, ecologically relevant flows from percentiles of the historical daily flows were analyzed seasonally using the approach of DePhilip and Moberg [42]. Overlaying key life history requirements for each group on representative hydrographs for each habitat type, relationships between species groups and seasonal and inter-annual stream flow patterns were explained in terms of flow needs by season. Thresholds were selected to approximate different ecologically relevant flows; Q5–Q10 corresponds to flood levels, Q10–Q75 represents high flows, Q75–Q95 represents low flows, and Q95–Q100 represent the extreme low flows at the site [42]. All daily data of the 46 years were arranged in descending order and their percentiles/exceedance calculated and mapped in Excel where the 5th, 10th, 75th and 95th percentile values are linear interpolations.

2.4.5. Irrigation Area Mapping and NDVI Analysis

Irrigation practices, sand mining and other economic activities were analyzed using the information obtained from farmers' discussion and field ground truth data collection using GPS and mapping aided by Google Earth. Discussion with farmers was organized by the local agriculture office experts. A focus group discussion was undertaken for each irrigation and sand mining work. Five farmers who initially started irrigation and 10 individuals who were collecting sand were contacted for discussion using a prepared checklist. The vegetation conditions in the catchment area were also analyzed.

Normalized difference vegetation index (NDVI) from MODIS satellite data was calculated in google earth engine, cloud computing platform, to link the impact of vegetation cover change on both low- and high-flow trend.

3. Results and Discussion

3.1. Precipitation

Data performance from CHRIPS checked by the Pearson's correlation test with the observed data in SPSS was found to be a correlation coefficient of 0.51 for Debretabor station and a correlation coefficient of 0.44 for Wanzaye station, significant at a 0.01 level (Figure 4). The correlation of the precipitation (PCP) and flow data of Gumara also tested with Pearson correlation test in SPSS and was found to have an r of 0.49; correlation significant at a 0.01 level. The annual precipitation trend of Gumara catchment from the areal estimate for the whole dataset showed a slight increase (Figure 5). The annual PCP trend tested with Mann–Kendal's test was found to be significant at the 0.05 level.

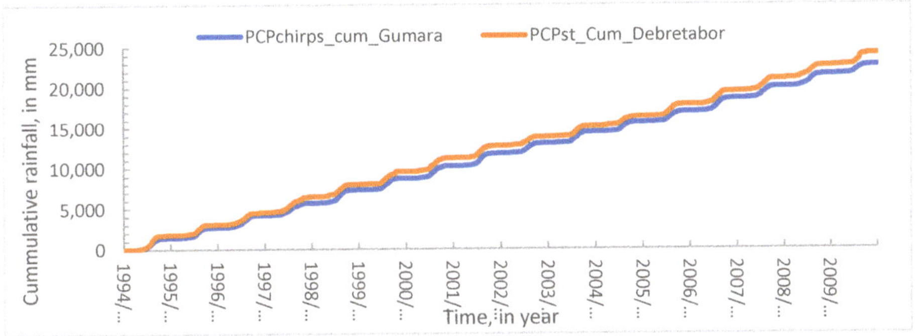

Figure 4. Cumulative precipitation measured at rain gauge stations (PCPst_Cum_Debretabor) and CHIRPS Satellite Data (PCPchirps_Cum_Gumara) versus time for Gumara river catchment (1994–2018).

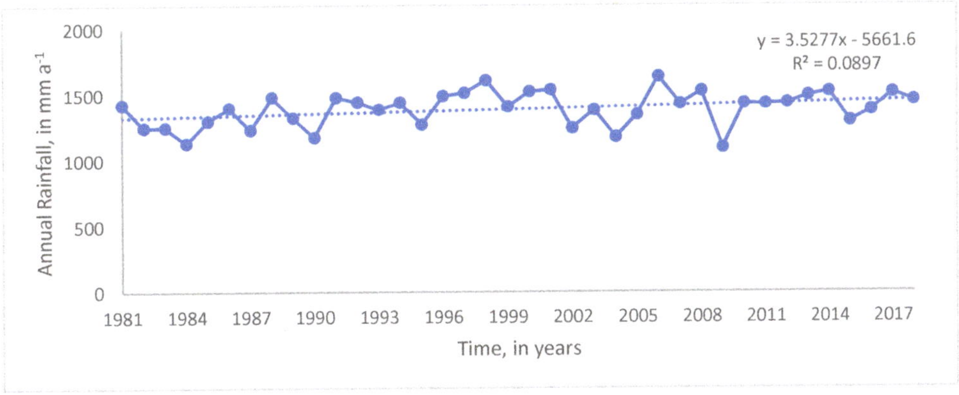

Figure 5. Annual precipitation of Gumara catchment (1981–2018) predicted with "Climate Hazards Group InfraRed Precipitation with Station Data" (CHIRPS). The trend line was found to be increasing at a rate of 4 mm per year.

3.2. Flow Indices

3.2.1. Monthly Flows

The monthly flow analysis indicated that low flow occurs from December to June, the lowest being in March, with a median flow of 4.0 m^3 s^{-1} and a standard error (SE) 0.55 and high flows occur from July to November; the highest being in August, with a median flow of 236 m^3 s^{-1} with a standard deviation of 8 m^3 s^{-1} (Table 1 and Figure 6). A high coefficient of dispersion (COD) was found from November to May during the dry season (winter and spring); and a low COD was observed from June to October during the wet season (summer and autumn).

All monthly flows have a decreasing trend (Figure 6). Flows in the dry season in March and April, decreased in time likely due to pump irrigation and the expansion of eucalyptus trees.

Table 1. Monthly Median Flows (m^3 s^{-1}) of Gumara River at the outlet (1973–2018). See Figure 1 for the location of the outlet; Q25 is flow that is exceeded 25% of the time, Q50 is the median and Q75 is flow exceeded 75% of the time.

Months	Median Flow (Q50), m^3 s^{-1}	Coefficient of Disp.; (Q75–Q25)/Q50	Months	Median Flow (Q50), m^3 s^{-1}	Coeff. of Disp. (Q75–Q25)/Q50
January	9.7	1.3	August	236	0.3
February	5.9	1.7	September	151	0.4
March	4.0	1.7	October	65.9	0.9
April	4.2	1.3	November	31.2	1.4
May	4.7	1.7	December	16.5	1.5
June	13.3	1.0			
July	123.6	0.5			

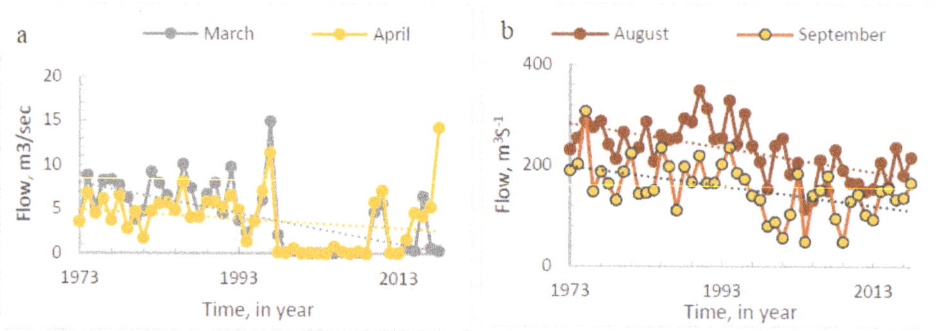

Figure 6. Monthly flow of Gumara 1973–2018 in (**a**) dry season (March and April) and (**b**) wet season (August and September). Both the dry and wet season monthly flows have a decreasing trend.

3.2.2. Low and High Flows

Low Flow

The Mann–Kendall trend test indicated that the 1-Day and 90-Day low flow decreased significantly over the study period at $p = 0.01$. Quantitatively, 1-Day low flow decreased from 1.55 m^3 s^{-1} in 1973 to 0.16 m^3 s^{-1} in 2018 and 90-Day (seasonal) low flow decreased from 4.88 m^3 s^{-1} in 1973 to 2.04 m^3 s^{-1} in 2018. The decrease in low flow after 1997 was verified in the discussion with farmers and district experts living and working in the study area. According to the discussants, pump irrigation started in 1997 using the pumps supported by German International Cooperation (GIZ). In the first year, pump irrigation was started by 25 farmers who were cultivating maize, then it reached maximum to all households in 2005. The delineation using Google Earth indicated that irrigated area was 29 km^2 in

2019 (Figure 7). The average historical low flow before 1997 was 3.03 m^3 s^{-1} or 3,141,504 m^3 of water per year. The net irrigable area of the lower Gumara is 15.25 km^2, distributed 6 km^2 Onion, 0.2 km^2 Tomato, 0.82 km^2 Garlic, 0.16 km^2 Pepper, 2.0 km^2 Tef, 6 km^2 Maize, and 0.07 km^2 Lentil (Annual report 2018/19, Dera District Agriculture Office). The irrigation water requirement of Onion is 288,300 m^3 km^{-2}, Tomato 168,900 m^3 km^{-2}, Pepper 127,100 m^3 km^{-2}, Garlic 144,200 m^3 km^{-2}, Tef 84,150 m^3 km^{-2}, Lentil 308,300 m^3 km^{-2}, and Maize 167,450 m^3 km^{-2} [43]. This amounts to a demand of 3,096,741 m^3 of water per year for 15.25 km^2, which is 99% of the value of the historical low flows before 1997.

There was an abrupt decrease of low flow in 1997 which remained at a bare minimum onwards. The coefficient of dispersion (COD) values are greater for all duration of flow, indicating greater variability in low flow (Table 2 and Figure 8). The mean decadal low flows were 3.02, 3.19, 1.96, 0.002, and 0.029 m^3 s^{-1} for 1973–1980, 1981–1990, 1991–2000, 2001–2010 and 2011–2018, respectively (Table 3).The extreme decrease in low flow components impacts the predators of fish, reducing their mobility and ability to access prey concentrated in smaller pools.

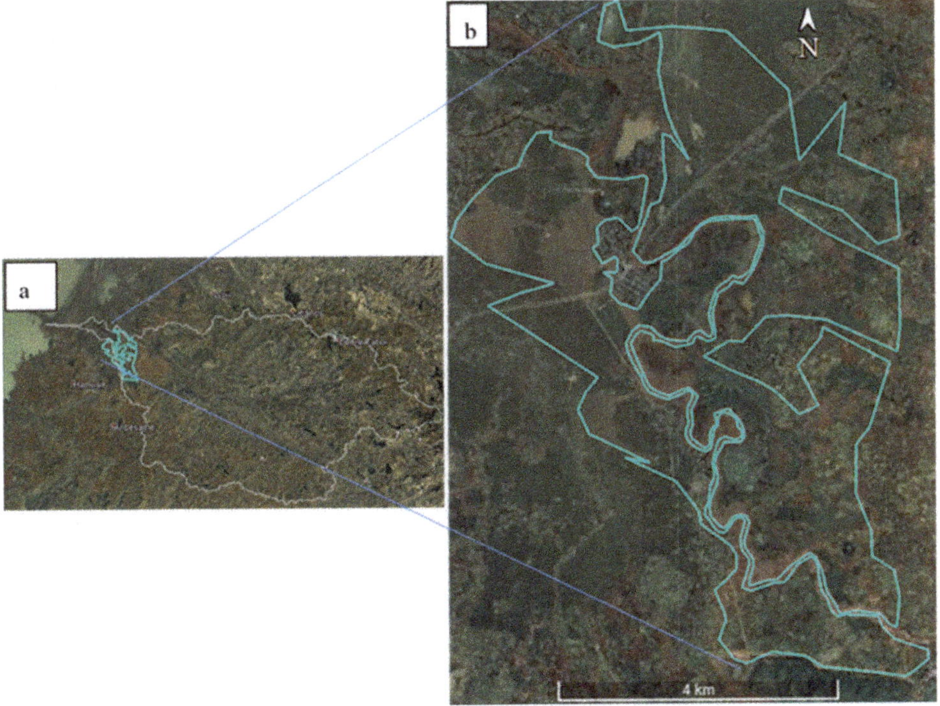

Figure 7. Gumara watershed (**a**) and Pump irrigation sites between the bridge and Wanzaye town (**b**); taken from Google Earth of 2019 image. The blue line indicates the total irrigation area (29 km^2) where the net irrigated area is 15.25 km^2.

High Flow

The maximum flows, similarly to the low flows, decreased over the time periods: 1-Day r^2 of 0.53 with significant trend at p = 0.01 with Mann–Kendall's test and 90-Day r^2 of 0.32 (p = 0.01). Quantitatively, 1-Day high flow decreased from 335 m^3 s^{-1} in 1973 to 266 m^3 s^{-1} in 2018 and 90-Day (seasonal) high flow decreased from 188 m^3 s^{-1} in 1973 to 185 m^3 s^{-1} in 2018. The 1-Day maximum is a good indicator for large flood decrease as it indicates individual peaks rather than averages as the other durations do (Table 2 and Figure 8). The mean decadal high flow ranged from 432 m^3 s^{-1} to 261 m^3 s^{-1} between 1973 and 1980 and 2001 and 2010, respectively (Table 3).

Table 2. Minimum and Maximum Flow ($m^3 \ s^{-1}$) of Different Duration in Gumara River at the 'outlet'. These are the common variables in the Ecological Limits of Hydrologic Alteration (ELOHA) analysis. The abbreviations are listed in Table 1.

Duration	Median (Q50)	Coeff. of Disp.; (Q75-Q25)/Q50	Duration	Median (Q50)	Coeff. of Disp.; (Q75-Q25)/Q50
1-day maximum	335	0.49	1-day minimum	1.62	1.89
3-day maximum	316	0.41	3-day minimum	1.70	1.85
7-day maximum	293	0.41	7-day minimum	1.71	1.91
30-day maximum	248	0.32	30-day minimum	2.36	1.91
90-day maximum	183	0.37	90-day minimum	4.27	1.54

Table 3. Mean decadal (a) low- and (b) high-flows ($m^3 \ s^{-1}$) and decadal percentage changes of Gumara River at the 'outlet'. The mean decadal low flow showed a decreasing trend since 1973–1980 and reached nearly zero for 2001–2010 and 2011–2018.

Years	(a)	Percent Change in Low Flow	(b)	Percent Change in High Flow
1973–1980	3.02		432.5	
1981–1990	3.19	5.7	441.3	2.0
1990–2000	1.96	−38.7	384.0	−13.0
2001–2010	0.00	−99.9	261.0	−32.0
2011–2018	0.00	-	263.4	0.9

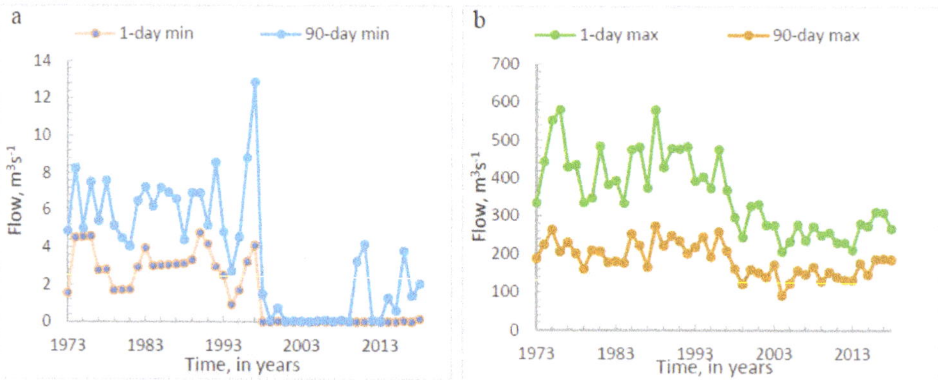

Figure 8. One-day and 90-day low (**a**) and high flow (**b**) of Gumara River at the outlet (1973–2018). The figure depicts (**a**) the lowest flow from each individual day of the year as 1-day duration low flow and the lowest flow of average 90 days as 90-day (seasonal) duration low flow and (**b**) highest flow from each individual day of the year as 1-day duration high flow; and highest flow of average 90 days as 90-day duration high flow.

From these results, we infer that the river is becoming disconnected earlier from the floodplain wetlands because of a decrease in "ecologically relevant" large floods. As studies indicated, this has an impact on fish breeding habitat shrinkage in a short period of time [44]. A decline in juvenile labeobarb abundance in the pool habitats of Gumara River occurred because of excess irrigation water abstraction, especially in the dry season months of March to May [44]. In addition, several studies found that in recent years, fish stocks declined rapidly, especially commercially important fish species like Labeobarbus' which are migratory fishes requiring wetland habitats for breeding [23,32,45].

We inferred that the decrease in large flood and in low flows during dry season is attributed to unmanaged pump irrigation, the expansion of plantations and soil conservation works being undertaken since 2010 through government mass mobilization program. This agrees with the other studies in lake Tana Basin [46]. In addition, a study on the hydrological impact of a Eucalyptus plantation found that the cumulative rainfall required to generate 3 mm runoff was higher after a threefold expansion of the plantation area [47].

The normalized difference vegetation index (NDVI) for Gumara catchment was checked by extracting NDVI data in Google Earth engine, cloud computing platform, and showed a significant increasing trend at a 0.05 level (Figure 9). An increase in vegetation is expected to decrease (direct) runoff due to increasing evapotranspiration.

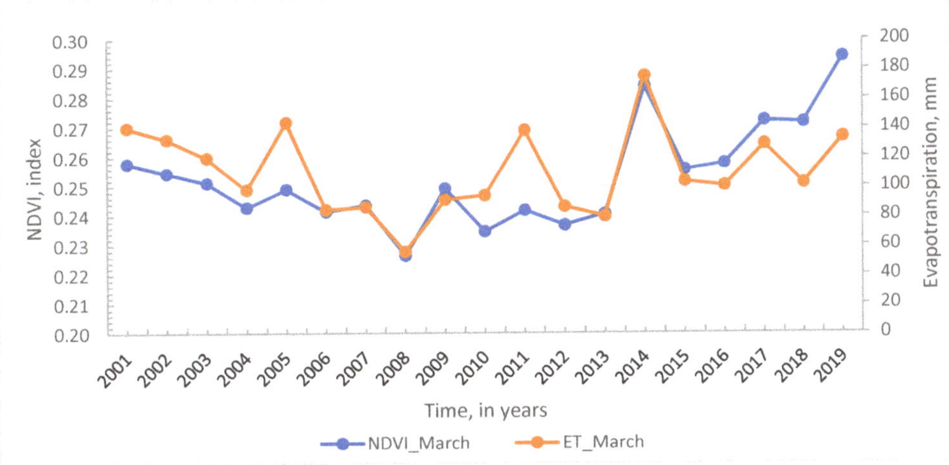

Figure 9. NDVI of Gumara River catchment in Dry season month of March between 2000 and 2019 and March Evapotranspiration over the Gumara catchment; satellite data of "MODIS Global Terrestrial Evapotranspiration 8-day Global 1 km" resolution was used for the estimate.

Therefore, using the satellite image data of MODIS, the dry season, the month of March, evapotranspiration over the Gumara was extracted and found to be increasing between 2001 and 2019 (Figure 9). The correlation of evapotranspiration with NDVI was 0.64 with Pearson's correlation test; significant at the 0.01 level. The evapotranspiration is in line with the vegetation increase in the Gumara Catchment. An increase in vegetation has increased the evapotranspiration, which, in turn, increased the amount of infiltration water need to saturate soils before runoff generation. This suggests that the hydrological process is highly influenced by tree plantations [47,48].

Research has shown that in the Ethiopian highlands where saturation excess runoff dominates, daily discharge is a function of daily amount of rainfall, not of the rainfall intensity [48–50]. For watersheds to start generating surface runoff after the dry monsoon phase, the soil needs to become saturated [48–50]. We divided the study period into three blocks before catchment management interventions, including vegetation expansions (1998 to 2011), during interventions (2012 to 2014) and after soil and water conservation and vegetation expansions (2015 to 2018). The average cumulative rainfall required for a 20 mm runoff depth increased from 371 mm in the first period before intervention to 442 mm after interventions (Figure 10). The result agrees with the finding of another recent study from the Ethiopian Highlands [47].

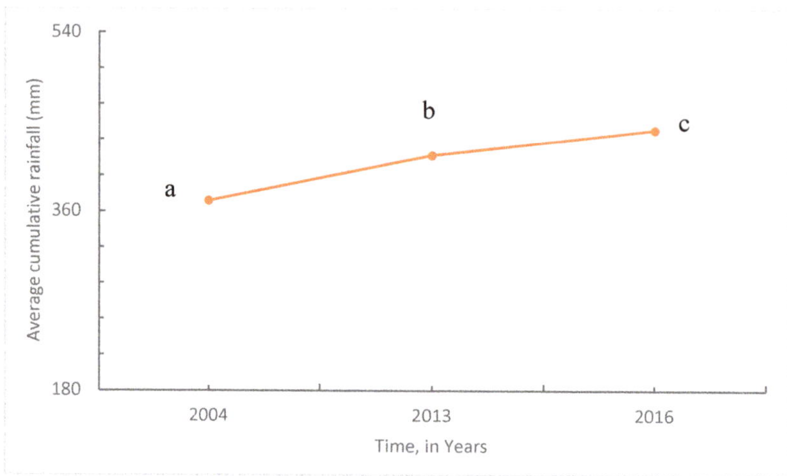

Figure 10. Average cumulative rainfall in three periods 1998–2011 (**a**), 2012–2014 (**b**) and 2015–2018 (**c**). The cumulative rainfall records considered here lie in the same day with 20 mm cumulative runoff depth records for each record year; that is the cumulative rainfall required for 20 mm run off generation.

3.3. Environmental Flow Components (EFC), Durations and Timing

Environmental Flow Components

The EFC analysis found four large floods during the last 50 years, i.e., 1975, 1976, 1981 and 1988 (Figure 11). Interestingly, no large flood was recorded after 1989 and small floods exhibited a decrease after 1997. This has a similar interpretation with low- and high-flow condition. This results in an early disconnection of floodplain wetlands from the river and the lake, which impacts fish migration, spawning/breeding, and the growth period for juveniles. High-flow pulse increased and shows a nearly uniform magnitude in the last decades. On the other hand, extreme low flow and low flow decreased and could have an impact on predator-prey relationships as species are concentrated in smaller pools (Figure 11).

Flow Duration and Timing of Environmental Flow Components

All environmental flow components showed increasing duration except for high flow pulses (Figure 12). Extreme low flow showed increase in trend where low flow in other ways decreased in recent years. According to the flow components analysis, large and small floods were not available after 1988 and 2001 respectively. High-flow pulse, which occurs at the beginning of the wet season, is almost undisturbed, which can give fish and other aquatic animals increased access to upstream areas. This flow component is not enough for complete fish reproduction, which need spawning and a growth period in the river and flood plain wetlands provided by small and large floods.

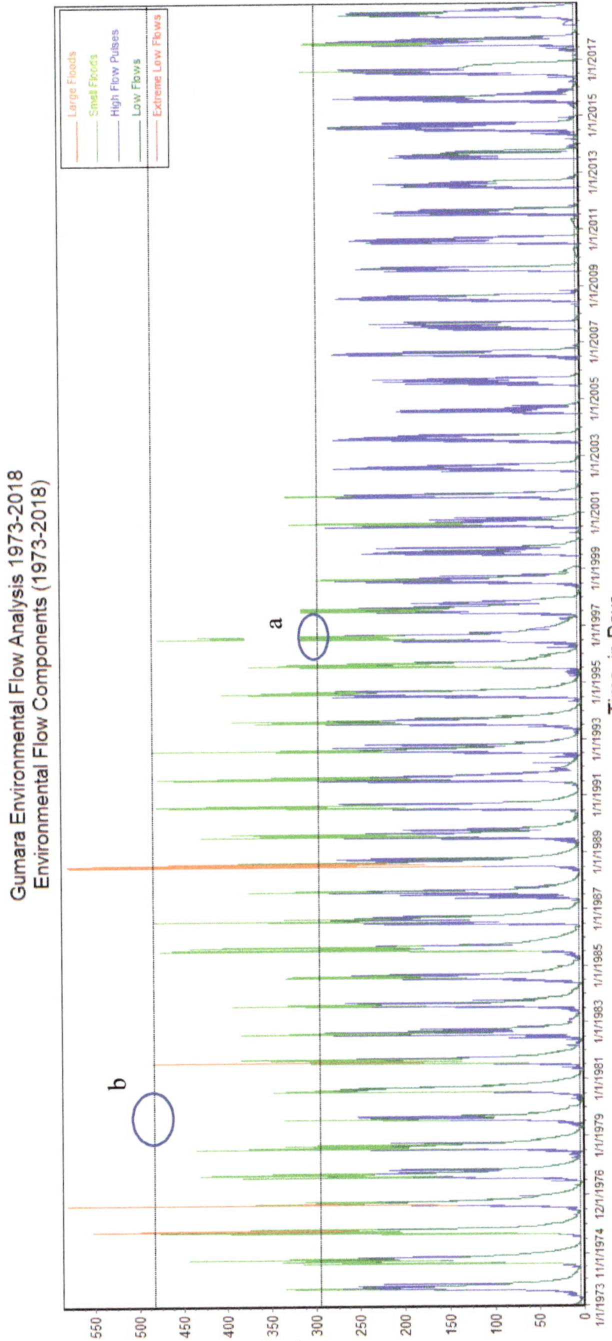

Figure 11. Environmental Flow Components of Gumara River; Extreme Low Flow, Low Flow, High Flow Pulse, Small Floods and Large Floods. The horizontal line (**a**) shows the small flood minimum peak flow and the horizontal line (**b**) shows large flood minimum peak flow.

The timing (the Julian day) of small flood occurrence is stable between 214 and 244 with a median of 228 and a COD of 0.039 but it is interrupted after 2001. Other environmental flow components are highly variable; for example, high-flow pulse moved from 194 Julian day in 1973 to about 230 Julian day in 2018, which is almost a month delay (Figure 12). This can cause a disruption of the reproduction cycle of fish and other aquatic animals which live both in the lake and river. High-flow pulses are a signal for these species to begin migrating into rivers to reach floodplain spawning areas.

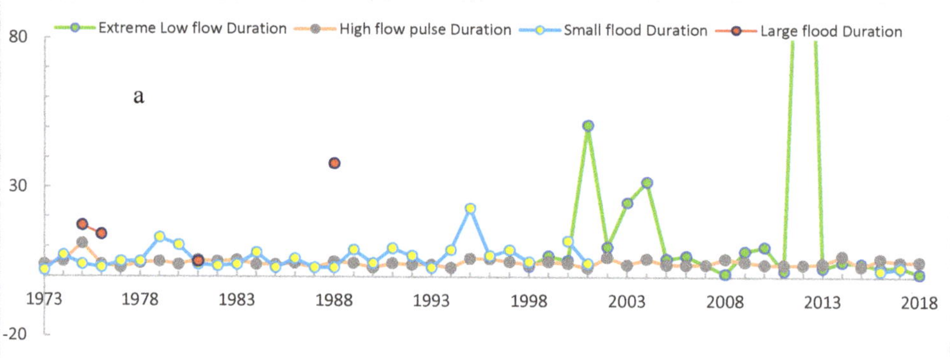

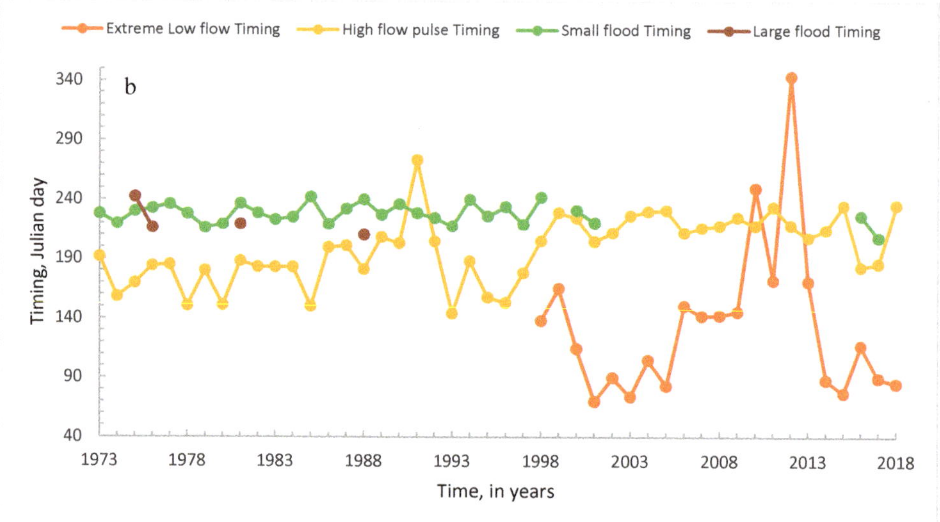

Figure 12. Duration (**a**) and Timing (**b**) of environmental flow components. Duration is the number of days a given flow component occurred and Timing is the Julian date when a given flow component occurred. Note. The Duration (**a**) in 2012 is 200 days.

3.4. Flow Components and Needs

The seasons considered in northwest Ethiopia are: the rainy season (Summer)—June, July and August; the beginning of the dry season (Autumn)—September, October and November; the dry season (Winter)—December, January and February; and end of dry season (Spring)—March, April and May. Figure 13 depicts the percentile flows for individual days of the calendar year over the duration of the discharge record for the bridge gauging station.

To relate flow regimes to ecological responses, we looked at fish spawning migration and reproduction with the percentile flows of Gumara River classified into different ecologically relevant flow components (Figure 13). This is in line with another similar study [42]. The definition of life histories of indicator fish species sensitive to hydrologic alterations in the study area is based on the literature [22,30–32,45,51–56] (Table 4). An overlay graph, Figure 13, depicts periods of fish spawning migration and reproduction with the percentile flows of Gumara river classified into different ecologically relevant flow components.

Table 4. Ecological condition of fish species in Lake Tana-Gumara River (review).

S.N.	Fish Species	Migration/Aggregation/ Period	Breeding Period/Catch	Habitat/Spawning Places/Location
1	**Labeobarbus spp.**	July–October	June–August (min in May, peak spawning in August)	Fast flowing, clear, highly oxygenated water, and gravel-bed streams or rivers;
	L. intermedius	from July–3rd week of September		
	L. tsanensis	from July–3rd week of September		
	L. brevicephalus	3rd week of August–3rd week of September		
	L. nedgia	1st week of September–1st week of October		
2	*Oreochromis niloticus*		June to October (peak in July); (3 months, June–September)	shallow littoral zone
3	*Clarias gariepinus*		April to July (peaked in June); max catch in Rainy season (peaked in June), min catch in Jan; short breeding period in July; high catch dry season (December-February); the breeding periods (1.5 months, June–July); peak in May	Largest aggregation in Gumara, abundant in the river mouth habitat; found mainly in the deeper open water area
4	*Varicorhinus beso*		max catch in August, min catch in Sep, Oct and Jan	dominated in the littoral
5	**Small barbs**			
	b. humilis		Between March and September	spawn in shallow riverine backwaters during the rainy season
	b. tanapelagius		March and September	
	b. pleurogramma			found only in the flood plain during the rainy season
6	**Large barbs or piscivorous barbs**	July to September	breeding period (4 months, mid-June to mid-October)	Gumara river, at Wanzaye. the 'large' piscivorous barbs migrate to affluent rivers for spawning

The overlaid graph, Figure 13, shows that spawning migration and reproduction begin in June and July for most fish species. Migration kicks off as freshest water or flow pulses reach the lake; high-flow pulses begin from 4.8 m^3 s^{-1} in June and small floods from 294 m^3 s^{-1} in July (Figure 13). However, some fish species, such as *B. humilis* and *B. tanapelagious*, stay in the littoral zone of the lake for reproduction [45].

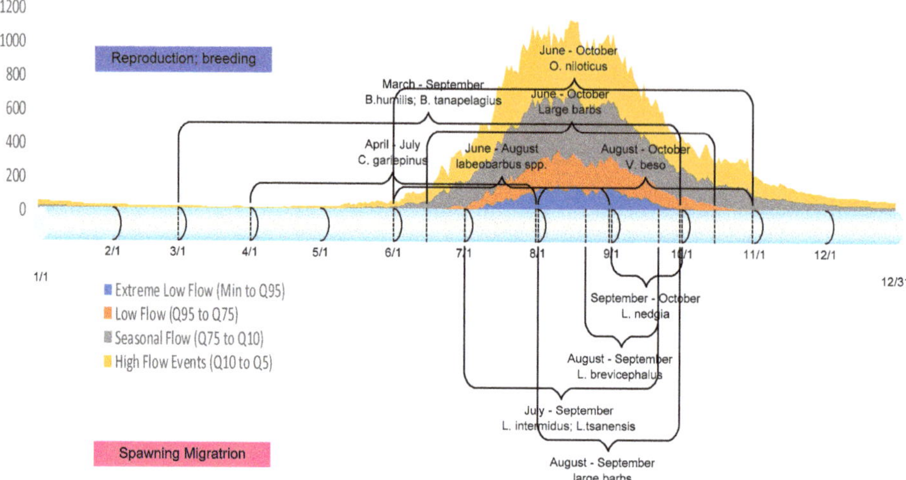

Figure 13. Flow Components vis-a-vis Spawning Migration Period (Below the line) and Reproduction Period (Above the line) of Fish Species in Lake Tana-Gumara River. Spawning migration and reproduction begin in June and July for most of the fish species. Migration kicks off as freshest water or flow pulses reach the lake, i.e., between high flow pulse begin from 4.8 m^3 s^{-1} in June and small flood of 294 m^3 s^{-1} in July.

As studies indicated, among commercially important migratory fishes of Lake Tana like *C. gariepinus* (cat fish), the major breeding season extends from April to July [51] (Table 4 and Figure 13). *C. gariepinus* requires the seasonal flow to emigrate from the Lake to the flood plain wetlands of Shesher and Welala via Gumara River to start the spawning. As shown in Figure 11, the high flow pulse (seasonal flow) has been delayed by one-month. This likely leads to a corresponding delay in the beginning of the spawning period of species like *C. gariepinus* (Figure 13).

This study has comparable results with recent studies globally which developed analytical connections between flow alterations and ecological responses (in testing the ELOHA framework) and suggested restoration possibilities [57–59]. Hence, results from this study indicate that the Gumara River and associated wetlands need restoration of ecologically relevant environmental flow components (large flood, small flood, high flow pulse, low flow and extreme low flow) to reverse the deterioration of the aquatic ecosystems in the river-wetland-lake interconnections. This will help to restore the aquatic ecosystem through regulating water resources use and appropriate conservation works in the upper watershed.

4. Conclusions

The results of this study indicate that low- and high-flow regimes of Gumara River have decreased over time. The low flow decrease was abrupt since 1997. Large floods also disappeared since 1988. One-day low flows decreased from 1.55 m^3 s^{-1} in 1973 to 0.16 m^3 s^{-1} in 2018, and 90-Day (seasonal) low flow decreased from 4.88 m^3 s^{-1} in 1973 to 2.04 m^3 s^{-1} in 2018. The decrease in flows in time is attributed to both water abstractions, catchment management interventions and rainfall variability. The cumulative rainfall required to generate runoff has increased over the study years. This flow reduction results in early disconnection of floodplain wetlands from the river, which, in turn, affects the breeding cycle of migratory fish species in the floodplain wetlands. Hence, the results of this study indicate that the Gumara River and associated wetlands need restoration of ecologically relevant flows (large flood, small flood, high-flow pulse, low-flow and extreme low-flow) to reverse the deterioration of the aquatic ecosystems in the river-wetland-lake interconnections. This will help to restore the aquatic ecosystem through regulating water resources use and appropriate conservation work in the upper watershed. The environmental flow management framework developed by NBI and the Ecological limits of hydrologic alteration (ELOHA) have helped to guide the study of the environmental flow components dynamics of the Gumara River. Finally, these results serve as the hydrological foundation for continued studies in the Gumara catchment, with the eventual goal of quantifying environmental flow requirements.

Author Contributions: Conceptualization, W.B.A., S.A.T. and M.E.M.; Data curation, W.B.A.; Formal analysis, W.B.A., S.A.T., T.A.N., D.A.M., M.V.C., K.W. and M.E.M.; Funding acquisition, W.B.A., S.A.T. and K.W.; Investigation, W.B.A.; Methodology, W.B.A., S.A.T., M.M.M., A.W. and M.E.M.; Project administration, S.A.T.; Resources, S.A.T., M.M.M., A.W. and K.W.; Supervision, S.A.T., M.M.M., A.W. and M.E.M.; Validation, T.S.S., K.W. and M.E.M.; Writing—original draft, W.B.A.; Writing—review and editing, W.B.A., S.A.T., M.M.M., M.G.D., T.A.N., D.A.M., T.S.S., M.V.C., K.W. and M.E.M. All authors have read and agreed to the published version of the manuscript.

Funding: This research was funded by Amhara Design and Supervision Works Enterprise, Global Minds Fund of Laboratory of Applied Geology and Hydrogeology in Ghent University and the PIRE project: "Taming Water in Ethiopia—An Interdisciplinary Approach to Improve Human Security in a Water-Dependent Emerging Region" supported by the National Science Foundation (NSF) under Grant No. 154587.4.

Acknowledgments: The flow data were obtained from Ministry of Water, Irrigation and Electric. Abbay basin Authority provided stage and discharge data. The authors acknowledged financial and technical support by the International Center for Tropical Agriculture (CIAT), the USAID-funded Africa RISING program, and the Water, Land and Ecosystems (WLE) Program of the CGIAR. The authors acknowledged the three anonymous reviewers and Professor Jan Nyssen who have provided comments on the manuscript.

Conflicts of Interest: The authors declare that there is no conflict of interest.

References

1. Richter, B.D.; Baumgartner, J.V.; Powell, J.; Braun, D.P. A method for assessing hydrologic alteration within ecosystems. *Conserv. Biol.* **1996**, *10*, 1163–1174. [CrossRef]

2. McClain, M.E. Balancing water resources development and environmental sustainability in Africa: A review of recent research findings and applications. *Ambio* **2013**, *42*, 549–565. [CrossRef] [PubMed]

3. Pahl-Wostl, C.; Arthington, A.; Bogardi, J.; Bunn, S.; Hoff, H.; Lebel, L.; Nikitina, E.; Palmer, M.; Poff, L.N.; Richards, K.; et al. Environmental flows and water governance: Managing sustainable water uses. *Curr. Opin. Environ. Sustain.* **2013**, *5*, 341–351. [CrossRef]

4. Reitberger, B.; McCartney, M. *Concepts of Environmental Flow Assessment and Challenges in the Blue Nile Basin, Ethiopia, in Nile River Basin*; Springer: Berlin/Heidelberg, Germany, 2011; pp. 337–358.

5. Arthington, A.H.; Naiman, R.J.; McClain, M.E.; Nilsson, C. Preserving the biodiversity and ecological services of rivers: New challenges and research opportunities. *Freshw. Biol.* **2010**, *55*, 1–16. [CrossRef]

6. Caissie, D.; El-Jabi, N.; Hébert, C. Comparison of hydrologically based instream flow methods using a resampling technique. *Can. J. Civ. Eng.* **2007**, *34*, 66–74. [CrossRef]

7. Hughes, D.A.; Hannart, P. A desktop model used to provide an initial estimate of the ecological instream flow requirements of rivers in South Africa. *J. Hydrol.* **2003**, *270*, 167–181. [CrossRef]

8. Tharme, R.E. A global perspective on environmental flow assessment: Emerging trends in the development and application of environmental flow methodologies for rivers. *River Res. Appl.* **2003**, *19*, 397–441. [CrossRef]

9. Hughes, D. Providing hydrological information and data analysis tools for the determination of ecological instream flow requirements for South African rivers. *J. Hydrol.* **2001**, *241*, 140–151. [CrossRef]

10. King, J.M.; Tharme, R.E.; De Villiers, M. *Environmental Flow Assessments for Rivers: Manual for the Building Block Methodology*; Water Research Commission: Pretoria, South Africa, 2000.

11. Zalewski, M.; Negussie, Y.Z.; Urbaniak, M. Ecohydrology for Ethiopia—Regulation of water biota interactions for sustainable water resources and ecosystem services for societies. *Ecohydrol. Hydrobiol.* **2010**, *10*, 101–106. [CrossRef]

12. Poff, N.L.; Richter, B.D.; Arthington, A.H.; Bunn, S.E.; Naiman, R.J.; Kendy, E.; Acreman, M.; Apse, C.; Bledsoe, B.P.; Freeman, M.C.; et al. The ecological limits of hydrologic alteration (ELOHA): A new framework for developing regional environmental flow standards. *Freshw. Biol.* **2010**, *55*, 147–170. [CrossRef]

13. Poff, N.L.; Allan, J.D. Functional organization of stream fish assemblages in relation to hydrological variability. *Ecology* **1995**, *76*, 606–627. [CrossRef]

14. *NBI, Preparation of NBI Guidance Document on Environmental Flows: Nile E-Flows Framework Technical Implementation Manual*; HYDROC GmbH: Siegum, Germany, 2016.

15. Alemayehu, T.; McCartney, M.; Kebede, S. The water resource implications of planned development in the Lake Tana catchment, Ethiopia. *Ecohydrol. Hydrobiol.* **2010**, *10*, 211–221. [CrossRef]

16. Awulachew, S.B. *Water Resources and Irrigation Development in Ethiopia*; Iwmi: Colombo, Sri Lanka, 2007; Volume 123.

17. Setegn, S.G.; Rayner, D.; Melesse, A.M.; Dargahi, B.; Srinivasan, R. Impact of climate change on the hydroclimatology of Lake Tana Basin, Ethiopia. *Water Resour. Res.* **2011**, *47*. [CrossRef]

18. Belete, M.A. *Modeling and Analysis of Lake Tana Sub Basin Water Resources Systems, Ethiopia*; Univ. Agrar-und Umweltwiss: Rostock, Germany, 2014.

19. Duan, W.; He, B.; Takara, K.; Luo, P.; Nover, D.; Hu, M. Impacts of climate change on the hydro-climatology of the upper Ishikari river basin, Japan. *Environ. Earth Sci.* **2017**, *76*, 490. [CrossRef]

20. Zou, S.; Jilili, A.; Duan, W.; De Maeyer, P.; Van De Voorde, T. Human and Natural Impacts on the Water Resources in the Syr Darya River Basin, Central Asia. *Sustainability* **2019**, *11*, 3084. [CrossRef]

21. Abebe, W.B.; Douven, W.J.A.M.; McCartney, M.; Leentvaar, J. *EIA Implementation and Follow Up: A Case Study of Koga Irrigation and Watershed Management Project ETHIOPIA*; Unesco-IHE: Delft, The Netherlands, 2007.

22. Anteneh, W.; Getahun, A.; Dejen, E.; Sibbing, F.A.; Nagelkerke, L.A.J.; De Graaf, M.; Wudneh, T.; Vijverberg, J.; Palstra, A.P. Spawning migrations of the endemic Labeobarbus (Cyprinidae, Teleostei) species of Lake Tana, Ethiopia: Status and threats. *J. Fish Biol.* **2012**, *81*, 750–765. [CrossRef] [PubMed]

23. Dejen, E.; Anteneh, W.; Vijverberg, J. The decline of The Lake Tana (Ethiopia) fisheries: Causes and possible solutions. *Land Degrad. Dev.* **2017**, *28*, 1842–1851. [CrossRef]

24. Duan, W.; He, B.; Nover, D.; Yang, G.; Chen, W.; Meng, H.; Zou, S.; Liu, C. Water quality assessment and pollution source identification of the eastern Poyang Lake Basin using multivariate statistical methods. *Sustainability* **2016**, *8*, 133. [CrossRef]

25. Dagnew, D.C.; Guzman, C.D.; Zegeye, A.D.; Akal, A.T.; Moges, M.A.; Tebebu, T.Y.; Mekuria, W.; Ayana, E.K.; Tilahun, S.A.; Steenhuis, T.S. Sediment loss patterns in the sub-humid ethiopian highlands. *Land Degrad. Dev.* **2017**, *28*, 1795–1805. [CrossRef]

26. Abebe, W.B.; GMichael, T.; Leggesse, E.S.; Beyene, B.S.; Nigate, F. Climate of Lake Tana Basin. In *Social and Ecological System Dynamics*; Springer: Berlin/Heidelberg, Germany, 2017; pp. 51–58.

27. Atnafu, N.; Dejen, E.; Vijverberg, J. Assessment of the Ecological Status and Threats of Welala and Shesher Wetlands, Lake Tana Sub-basin (Ethiopia). *J. Water Resour. Prot.* **2011**, *3*, 540–547. [CrossRef]

28. Karlberg, L.; Hoff, H.; Amsalu, T.; Andersson, K.; Binnington, T.; Flores-López, F.; de Bruin, A.; Gebrehiwot, S.; Gedif, A.B.; Heide, F.Z.; et al. Tackling complexity: Understanding the food-energy-environment nexus in Ethiopia's Lake Tana sub-basin. *Water Altern.* **2015**, *8*, 710–734.

29. Nagelkerke, L.; Sibbing, F.A. The Barbs of Lake Tana, Ethiopia: Morphological Diversity and Its Implications for Taxonomy, Trophic Resource Partitioning, and Fisheries. Ph.D. Thesis, Wageningen Agricultural University, Wageningen, The Netherlands, 1997.

30. Goshu, G.; Tewabe, D.; Adugna, B.T. Spatial and temporal distribution of commercially important fish species of Lake Tana, Ethiopia. *Ecohydrol. Hydrobiol.* **2010**, *10*, 231–240. [CrossRef]

31. Shitaw, T.; Medehin, S.G.; Anteneh, W. Spatio-temporal distribution of Labeobarbus species in Lake Tana. *Int. J. Fish. Aquat. Stud.* **2018**, *6*, 562–570.

32. Ameha, A.; Assefa, A. The Fate of The Barbus of Gumara River, Ethiopia. *SINET Ethiop. J. Sci.* **2002**, *25*, 1–18. [CrossRef]

33. Aynalem, S.; Bekele, A. Species composition, relative abundance and distribution of bird fauna of riverine and wetland habitats of Infranz and Yiganda at southern tip of Lake Tana, Ethiopia. *Trop. Ecol.* **2008**, *49*, 199.

34. Zur Heide, F. *Feasibility Study for a Lake Tana Biosphere Reserve, Ethiopia*; Bundesamt für Naturschutz, BfN: Bonn, Germany, 2012.

35. Funk, C.; Peterson, P.; Landsfeld, M.; Pedreros, D.; Verdin, J.; Shukla, S.; Husak, G.; Rowland, J.; Harrison, L.; Hoell, A.; et al. The climate hazards infrared precipitation with stations—A new environmental record for monitoring extremes. *Sci. Data* **2015**, *2*, 150066. [CrossRef]

36. Kendall, M. *Rank Correlation Methods*; Charles Griffin: London, UK, 1975.

37. Mann, H.B. Non-parametric tests against trend. *Econom. J. Econom. Soc.* **1945**, *13*, 245–259.

38. Running, S.; Mu, Q.; Zhao, M. *MOD16A2 MODIS/Terra Net Evapotranspiration 8-Day L4 Global 500 m SIN Grid V006*; NASA EOSDIS Land Processes DAAC: Washington, DC, USA, 2017.

39. The Nature Conservancy. *Indicators of Hydrologic Alteration Version 7.1 User's Manual*; The Nature Conservancy: Virginia, VA, USA, 2009.

40. Williams, G.P. Bank-full discharge of rivers. *Water Resour. Res.* **1978**, *14*, 1141–1154. [CrossRef]

41. Nigatu, T.A.; Zimale, F.A.; Tilahun, S.A.; Abebe, W.B.; Behulu, Y.M.; Tegegn, B.A.; Steenhuis, T.S. Rating Curves for Infrequently Calibrated River Gauging Stations at the Rivers in Lake Tana Basin, Ethiopia. *Water* **2020**, in press.

42. DePhilip, M.; Moberg, T. *Ecosystem Flow Recommendations for the Susquehanna River Basin*; The Nature Conservancy: Harrisburg, PA, USA, 2010.

43. ADSWE. *Upper Rib Large Scale Irrigation Project, Volume IV: Irrigation Agronomy*; Amhara National Regional State, Bureau of Water, Irrigation & Energy Development: Bahir Dar, Ethiopia, 2018.

44. Dejen, E.; Vreven, E. Habitat Use and Downstream Migration of 0+ Juveniles of the Migratory Riverine Spawning Labeobarbus Species (Cypriniformes: Cyprinidae) of Lake Tana (Ethiopia). In Proceedings of the Fifth International Conference of the Pan African Fish and Fisheries Association (PAFFA5), Bujumbura, Burundi, 16–20 September 2013.

45. Gebremedhin, S.; Getahun, A.; Anteneh, W.; Bruneel, S.; Goethals, P. A Drivers-Pressure-State-Impact-Responses Framework to Support the Sustainability of Fish and Fisheries in Lake Tana, Ethiopia. *Sustainbility* **2018**, *10*, 2957. [CrossRef]

46. Enku, T.; Tadesse, A.; Yilak, D.L.; Gessesse, A.A.; Addisie, M.B.; Abate, M.; Zimale, F.A.; Moges, M.A.; Tilahun, S.A.; Steenhuis, T.S. Biohydrology of low flows in the humid Ethiopian highlands: The Gilgel Abay catchment. *Biologia* **2014**, *69*, 1502–1509. [CrossRef]

47. Mhiret, D.A.; Dagnew, D.C.; Alemie, T.C.; Guzman, C.D.; Tilahun, S.A.; Zaitchik, B.F.; Steenhuis, T.S. Impact of Soil Conservation and Eucalyptus on Hydrology and Soil Loss in the Ethiopian Highlands. *Water* **2019**, *11*, 2299. [CrossRef]

48. Enku, T.M.; Ayana, A.M.; Tilahun, E.K.; Mengiste, S.A.; Abate, M.; Steenhuis, T.S. Groundwater use of a small eucalyptus patch during the dry monsoon phase. *Bioloigia* **2020**, in press.

49. Liu, B.M.; Collick, A.S.; Zeleke, G.; Adgo, E.; Easton, Z.M.; Steenhuis, T.S. Rainfall-discharge relationships for a monsoonal climate in the Ethiopian highlands. *Hydrol. Process. Int. J.* **2008**, *22*, 1059–1067. [CrossRef]

50. Tilahun, S.A.; Ayana, E.K.; Guzman, C.D.; Dagnew, D.C.; Zegeye, A.D.; Tebebu, T.Y.; Yitaferu, B.; Steenhuis, T.S. Revisiting storm runoff processes in the upper Blue Nile basin: The Debre Mawi watershed. *Catena* **2016**, *143*, 47–56. [CrossRef]

51. Ameha, A.; Abdissa, B.; Mekonnen, T. Abundance, Length-Weight Relationships and Breeding Season of Claries Gariepinus (Toleostei: Clariidae) in Lake Tana Ethiopia. *SINET Ethiop. J. Sci.* **2006**, *29*, 17–176.

52. Wudneh, T. *Biology and Management of Fish Stocks in Bahir Dar Gulf, Lake Tana, Ethiopia*; WUR: Wageningen, The Netherlands, 1998.

53. Dejen Dresilign, E. Ecology and Potential for Fishery of the Small Barbs (Cyprinidae, Teleostei) of Lake Tana, Ethiopia. Ph.D. Thesis, Wageningen University, Wageningen, The Netherlands, 2003.

54. Tewabe, D. Spatial and Temporal Distributions and Some Biological Aspects of Commercially Important Fish Species of Lake Tana, Ethiopia. *J. Coast. Life Med.* **2014**, *2*, 589–595.

55. Nagelkerke, L.A.; Sibbing, F.A. The large barbs (Barbus spp., Cyprinidae, Teleostei) of Lake Tana (Ethiopia), with a description of a new species, Barbus osseensis. *Neth. J. Zool.* **2000**, *50*, 179–214. [CrossRef]

56. De Graaf, M.; Machiels, M.A.; Wudneh, T.; Sibbing, F. Declining stocks of Lake Tana's endemic Barbus species flock (Pisces, Cyprinidae): Natural variation or human impact? *Biol. Conserv.* **2004**, *116*, 277–287. [CrossRef]

57. Endreny, T.A.; Kwon, P.; Williamson, T.N.; Evans, R. Reduced Soil Macropores and Forest Cover Reduce Warm-Season Baseflow below Ecological Thresholds in the Upper Delaware River Basin. *J. Am. Water Resour. Assoc.* **2019**, *55*, 1268–1287. [CrossRef]

58. McManamay, R.A.; Orth, D.J.; Dolloff, C.A.; Mathews, D.C. Application of the ELOHA framework to regulated rivers in the Upper Tennessee River Basin: A case study. *Environ. Manag.* **2013**, *51*, 1210–1235. [CrossRef]
59. McClain, M.E.; Subalusky, A.L.; Anderson, E.P.; Dessu, S.B.; Melesse, A.M.; Ndomba, P.M.; Mtamba, J.O.; Tamatamah, R.A.; Mligo, C. Comparing flow regime, channel hydraulics, and biological communities to infer flow–ecology relationships in the Mara River of Kenya and Tanzania. *Hydrol. Sci. J.* **2014**, *59*, 801–819. [CrossRef]

Article

A New Water Environmental Load and Allocation Modeling Framework at the Medium–Large Basin Scale

Qiankun Liu [1], Jingang Jiang [2], Changwei Jing [3], Zhong Liu [4],* and Jiaguo Qi [5],*

[1] Ocean College, Zhejiang University, Zhoushan 310058, China; 11534012@zju.edu.cn
[2] Institute of Technical Biology & Agriculture Engineering, Hefei Institutes of Physical Science, Chinese Academy of Sciences, Hefei 230031, China; gangzg@ipp.ac.cn
[3] School of Science, Hangzhou Normal University, Hangzhou 310018, China; changweij@hznu.edu.cn
[4] Ningbo Scientific Research and Design Institute of Environmental Protection, Ningbo 315000, China
[5] Center for Global Change and Earth Observation, Michigan State University, 1405 South Harrison Road, East Lansing, MI 48824, USA
* Correspondence: lqk19890627@163.com (Z.L.); qi@msu.edu (J.Q.);
   Tel.: +86-139-578-72922 (Z.L.); +86-136-9327-8045 (J.Q.)

Received: 29 September 2019; Accepted: 12 November 2019; Published: 15 November 2019

Abstract: Waste load allocation (WLA), as a well-known total pollutant control strategy, is designed to distribute pollution responsibilities among polluters to alleviate environmental problems, but the current policy is unfair and limited to single scale or single pollution types. In this paper, a new, alternative, multi-scale, and multi-pollution WLA modeling framework was developed, with a goal of producing optimal and fair allocation quotas at multiple scales. The new WLA modeling framework integrates multi-constrained environmental Gini coefficients (EGCs) and Delphi-analytic hierarchy process (Delphi-AHP) optimization models to achieve the stated goal. The new WLA modeling framework was applied in a case study in the Xian-jiang watershed in Zhejiang Province, China, in order to test its validity and usefulness. The results, in comparison with existing practices by the local governments, suggest that the simulated pollutant load quota at the watershed scale is much fairer than the existing policies and even has some environmental economic benefits at the pollutant source scale. As the new WLA is a process-based modeling framework, it should be possible to adopt this approach in other similar geographic areas.

Keywords: total water pollutant control; pollutant load allocation; equity and efficiency; regional and site-specific scale; environmental Gini coefficient models; Delphi-analytic hierarchy process models

1. Introduction

Water pollution may cause serious environmental concerns, such as oxygen deficiency and toxic algal blooms, that are unsafe and unsuitable for daily life, industry, and even agriculture [1]. Total water pollutant control is an important policy instrument to constrain pollutant discharges in order to ensure water quality [2–4]. Proper allocation of discharge quotas or waste load allocation (WLA) based on a rational and fair basis is critical for water pollution control [5–7]. However, challenges exist in practice as WLA is directly related to the economic benefit allotment and coordination among different pollution contributors. Economic development, which is viewed as a major cause of water pollution, and environmental protection, as well as social benefits, must be balanced in principal when allocating pollution quotas [8] through a fair and rational WLA system.

The United States of America was one of the first countries to adopt a total water pollutant control policy and developed the technical guidelines for WLA in 1972 [9]. These guidelines were based on various comparative analyses to determine optimal practical procedures that specifically consider cost, feasibility and effectiveness. Ronald et al. [10], for example, compared and evaluated eight different optimization formulations from 25 WLA methods and then eliminated one of them due to its excessive costs and overly conservative load estimates. The WLA concept was introduced to China in the 1980s and subsequently attracted considerable interest from many scholars and decision makers to assess its validity and practicality, and curtail China's pressing water pollution issues [11,12]. At the moment, WLA guidelines are applied at either region or site-specific scales [13–16], with a primary focus on single pollutant industrial point source (PS) pollution [17,18].

A notable drawback of existing WLA, when applied at town or village scales in China, is its inability to practically identify individual polluters, the terminal implementers of WLA (Figure 1). This is because the current WLA is pollutant-specific, without specific targeted sources of pollutions that can, in fact, come from many sources such as domestic sewage, agricultural non-point sources (NPS), and large-scale livestock breeding farms [19]. This ambiguity creates unfair and unpractical WLA policies, leading to ineffective practices. Therefore, there is a clear need to re-think the existing WLA and devise alternative fair allocation guidelines that can integrate multiple sources (either NPS or PS) at multiple scales, in order to effectively reduce water pollutions.

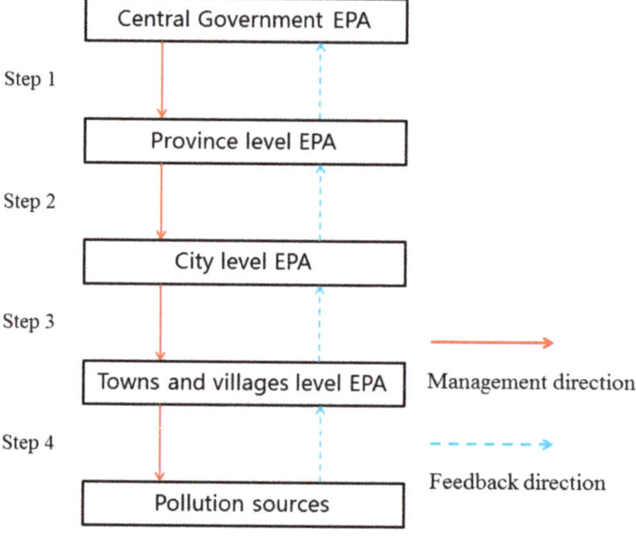

Figure 1. The cascade pollution discharge management structure in China: The solid red lines are the management decision making and the blue dashed lines are the cumulative feedback.

Another issue, when applied at regional scale, is related to the diverse nature of pollution contributors. Different administrative units and polluters design their own implementation plans based on their economic situation, natural resource availability, and administrative structures [6]. These units are likely to have different levels of pollution contributions [20], but the existing regional WLA policy treats them with equal pollution responsibility.

Similarly, unfairness exists at the pollution source scale, due to the nature of the pollutions (for example, industrial factories vs. residential vs. confined feeding) and the heterogeneity in economic status, social benefits, and organizational structures. These differences and their socioeconomic characteristics should be considered in order to design an effective WLA system [21]. Few WLA frameworks, to our knowledge, take into account of these diversity and heterogeneity issues among pollution sources, leading to unfair and, thus, inefficient and ineffective pollution reduction policies.

In this study, a new multi-scale and multi-sector optimal WLA framework for water pollution control was developed by integrating multi-constraint environmental Gini coefficients (EGCs) [22] and Delphi-analytic hierarchy process (Delphi-AHP) [23] models to account for the above unfairness issues. The new cascade WLA approach simultaneously allocates waste load reduction quotas at both the regional scale and site-specific scales. It addresses fairness issues and considers multiple pollution sources to account for socioeconomic benefits. The new WLA method was tested and validated in a case study in the Xian-jiang watershed, one of the most seriously water-polluted watersheds within the Yangtze River Delta, China. The results suggested that the new WLA approach provided a fair and optimal pollution discharge strategy for each pollution source in each district, implying that water quality targets can be practically achieved through policy interventions.

2. Multi-Scale WLA Optimization Modeling

The technical approach to integrate optimization procedures and multi-constraint criteria with socioeconomic variables is illustrated in Figure 2. Independent from the geographical characteristics (e.g., local climate, topography, and soil properties), the new WLA modeling framework integrates all pollution sources and the socioeconomic status in a watershed. Therefore, it has a broad applicability in other watersheds. The regional scale WLA (upper box in Figure 2) first uses population, GDP (gross domestic product), and total land use area data as inputs to the EGC model for allocating pollution load quotas at the administrative level (county or district level, for example). Then, the administrative-level WLA allocations are cascaded to a pollution source or a site-specific level (the lower box in Figure 2), which considers in situ discharges, population, costs associated with pollution reduction, the level of technical challenges, and pollutant discharge per unit GDP or income in the Delphi-AHP model. This model finally allocates quotas to different sectors (industrial plants, agriculture, livestock farms, and residential sewage). This integrated modeling framework was coded and implemented in the MATLAB (R2012b, ver. 8.0) environment.

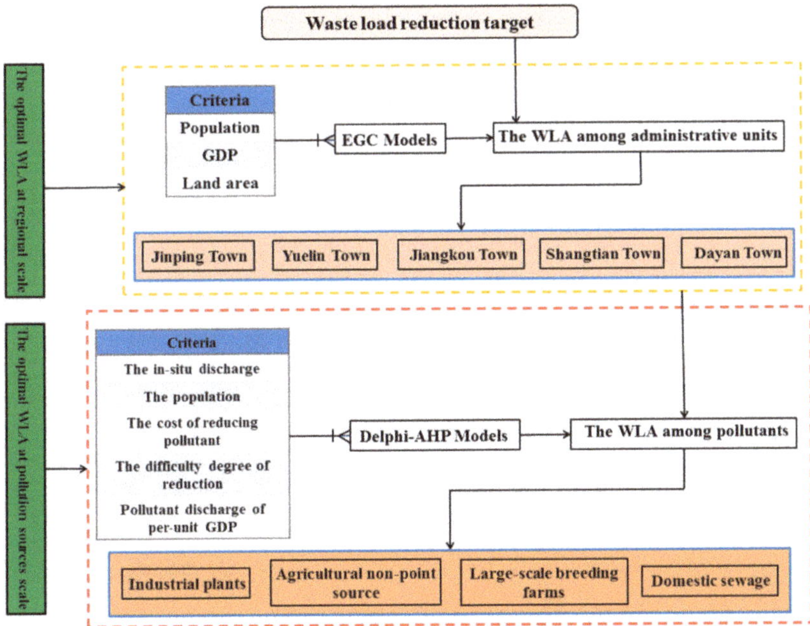

Figure 2. An integrated, multi-scale, and multi-sector optimal waste load allocation (WLA) framework in the Xian-jiang watershed, China.

2.1. Regional-Scale EGC Modeling

2.1.1. Gini Coefficient

The Gini coefficient, developed by the Italian economist Gini in the early 20th century, is an indicator to measure the extent of equality in wealth distribution and is calculated based on the Lorenz curve, using the trapezoidal area method [24,25]. The lower the Gini coefficient, the more allocation equality the society has; conversely, lower equality is indicated by a higher Gini coefficient.

2.1.2. Gini Coefficient in Environment

In recent years, the Gini coefficient has been adopted in the field of environmental inequality. From the environmental viewpoint, pollutant discharge quotas as a public issue can take advantage of the spirit of the Gini coefficient to express the fairness of various regions' emission intensities, which emphasizes the equal right in sharing the pollution discharge quota among implementers. Therefore, the EGC model was introduced to measure the inequality in the allocation of this 'resource' by comparing neighborhoods on a regional basis [26,27]. Furthermore, the EGC expands the original Gini coefficient from a single-criterion basis to a multi-criteria system. A lower EGC value suggests a more equal allocation of water pollution loads, or means that the distribution is appropriate for the region's actual social and economic development [28].

The selection of appropriate evaluation indices is crucial in the regional allocation of waste loads using EGC models and regional perspectives, such as environmental resources, local economy, and social conditions, are important in WLA quota allocation. In this study, the region's total population, GDP and land area were selected to be typical control targets of EGCs based on the following considerations [13]: First, population is a social indicator. As a public resource, each person has an equal right to enjoy the water environment capacity (WEC). This means the waste load quota should be allocated proportionally to the district populations. Second, GDP, an important economic indicator, represents incomes and, thus, the financial capacity of a local district. Third, the total land area as an indicator of natural resources is both a contributor to pollutions through, for example, NPS, but also functions as a purification medium through, for example, wetlands. Further, more land area often means that the district has more potential for industrial expansion, population growth, and economic development.

2.1.3. EGC Optimization for WLA

The optimization of WLA using EGCs at the regional scale can be realized by the following steps:

1. Determine the total amount of pollution load that has exceeded the WEC and that needs to be allocated within the watershed to meet the specified water quality goals and ensure sound water functions.
2. Compute the EGCs of the three indicators (population, GDP, and land area) under the current waste load discharge in each district as the initial values of the optimization (Equation (1)).

$$G_{0(j)} = 1 - \sum_{i=1}^{n} \left(X_{j(i)} - X_{j(i-1)} \right) \left(Y_{0(i)} + Y_{0(i-1)} \right) \qquad (1)$$

where $G_{0(j)}$ is the initial EGC of criterion j; $X_{j(i)}$ is the cumulative percentage of criterion j of the ith district; $Y_{0(i)}$ is the cumulative percentage of current pollution discharge of the ith district; and n is the number of administrative units within the watershed.

3. Compute the EGCs of the three indicators (population, GDP, and land area) under the current waste load discharge in each district as the initial values of the optimization (Equation (2)).

$$\begin{cases} G_{(j)} = 1 - \sum_{i=1}^{n} \left(X_{j(i)} - X_{j(i-1)} \right)\left(Y_{(i)} + Y_{(i-1)} \right) \\ Y_{(i)} = Y_{0(i)} - w_i \end{cases} \tag{2}$$

where $G_{(j)}$ represents the EGC of criterion j after reduction; $Y_{(i)}$ represents the cumulative percentage of waste load discharge after reduction in the ith district; and w_i represents pollution load removals of the ith district.

The multi-constrained minimum EGC models can be implemented to obtain the optimal WLA scheme, i.e., to find a set of w_i optimal solutions based on fairness. The optimization process can be described by the optimization objective function, as follows:

$$\text{Min } F = \sum_{j=1}^{3} G_j \tag{3}$$

which represents the minimum summation of EGCs for each criterion j, and by the constraint function of the total amount of removals, which is expressed as follows:

$$W = \sum_{i=1}^{n} w_i \tag{4}$$

The sum of the reduced pollution loads in each district should be the total removals (W) in the whole basin.

The constraint function for each EGC is as follows:

$$G_{(j)} \leq G_{0(j)} \tag{5}$$

The EGC for every criterion j should get smaller or be at least equal to the initial EGC so that more equity can be achieved after optimization.

The constraint of the removal rate in each district is stated as follows:

$$\begin{cases} P_i = \dfrac{w_i}{w_{0(i)}} \\ P_{i0} \leq P_i \leq P_{i1} \end{cases} \tag{6}$$

where P_i is the pollution load removal rate for the ith district; P_{i0} and P_{i1} are the lower and upper limits of the removal rate for the ith district, respectively; and $w_{0(i)}$ is the current waste load discharge of the ith district. An appropriate upper limitation of 20% and lower limitation of 1% were set in this study to make sure each district had a waste load removal, but not one that was beyond its capability.

The multi-constrained optimization model equations can be solved using the Monte Carlo simulation method [29] to find the optimal solution vector, w_i, for WLA at the regional scale and to avoid causing conflicts of interest among districts.

2.2. Multi-Sector Scale: The Delphi-AHP Models

The AHP [23] model, an important approach to multi-criteria decision making (MCDM), can help decision-makers address a complex problem, while permitting the prioritization of alternatives. It involves four basic steps, as follows: (a) modelling a complex problem as a hierarchy, (b) the valuation of weights for each criterion, (c) the aggregation of weights into overall priorities for the alternatives, and (d) consistency test [23]. Since its introduction, the AHP has been widely used in diverse fields, such as manufacturing systems, supplier selection, strategy selection, and many others [30–33].

The hierarchical modelling of the problem and the ability to simultaneously adopt qualitative and quantitative judgements are its major strengths [23], making it appropriate for our study.

2.2.1. Hierarchical Structure Construction

Considering the overall benefits of environmental, economic, and social factors, as well as the technological level in the districts, a hierarchical structure of the problem was designed based on the evaluation criteria system, combining qualitative and quantitative analyses (Figure 3). The need to allocate the total pollutant discharge was selected as the overall goal (Level 1). The indicators selected at level 2 include the in situ discharge (b_1), the population size (b_2), the pollutant reduction cost (b_3), the technical difficulties of pollutant reduction (b_4), and the discharge per unit of GDP (b_5) for each pollution source.

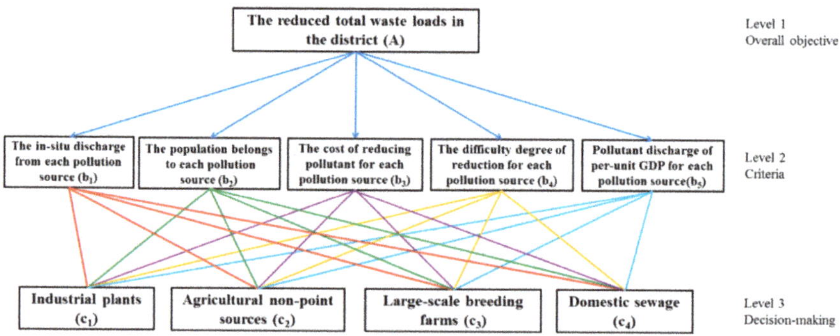

Figure 3. The hierarchy analysis framework of optimization criteria.

The main considerations at Level 2 are as follows: (1) The value b_1 not only reflects the current status of pollution source discharge, but also indicates the shares from various sectors of total pollution in a district; (2) b_2, as a typical social criterion, was chosen to represent the demographic distribution among different pollution sources; (3) b_3 and b_4 are important criteria of economic efficiency in the WLA between various sectors (here, we argue that WLA should be inclined to sectors with the features of easy operation and low cost in waste load reduction); and (4) b_5 reflects the differences in technical management levels, which can effectively promote technological innovation in pollution reduction, production efficiency, and low emissions. The main pollutant sources, as WLA receptors in the district, constitute level 3 as the final decision-makers. This optimization is to ensure that the removal quotas were allocated in an equitable and highly economically efficient way, which is very important in water resource management in China.

2.2.2. Pairwise Comparison Matrix Construction

The construction of the pairwise comparison, $A = \left(a_{ij}\right)_{n \times n}$, as a positive reciprocal matrix is as follows:

$$A = \begin{bmatrix} 1 & a_{12} & \cdots & a_{1n} \\ a_{21} & \cdots & a_{ij} & \cdots \\ \cdots & a_{ji} & \cdots & \cdots \\ a_{n1} & \cdots & \cdots & 1 \end{bmatrix}_{n \times n} \tag{7}$$

where a_{ij} is the comparison value between object i and j; $a_{ij} = 1/a_{ji}$, $a_{ii} = 1$.

The value a_{ij} estimates the relative importance among pairs of i and j at a given level using a linear scale of the integers 1–9 with their reciprocals, as advanced by Saaty [34] (Table 1), which was regarded as the best scale to represent weight ratios. To minimize the impact of individual subjectivity, the Delphi method, which has been widely used, especially for natural science and technology fields [35–37], was used in building the above matrix. According to the expert selection principle, we adopted three steps in this study, as follows: First, we selected an evaluation team of 15 local experts in the field of water environment, who are knowledgeable about the studied watershed. The team was composed of five experts with PhD degrees, and ten engineers from environmental protection authorities in Ningbo city. Second, we asked the experts fill out a table that was designed based on the grading standard of Table 1 to evaluate the importance of mutual indicators. Finally, the survey results of all the experts are integrated in the pairwise comparisons using the geometric mean method. An analysis of the expert's opinions allowed us to minimize the individual subjectivity effects in the process of qualitative evaluation [38].

Table 1. The 1 to 9 fundamental scale.

Intensity of Importance	Definition
1	Equal importance
2	Weak
3	Moderate importance
4	Moderate plus
5	Strong importance
6	Strong plus
7	Very strong or demonstrated importance
8	Very, very strong
9	Extreme importance

Suppose the pairwise comparison contains n elements and p experts, then a_{ij} can be calculated using the following:

$$a_{ij} = \left(\prod_{r=1}^{p} a_{ij}^{r} \right)^{\frac{1}{p}},$$

(8)

where a_{ij}^{r} represents the importance of element i relative to j, determined by the rth expert.

2.2.3. Weight Ratios Calculation

To construct the comparison matrix $B = \left(b_{ij} \right)_{5\times5}$, the priorities $\vec{P} = \begin{bmatrix} P_1 & P_2 & \cdots & P_5 \end{bmatrix}^T$, which reflect the mutual importance of all indicators at the criteria level, can be derived from the eigenvalue method described in Equation (9), as follows:

$$\begin{cases} \overline{M_i} = \sqrt[5]{\prod_{j=1}^{5} b_{ij}} \\ P_i = \dfrac{\overline{M_i}}{\sum_{i=1}^{5} \overline{M_i}} \end{cases} \quad (i = 1, 2, \cdots, 5; \ j = 1, 2, \cdots 5)$$

(9)

where $\overline{M_i}$ is the geometric mean of the ith line in the pairwise matrix B and P_i is the weight of the ith criterion in level 2.

The local comparison matrices $(C_n = \left[\left(c_{ij} \right)_{4\times4} \right]_n, \ n = 1, 2, \cdots, 5)$ at the decision-making level were constructed with respect to each element in the preceding level. The local priorities of $L_i = \begin{bmatrix} \vec{L_1} & \vec{L_2} & \cdots & \vec{L_5} \end{bmatrix}_{4\times5}$ can be calculated according to the method proposed above. Finally, with an additive aggregation, with normalization, the sum of the local priorities to unity was adopted to determine the global priority, $\vec{W} = \begin{bmatrix} w_1 & w_2 & w_3 & w_4 \end{bmatrix}^T$, at the decision-making level for the WLA among pollution sources, expressed by Equation (10), as follows:

$$\vec{W} = \begin{bmatrix} \vec{L_1} & \vec{L_2} & \cdots & \vec{L_5} \end{bmatrix} \times \vec{P}$$

(10)

where $\vec{W}$ represents the global priorities of the pollution sources; L_i is the local priorities of the pollution sources with respect to the ith criteria; and $\vec{P}$ represents the weights of the criteria in level 2.

2.2.4. Consistency Test

As priorities make sense only if derived from consistent or near consistent matrices, the evaluation of each matrix must go through consistency verification to ensure the preferable credibility of the results. Saaty [39] proposed a consistency index (CI) (Equation (11)), which is related to the eigenvalue method, as follows:

$$CI = \frac{\lambda_{max} - n}{n - 1} \tag{11}$$

where n = the dimension of a comparison matrix; and λ_{max} = the maximal eigenvalue of the matrix.

Having obtained the CI, it was then substituted into Equation (12) to calculate the consistency ratio (CR), as follows:

$$CR = \frac{CI}{RI} \tag{12}$$

where RI is the random index (the average CI of 500 randomly filled matrices), which is depicted in Table 2 [40].

Table 2. Random indices form.

n	1	2	3	4	5	6	7	8	9
RI	0	0	0.58	0.89	1.12	1.26	1.36	1.41	1.46

A perfect consistency should not be expected when working with the AHP [18]. When RI < 0.1, the consistency of the comparison matrix is satisfied or the priorities should be slightly revised.

3. Study Area

The Xian-jiang watershed lies in Ningbo city in the Zhejiang province, China, and is a typical coastal watershed within the Yangtze River Delta (Figure 4). It has a drainage area of 306.70 km^2 with main rivers and 5 town-level administrative divisions (Figure 4). The watershed has experienced serious water pollution problems [41,42], resulting from industrial plants, domestic sewage, agricultural run-off, and large-scale livestock breeding. The primary pollutants are chemical oxygen demand (COD), ammonia nitrogen (NH$_3$-N), and total phosphorus (TP), providing a good case study for multi-scale and multi-sector WLA analysis.

This watershed was set to reach a water quality of grade II and III for the upper and lower reaches, respectively, by 2020, according to the standards set by Chinese EPA [43] as part of the 13th National Five-Year Plan of China. In this study, a steady-state 1-D water quality response model coupled with the matrix algorithm and the section control method was integrated within the WLA framework to calculate the pollutants that need to be reduced in order to meet the stated targets.

The watershed was divided into 1350 uniform computational elements for model calculation. The flow velocity in each element was obtained using the Manning hydrodynamic equation [44], and then the concentration of a specific pollutant at x distance was calculated based on the water balance equation [45], expressed as follows in Equation (13):

$$\frac{\partial c}{\partial t} + u\frac{\partial c}{\partial x} = E\frac{\partial^2 c}{\partial x^2} - k_1 c \tag{13}$$

where c is the pollutant concentration (mg/L); u is the flow velocity (m/s); t is the time of river water flowing through from the headwater to somewhere (s); x is the distance that river water flows through in time t (m), x(t) = ut and k_1 is the pollutant degradation coefficient (d^{-1}); and E is the longitudinal dispersion coefficient.

The computed pollutant concentrations of each element were applied to the section control method (Equations (14)) to assess the water environmental capacity of each element.

$$W_i = Q_i(C_{Si} - C_{0i}) + Q_i C_{Si}\left(e^{k_i t_i} - 1\right) + \sum_{l=1}^{n} q_l C_{si} + C_{si} \sum_{l=1}^{n} q_l\left(e^{k_i t_i} - 1\right) \tag{14}$$

where W_i, C_{si}, and C_{0i} represent the water environmental capacity, the target pollutant concentration, and the actual concentration of the pollutant of element i, respectively; k_i is the degradation coefficient of the pollutant in element i; and t_i is the time period used to flow through element i.

Eventually, the pollution loads that need to be reduced are 340.16, 25.11, and 11.41 tons for COD, NH_3–N, and TP, respectively, in order to meet the stated water quality targets by year 2020. Detailed descriptions of the simulation process and model validation are available in Liu et al. [19], and are not presented here for the sake of brevity.

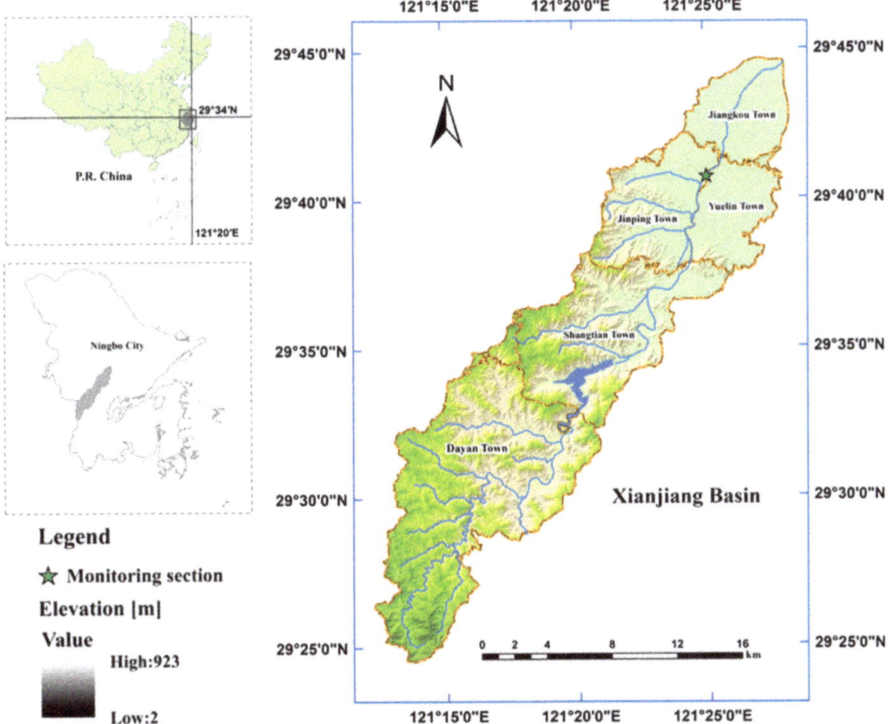

Figure 4. The geographic characteristics of Xian-jiang watershed within Ningbo City, China.

4. Results and Discussion

4.1. Regional-Scale Allocation Results

4.1.1. Optimal WLA Results in Districts

The population, GDP, and land area data of each town in the Xian-jiang watershed were obtained from the Towns Agency of Statistics of Ningbo City (2015) [46], while the in situ discharges of pollutants (2015) for each town in the basin were calculated (Table 3). The EGC for each criterion under the constraints was optimized, and the sum of all criteria's EGCs was minimized to obtain an optimal solution vector; namely, the reduced waste loads of the five towns considered.

Table 3. The criteria statistics and in situ discharges of pollutants of districts in the Xian-jiang watershed (2015). COD, chemical oxygen demand; NH$_3$-N, ammonia nitrogen; TP, total phosphorus; GDP, gross domestic product.

Districts	Criteria			Pollutants In Situ Discharge (t)		
	Population	GDP (Ten Million Chinese Yuan)	Land Area (km^2)	COD	NH$_3$-N	TP
Jinping Town	112,209	891.94	41.82	2664.71	201.41	50.68
Yuelin Town	49,262	1131.02	27.09	1696.88	97.93	29.66
Dayan Town	13,591	133.28	127.53	393.79	28.83	11.76
Jiangkou Town	29,885	367.74	31.10	1599.39	128.92	25.94
Shangtian Town	19,071	234.46	73.53	411.91	38.07	12.73

As shown in Figure 5, the targeted pollutant removals and proportion of COD discharge in Jiangkou, Jinping, Yuelin, Dayan, and Shangtian were 198.09 t (12.39%), >72.91 t (2.74%), >43.25 t (2.55%), >15.91 t (4.04%), and >10.00 t (2.43%), respectively. Jiangkou (14.96 t) and Jinping (5.49 t) were the two districts with the largest NH$_3$-N reduction loads, which accounted for 11.61% and 2.73% of the total removal rates, respectively. A total of 11.41 t of TP pollutants needed to be reduced, of which the largest removal (proportion) was in Jiangkou, with 5.24 t and 20.22%, followed by Jinping and Dayan with 3.97 t (7.84%) and 0.63 t (5.36%), respectively. The removal rate of the remaining districts (Shangtian and Yuelin) were relatively low, and were both less than 4.00%, with removals of 0.50 t and 1.06 t, respectively.

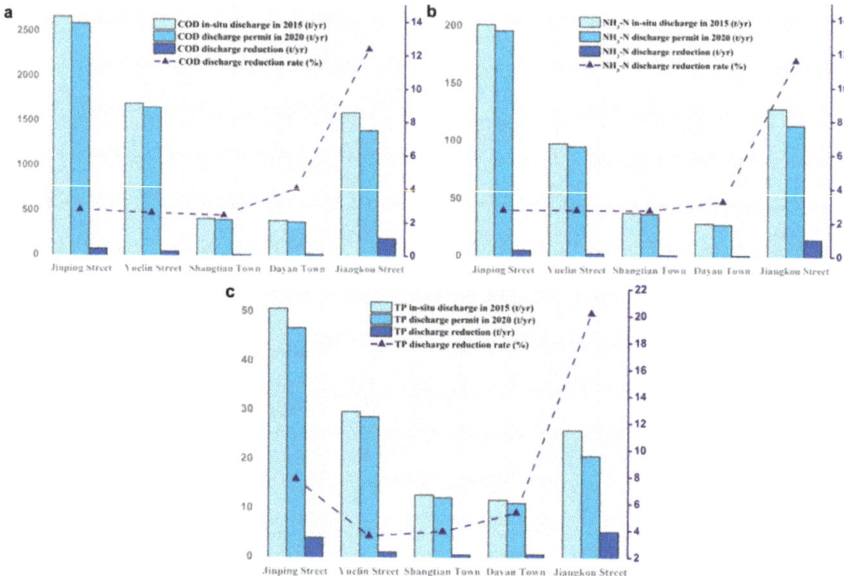

Figure 5. The in situ discharge (2015) and allocation of pollutant discharge quotas after optimization in the Xian-jiang watershed: (**a**) chemical oxygen demand (COD); (**b**) ammonia nitrogen (NH$_3$-N); and (**c**) total phosphorus (TP).

The regional allocation results revealed that Jiangkou was significantly higher than other districts in terms of both load removals and proportion of the three pollutants. Part of the reason was the high in situ pollutant discharges in the district and part was the modes of social economic developments. Conversely, as the town with the lowest pollutant discharge per unit of GDP and population, Shangtian was granted a lesser quota both in removal and rate, as predicted by the model.

Interestingly, the towns with the largest pollutant discharges were not necessarily those with the highest proportions of removal quota because the allocation of a pollutant discharge quota at the regional scale considered all factors, including the region's economic efficiency and social equality, not just the magnitude of the in situ discharge of pollutants. For example, although Jinping and Yuelin exceeded the in situ COD discharges of Jiangkou, it was assigned a lesser pollutant removal and proportion. This is probably due to the backward modes of social economic development in Jiangkou, such as less-developed domestic sewage networks and large areas of extensive agriculture, which resulted in higher discharges of pollutants per unit of GDP and population, compared with the other two towns.

4.1.2. EGCs Before and After Optimization

The EGC optimization models were used to calculate the EGC of each criterion among the districts (Table 4) and draw the Lorenz curves before and after the optimization for the three pollutants.

Table 4. EGC$_S$ of multiple criteria in the Xian-jiang watershed.

Pollutants	Criteria	EGC (P [1])	EGC (G [1])	EGC (L [1])	Total
	Before optimization	0.162	0.215	0.583	0.960
COD	After optimization	0.152	0.201	0.576	0.929
	Decrease	0.010	0.014	0.007	0.031
	Before optimization	0.146	0.271	0.569	0.986
NH$_3$-N	After optimization	0.129	0.258	0.569	0.956
	Decrease	0.017	0.013	0.000	0.030
	Before optimization	0.141	0.217	0.521	0.879
TP	After optimization	0.126	0.197	0.519	0.842
	Decrease	0.015	0.020	0.002	0.037

[1] P, G, and L indicate the abbreviations for population, GDP, and land area, respectively. EGC, environmental Gini coefficient.

Take the COD as an example, the EGC of population vs. COD in situ discharge is the smallest, at 0.162, and is within the reasonable range in WLA equity at the regional scale, indicating that the current distribution of COD discharge among districts is balanced according to population. The GDP-based EGC for COD in situ discharge among districts is 0.215, which keeps the COD discharge to the districts relatively reasonable in relation to local economy, but still has the potential for optimization. In particular, the greatest distribution inequity of COD discharge at regional scale occurs in the land area, with an EGC of up to 0.583, exceeding the warning sign of equity (0.400) [6], suggesting that the current COD load discharge in the five districts does not match well with the land area indicator.

Interestingly, considering the geographical pattern of pollution sources and function regionalization in the watershed using ArcGIS (ver. 10.2) (Figure 6), it can be further inferred that the high EGCs corresponding to land area are mainly imputed to Dayan town. This accounts for nearly half of the total land area (41.95%), but accommodates only little pollution discharge from pollution sources, making up only 5.82% (COD), 5.82% (NH$_3$-N), and 8.99% (TP) of the total pollutant discharges in the entire watershed. The situations of the other pollutants are similar to that of COD, having the same distribution pattern of EGCs for the equivalent criteria (Table 4).

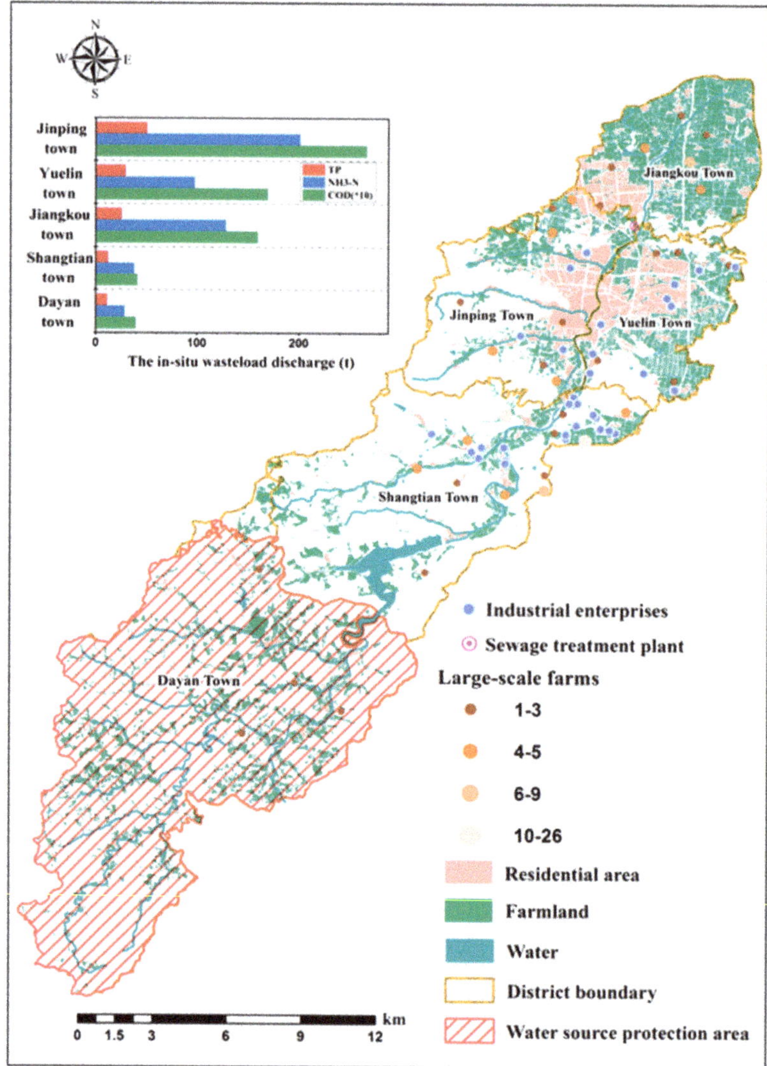

Figure 6. The administrative divisions, pollution source distribution, and geographical patterns of the Xian-jiang watershed.

Table 4 and Figure 7 highlight the decrement and amplitude of EGCs corresponding to the three criteria after optimization. Note that the EGCs of COD, NH_3-N, and TP after optimal WLA at the regional scale were all less than those of initial pollutant discharge, and the Lorenz curves after optimal allocation were closer to approaching the line of absolute equality. The results revealed that the optimal allocation of the removals at the regional scale brought more accordant and equitable responsibilities in relation to the districts' respective shares of socioeconomic development, and the optimal allocation coordinated the pollutant discharge quota with the natural environment. In other words, a more equitable pollutant discharge quota at the regional scale was achieved after WLA optimization using the EGC models.

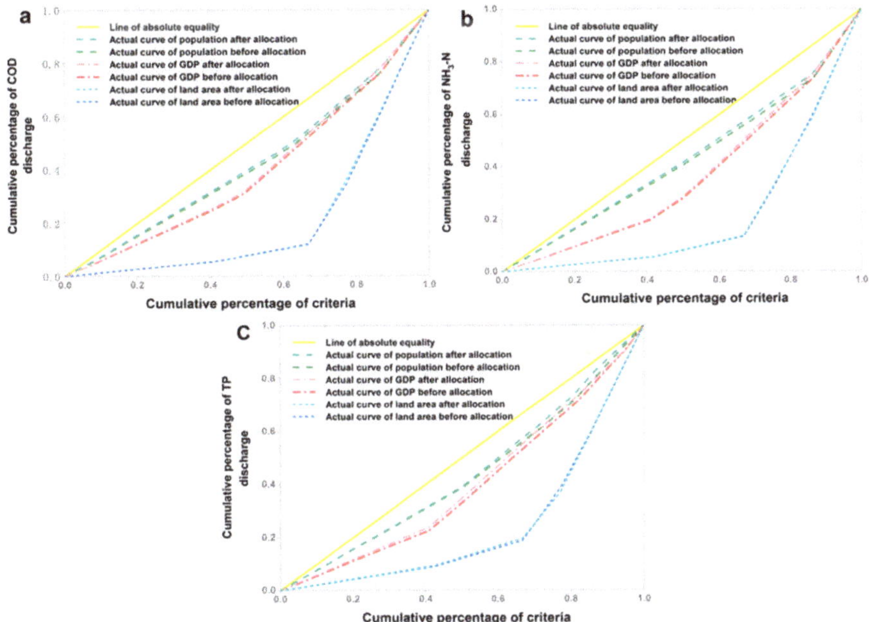

Figure 7. Lorenz curves of pollutant discharge based on the three criteria before and after allocation: (**a**) COD; (**b**) NH$_3$-N; and (**c**) TP.

In view of the overall trends, as depicted in Figure 7, it is noteworthy that the EGCs and the Lorenz curves do not change very much after the optimization. Two reasons can be identified as the cause, as follows: (1) The pollution discharge had been distributed to each district relatively equally according to the criteria, i.e., population and GDP, before optimization; and (2) there are upper and lower limits on the allocation of pollution removal to each district in order to be compatible with the local socioeconomic conditions. Considering the constraints in the EGC optimization models, the shape of the Lorentz curve can only be adjusted gradually to avoid an out-of-range affordability of removals for local managers, due to the tenacious struggle to decrease EGCs.

4.1.3. Factor Analysis

A contribution coefficient method was used to further determine the regional inequality factors referring to waste load in situ discharge vs. the criteria. The contribution coefficient, taking 1 as the threshold, is the contribution ratio of regional evaluation criteria to in situ pollutant discharge in a certain district, expressed as Equation (15) [47], as follows:

$$CC_j = \left(M_{ij}/M_j\right)/\left(W_i/W\right) \tag{15}$$

where CC_j represents the population contribution coefficient (PCC), the green contribution coefficient (GCC), and the land area contribution coefficient (LACC) with respect to the criteria j of population, GDP, and land area, respectively; M_{ij} is the magnitude of criterion j in the ith district, and M_j is the sum of criteria j in the watershed; W_i is the pollutant in situ discharge of the ith district; and W is the total pollutant discharges in the watershed.

Table 5 provides the CCs of the three criteria at each district in the watershed. As shown, in 2015, the PCCs of COD, NH$_3$-N, and TP pollutants for Jiangkou were lowest in the region, with 0.56, 0.51, and 0.67, respectively. This is possibly due to the relatively undeveloped economic and living conditions, with some deficiencies in domestic sewage treatment systems. Moreover, the district with well-developed river systems and fairly fertile soil, an area in the plain river network of the lower Xian-jiang watershed, is the primary rice-growing area in the watershed (Figure 6). Paddy fields make up as high as 83.2% of the whole area, leading to serious agricultural NPS pollution. In short, all these factors produced an increasingly incisive contradiction between the local population and the water environment, making this an unfair-factor district in terms of EGCs vs. population.

Table 5. Population contribution coefficients (PCCs), green contribution coefficients (GCCs), and land area contribution coefficients (LACCs) of districts for the three pollutants in the Xian-jiang watershed (2015).

Districts	PCC			GCC			LACC		
	COD	NH_3-N	TP	COD	NH_3-N	TP	COD	NH_3-N	TP
Jinping Town	1.40	1.23	1.29	0.82	0.79	0.83	0.35	0.34	0.36
Yuelin Town	0.88	1.11	0.97	1.64	2.07	1.81	0.36	0.45	0.40
Dayan Town	1.04	1.04	0.68	0.83	0.83	0.54	7.28	7.28	4.71
Jiangkou Town	0.56	0.51	0.67	0.56	0.51	0.67	0.44	0.40	0.52
Shangtian Town	1.27	1.11	0.87	1.40	1.11	0.87	4.01	3.18	2.51
Mean	1.03	1.00	0.89	1.05	1.06	0.94	2.49	2.33	1.70

Yuelin town, with the highest economic level in the basin, contributed 41% to the total GDP and received the largest GCCs for COD (1.64), NH_3-N (2.07), and TP (1.81). This indicated that the contribution rate of GDP to the entire region was more than that of pollution discharge in this district. It further revealed the advancements in the cleanliness of production processes and sewage treatment efficiency in this district. Although Yuelin exceeded two-fold the in situ pollutant discharges of Shangtian and Dayan towns, it was assigned to a lesser pollutant removal. In addition, the GCCs of Shangtian for COD and NH_3-N also surpassed the threshold of the green contribution factor (1.00). Moreover, the GCCs of COD and NH_3-N in Jiangkou, and TP in Dayan, both belonging to the less-developed areas, were less than the green contribution standard, with low values of 0.56, 0.51, and 0.54, respectively. It could be speculated that GDP output in these regions is characterized by high pollution and low efficiency, which are the main factors leading to the unfairness of WLA at the regional scale based on the GDP index.

Dayan town, referring to the criterion of land area, presented the largest LACCs among the districts for COD (7.28), NH_3-N (7.28), and TP (4.71), owing to its unique geographical location. In spite of vast expanses and richness in natural resources, the district is completely subject to the water source conservation area, leading to strict restrictions on local resident size and density, industrial development, and agricultural scale for the protection of drinking water security (Figure 6). In contrast, the LACCs of the three pollutants in Jinping and Yuelin towns were all below the value of 0.50, revealing a heavy discharge of pollutants into the river per unit of land area in these districts.

Different from the PCC and GCC, the LACC is not as high as it could be when there is a low utilization of land resources. Thus, the high LACCs of Dayan town should be lowered accordingly. However, our results found that its specific geographical location (water source conversation area) makes it hard for the area to be adjusted by human intervention, such as demographic migration, industrial distribution adjustment, and land use and land cover conversion. The human-induced adjustment may result in the destruction of the original ecological environment and directly threaten the safety of drinking water, which would instead lead to an 'inequity' in WLA. Consequently, the high LACCs are acceptable in this particular water function division. At the same time, we urge the optimal allocation of removals at the regional scale with the assistance of geographic information system (GIS) technology.

Furthermore, to mitigate the polluted water more efficiently, targeting a specific area instead of a whole watershed has been recommended as a cost-effective method in many previous studies [48–51]. Hence, the spatial zonation, in which the PCC and GCC were synthetically considered, into critical source areas (CSA) (PCC < 1, GCC < 1), improving areas (PCC < 1, GCC > 1 or GCC < 1, PCC > 1), and safety areas (PCC > 1, GCC > 1) among the five districts was performed by ArcGIS (ver. 10.2) (Figure 8).

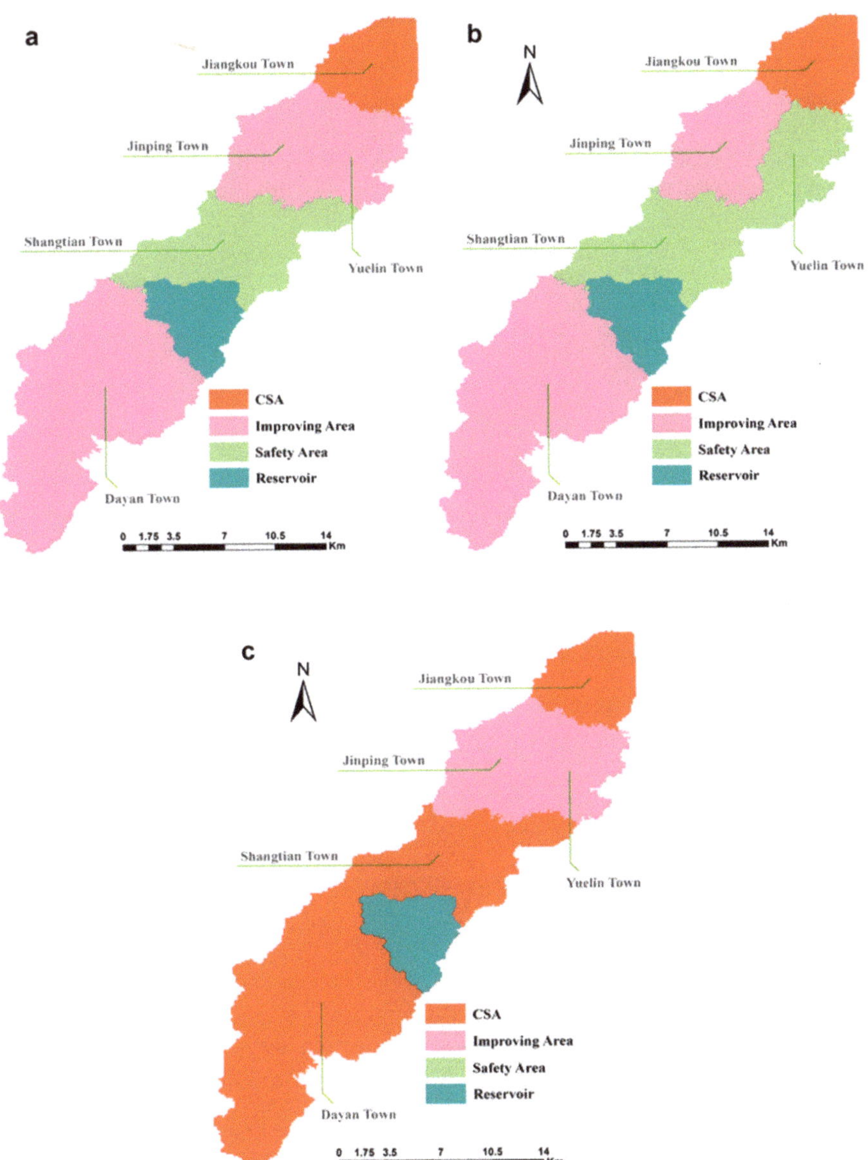

Figure 8. The spatial pattern of WLA unfair-factor districts: (**a**) COD; (**b**) NH_3-N; (**c**) TP.

As seen in Figure 8, both COD and NH_3-N have a similar pattern of unfair-factor districts, except Yuelin town. Among them, only Jiangkou town was identified as a CSA of unfair factors in the watershed, revealing that this area faces an acute contradiction between population, economic profit, as well as the water environment under per-unit pollutant discharges, and should be managed and controlled preferentially for the two pollutants. Hence, the EGC optimal models handed out the most discharge removals and highest proportions of the three pollutants to this district. Meanwhile, the fairness of the optimization allocation results of using EGCs models at regional level was further verified. Conversely, Shangtian town was characterized by having both PCC and GCC values of more than 1.00, and therefore belongs to the safety area, which exhibited a high-efficiency economic output with low consumption of WEC, while there was less impact of population scale on the water environment. In other words, with limited funds and manpower, environmental protection agencies can directly skip these regions in the total pollutant control process. Water environmental management in improving areas, which either have a high level of economic development but a sharp conflict between population and environment or are under a less stressful environment from per-unit population but have low environmental and economic benefits, should improve their deficiencies while maintaining the status level of superior contribution coefficients.

In particular, for TP, a significant difference was found for the grade distribution of unfair-factor districts compared with COD and NH_3-N, in that all the five districts were identified as CSAs or improving areas, and CSAs accounted for a higher proportion of administrative units than with the other two pollutants, suggesting that the in situ allocation of the TP discharge quota at the regional scale should preferentially be improved.

4.2. Results of Pollutant Source Scale Allocation

4.2.1. Pairwise Comparisons and Synthesis of the Weights

Jinping town was identified as the inner city of the watershed with multi-sector pollution, and was selected as a typical region to perform the optimal WLA at the pollution source scale for the three pollutants using the coupled Delphi-AHP algorithm. The partial parameters of the criteria at level 2 can be achieved through a quantitative calculation using the 2015 statistical yearbook data of villages and towns in Ningbo city (Table 6) [46]. Furthermore, the above-mentioned Delphi method was adopted to evaluate the indicators with quantizing difficulties (b_3 and b_4).

Table 6. The partial indicators of alternatives at the decision-making level, with respect to the elements of the criteria level (2015).

Pollution Sources	In Situ Discharge of Pollutants (t)			Population Scale			Pollutant Discharge Per Unit of GDP [1]		
	COD	NH_3-N	TP	COD	NH_3-N	TP	COD	NH_3-N	TP
Industrial plants	16.15	0.80	0.00	12,340	12,340	0	1.03	0.05	0
Agricultural NPS	158.33	6.61	5.66	1390	1390	1390	25.28	1.06	0.90
Large-scale breeding farms	82.57	8.00	6.45	360	360	360	13.17	1.28	1.03
Domestic sewage	1665.89	159.67	23.53	82,514	82,514	82,514	5.61	0.54	0.08

[1] Pollutant discharge per-unit of GDP: kg per 10,000 yuan. NPS, non-point sources.

The pairwise comparisons at the criteria level and the decision-making level were constructed on the basis of the above-mentioned statistical results. Subsequently, the respective local priorities of the three pollutants, namely, the importance weights of the attributes in each level, were solved using the principal eigenvector method [39], which was fully illustrated in Section 2.2.3. The maximum eigenvalues and the consistency test of comparison matrices are depicted in Table 7. The consistency test confirmed that the obtained pairwise comparisons were satisfactory in relation to consistency requirements (CR < 0.1), since they revealed a small inconsistency, inducing only a small distortion.

Table 7. The maximum eigenvalues and consistency parameters of the three pollutants.

Pollutants	Pairwise Comparison Matrices	Maximum Eigenvalue (λ_{max})	Consistency Index (CI)	Consistency Ratio (CR)
	B	5.057	0.014	0.013
COD	b1-C	4.158	0.053	0.058
	b2-C	4.165	0.055	0.061
	b3-C	4.051	0.017	0.019
	b4-C	4.051	0.017	0.019
	b5-C	4.135	0.045	0.050
NH$_3$-N	b1-C	4.158	0.053	0.058
	b2-C	4.165	0.055	0.061
	b3-C	4.051	0.017	0.019
	b4-C	4.051	0.017	0.019
	b5-C	4.102	0.034	0.038
TP	b1-C	3.029	0.015	0.025
	b2-C	3.065	0.032	0.056
	b3-C	3.009	0.005	0.008
	b4-C	3.009	0.005	0.008
	b5-C	3.025	0.012	0.021

Our preliminary results (Figure 9) revealed that the importance sequence of the five criteria at level 2 was b_5 (0.41), > b_1 (0.25), > b_3 (0.14) = b_4 (0.14), > b_2 (0.06). The criterion b_5 (pollutant discharge per unit of GDP), as the most important factor judged by experts in the field in Section 2.2.2, reflecting the environmental economic benefits of alternative pollution sources, captures 41% of the entire sum of weights, making it the most determinant criteria. This is understandable since economic development is the paramount issue that governments care about, especially for developing countries such as China, despite this factor's negative impacts on the environment, leading to the extreme importance of the tradeoff between economic development and environmental protection. The relative preference for b_5 can effectively promote technical innovation and guide waste load discharge in a low-level and high-efficiency direction.

Specific to each sector, taking COD as an example, domestic sewage at the decision-making level is the primary weight in influencing and shaping WLA among pollutant sources, with respect to b_1 (0.65) and b_2 (0.58) (Figure 9). This can be explained by the fact that the magnitude of domestic pollution, with respect to both the in situ pollutant discharge and the population scale, is much higher than the other sectors in the district. Agricultural NPS (c_2) and large-scale livestock farming (c_3) were identified as determinants in b_3 (c_2 with a priority of 0.47), b_4 (c_3 with a priority of 0.47), and b_5 (c_2 with a priority of 0.58).

In particular, the industrial PS discharge has the smallest weight ratios among the alternatives, in correspondence with all the criteria except b_2 (large-scale livestock breeding source, 0.04), with priorities of 0.04 (b_1), 0.07 (b_3), 0.07 (b_4), and 0.04 (b_5), respectively. One possible explanation is that not only does industrial PS have the lowest waste load discharge, but more crucially, it has the highest GDP output per unit of pollutant discharge, revealing a high environmental economic benefit far ahead of any other sectors. In addition, the cost and difficulty degree of pollution removal for industrial PS, specifically considering costly wastewater treatment facilities, exorbitant operating costs, and its economic contribution to local employment and taxation, were much higher compared to the other sectors, such as agricultural and large-scale farming sources.

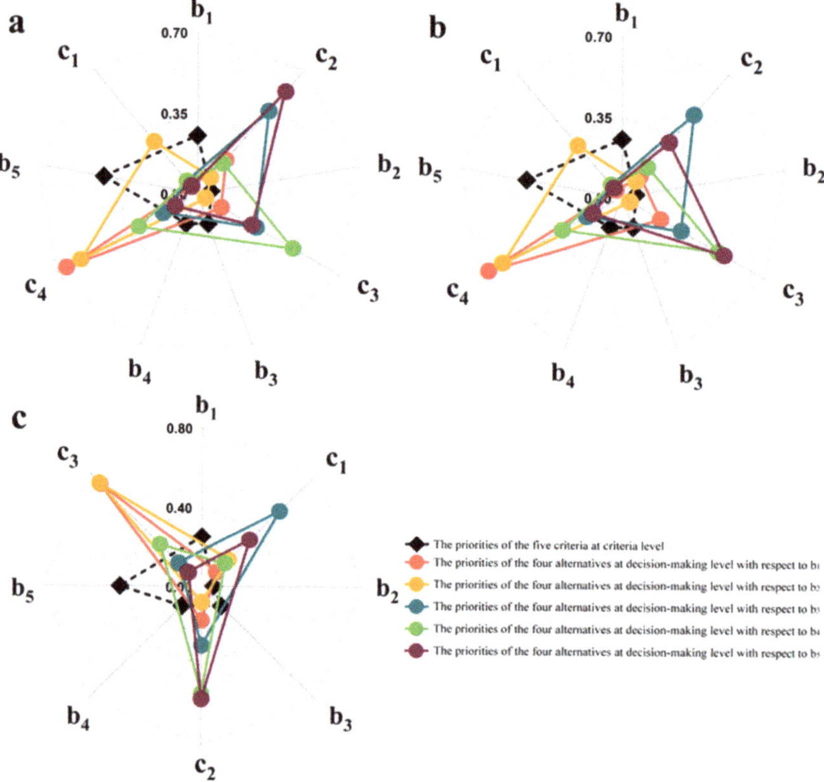

Figure 9. The local priorities of elements at criteria and decision-making levels: (**a**) COD; (**b**) NH$_3$-N; and (**c**) TP.

Subsequently, the results of the local priorities across all criteria were aggregated using Equation (10) to obtain the global priorities of the three pollutants, which presented an overall preference rating of WLA among each sector in the district (Table 8). The results indicated that the pollution source for COD, NH$_3$-N, and TP that possessed the highest weights was agricultural NPS (COD: 0.38) and large-scale breeding farms (NH$_3$-N, TP: 0.36 and 0.40), and the next sector was domestic sewage with priorities of 0.30 (COD), 0.32 (NH$_3$-N), and 0.33 (TP), respectively. Conversely, industrial PS was assigned with the least weight for COD (0.07) and NH$_3$-N (0.07), and agricultural NPS for TP with a priority of 0.27, referring to regional waste load removals. This implies that industrial PS was relatively unimportant in WLA at the pollutant source scale.

Table 8. The global priorities across the alternatives at the decision-making level, with respect to the overall objective.

Pollutants	The Normalized Global Priorities			
	C_1	C_2	C_3	C_4
COD	0.07	0.38	0.25	0.30
NH$_3$-N	0.07	0.25	0.36	0.32
TP	0.00	0.27	0.40	0.33

4.2.2. The Optimal WLA among Pollution Sources

WLA was completed for the pollution sources in the district according to the global priorities at the decision-making level (Figure 10). The results revealed that agricultural NPS had the largest removal of COD, with up to 27.71 t, and the large-scale livestock breeding source was the biggest contributor of waste load removals for NH_3-N (1.98 t) and TP (1.59 t).

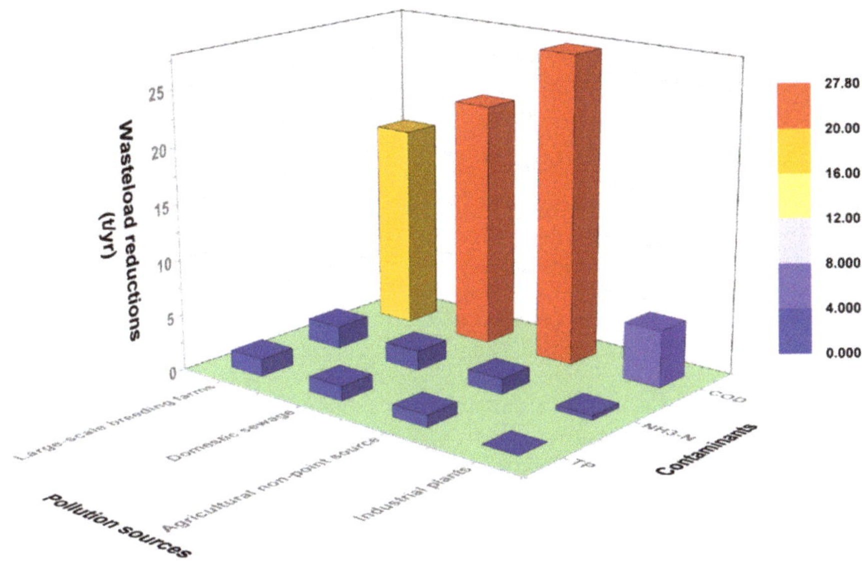

Figure 10. The WLA of the three pollutants among pollution sources in Jinping town.

For COD, the agricultural source had the second biggest in situ discharge, which was slightly inferior to domestic sewage, and, concurrently, had a lower cost of discharge reduction compared to the other sectors. Moreover, it was noticed that the pollutant discharge per unit of GDP (25.28 kg/ten thousand yuan) of agricultural NPS was much greater than that of other pollution sources, being 1.92, 4.51, and 24.54 times greater that of large-scale livestock breeding (13.17 kg/ten thousand yuan), domestic sewage (5.61 kg/ten thousand yuan), and industrial plants (1.03 kg/ten thousand yuan), respectively (Table 6). Such results suggest that the COD discharge from agricultural NPS not only has a significant impact on the water environment, but more seriously, it leads to economic inefficiency in water resource utilization. China has to feed 20% of the world's population with 7% of the world's cropland [52], leading to the excessive use of fertilizer and manure to increase food production. Conversely, less developed agricultural technology and extensive cultivation result in lower fertilizer application efficiency and a larger loss of nutrients [53]. Therefore, controlling COD discharge from agricultural sources has become the most critical and preferred way for environmental management sectors to achieve the goal of clean water in this district, such as through the gradual reduction of fertilizer usage, the establishment of ecological agriculture, and the control of water and soil loss.

Large-scale livestock breeding, different from COD, has the second NH_3-N and TP discharges in size, following domestic sewage, and has the most pollutant discharge per unit of GDP; namely, the lowest environmental and economic performance generated by per-unit pollutant discharge. On the other hand, synthetically considering its low removal difficulty and cost relative to sewage treatment facilities for industrial plants and underground sewer networks for domestic sewage, the Delphi-AHP optimal models expressed a WLA preference for the large-scale breeding source, thereby making it the key sector for removals of NH_3-N and TP in the district.

Industrial enterprises above the designated size were the sector with the least removal for COD and NH_3-N, with only 5.1 t and 0.38 t, respectively. This is mainly due to its minimal waste load discharge and, furthermore, its maximal economic performance under per-unit pollutant discharges, as compared with the other three sectors in the system. Our optimal results confirmed the viewpoint, as noted by Sun et al. [13], that the allocated sectors of equal pollution discharge with higher economic efficiency should get more shares of the waste load discharge quota.

5. Conclusions

Local EPAs used to allocate waste removal directly to districts or pollution sources within their jurisdictions through administrative orders, based primarily on their past experiences [54]. However, a common consequence of these WLA allocations is unfairness that leads to disputes and blaming each other for shared waste responsibilities, owing to the fuzzy allocation basis and biased subjectivity. The developed WLA framework allocates waste load removal simultaneously, at multiple scales and among different sectors, considering both the principles of equity and efficiency for the specific implementers. This is very valuable for decision-makers in providing critical information (i.e., the best compromise solutions for WLA) and practical guidance on water pollution control. The new modeling framework, based on the premise of equality, minimizes environmental costs while maximizing economic efficiency, which is extremely important for communities in developing countries.

The results revealed that the removal and proportions of pollutants are significantly associated with the region's actual socioeconomic development modes, which confirms the viewpoint that socioeconomic factors will have significant impact on water management in the future, as noted by Reynard et al. [55]. Inadequate sewage networks, the lack of wastewater treatment technologies, and intensive land cultivation resulted in high pollutant discharges per unit of GDP and population. Thus, for local authorities, these are the key targeted regions for the total waste load control. Decision-makers should encourage advanced regions in their continuous efforts to improve the economic efficiency of water environmental resources, while impelling relatively less developed areas to economically transition and gradually phase out inefficient sectors.

The distribution of unfair districts (Figure 8) provided information on how different waste load removal priorities should be considered when different districts are targeted for a specific pollutant. Meanwhile, for different pollutants, the urgency also varied throughout the administrative units across the watershed. This suggested that the current one-size-fits-all allocation strategy of waste removal adopted by environmental management sectors should be changed, and instead of preferentially reducing total pollution loads, they should focus on key pollutants (TP in our study) and regions (CSAs). This can help decision-makers save significant costs under the conditions of limited capital and energy.

The WLA results at the pollution sources scale suggest agricultural NPS as the sector with the largest removal quota of COD, as they are the large-scale breeding source for NH_3-N and TP. There are certain characteristics that high-reduction sectors tend to share, as follows: (1) These sectors are the major contributors of waste load discharge in the district. (2) Moreover, they have the most pollutant discharge per unit of GDP, implying the lowest environmental and economic benefits in relation to water resource utilization, and (3) at the same time, they feature lower removal costs and operational challenges. Therefore, the pollution sources with the above-mentioned characteristics should be the top priorities in the reduction of surplus waste load discharge.

It is also noteworthy that most previous studies focused primarily on the WLA of removals among PS pollution. Conversely, our results highlight the industrial pollution source as the last option for reduction from an environmental-economic benefit perspective. Instead, the often overlooked types, such as agricultural NPS and domestic sources, deserve more attention, especially in extensive rural areas.

Jinping town was selected as a typical example to demonstrate the optimal allocation of removals among pollution sources. Further efforts should include comparative analyses of districts with various

Water **2019**, *11*, 2398

landscape features to explore the optimization of WLA at the pollution source scale, under different regional characteristics. The multi-scale and multi-sector WLA method developed in this study can provide an important reference for similar research in other watersheds, especially for extensive developing areas that are subjected to multi-source pollution and serious surface water deterioration.

Author Contributions: Conceptualization, Q.L. and J.Q.; Data curation, Q.L.; Formal analysis, Q.L.; Funding acquisition, Z.L. and J.Q.; Investigation, Q.L. and J.J.; Methodology, Q.L. and J.J.; Project administration, Z.L. and J.Q.; Software, Z.L. and J.Q.; Supervision, Z.L.; Validation, Q.L. and J.J.; Visualization, Q.L.; Writing—original draft, Q.L.; Writing—review & editing, C.J. and J.Q.

Funding: This research was funded by Asian-Pacific Network for Global Change Research, grant number ARCP2015-06CMY-Wu, and USDA project through the AgBioResearch at Michigan State University, grant number MICL02264.

Acknowledgments: The authors would like to specially thank the Ningbo Scientific Research and Design Institute of Environment Protection for providing the valuable census data of pollution sources. The socio-economic statistical data were provided by the township governments of Ningbo city. We also gratefully acknowledge the administrative support by the Ningbo Environmental Protection Bureau.

Conflicts of Interest: The authors declare no conflict of interest. The sponsors had no role in the design, execution, interpretation, or writing of the study.

References

1.	Carpenter, S.R.; Caraco, N.F.; Correll, D.L.; Howarth, R.W.; Sharpley, A.N.; Smith, V.H. Nonpoint pollution of surface waters with phosphorus and nitrogen. *Ecol. Appli.* **1998**, *8*, 559–568. [CrossRef]

2.	Ge, M.; Wu, F.P.; You, M. A provincial initial water rights incentive allocation model with total pollutant discharge control. *Water* **2016**, *8*, 525. [CrossRef]

3.	Yandamuri, S.R.; Srinivasan, K.; Murty Bhallamudi, S. Multiobjective optimal waste load allocation models for rivers using nondominated sorting genetic algorithm-II. *J. Water Resour. Plan. Manag.* **2006**, *132*, 133–143. [CrossRef]

4.	Ghadouani, A.; Coggins, L.X. Science, technology and policy for water pollution control at the watershed scale: Current issues and future challenges. *Phys. Chem. Earth A/B/C* **2011**, *36*, 335–341. [CrossRef]

5.	Petrosjan, L.; Zaccour, G. Time-consistent Shapley value allocation of pollution cost reduction. *J. Econ. Dyn. Control* **2003**, *27*, 381–398. [CrossRef]

6.	Zhang, Y.; Wang, X.Y.; Zhang, Z.M.; Shen, B.G. Multi-level waste load allocation system for Xi'an-Xianyang section, Weihe River. *Procedia Environ. Sci.* **2012**, *13*, 943–953. [CrossRef]

7.	Mao, Z.P.; Li, H.E. Solution to the optimal allocation of reduced pollutant in the total amount control. *Water Resour. Water Eng.* **1999**, *10*, 25–30. (In Chinese)

8.	Jørgensen, S.; Zaccour, G. Incentive equilibrium strategies and welfare allocation in a dynamic game of pollution control. *Automatica* **2001**, *37*, 29–36. [CrossRef]

9.	Driscoll, E.D.; Mancini, J.L.; Mangarella, P.A.; Pawlow, J.R. Biochemical oxygen demand/dissolved oxygen. In *Technical Guidance Manual for Performing Waster Load Allocations. Book 2, Streams and Rivers*, 1st ed.; USEPA: Washington, DC, USA, 1983; pp. 108–110. ISBN EPA-440/4-84-020.

10.	Ronald, C.; Irene, K. An evaluation of eight waste-load allocation methods. *JAWRA J. Am. Water Resour. Assoc.* **1985**, *21*, 833–839. [CrossRef]

11.	Meng, C.; Wang, X.; Li, Y. An optimization model for waste load allocation under water carrying capacity improvement management, a case study of the Yitong River, Northeast China. *Water* **2017**, *9*, 573. [CrossRef]

12.	Gu, R.; Dong, M. Water quality modeling in the watershed-based approach for waste load allocation. *Water Sci. Technol.* **1998**, *38*, 165–172. [CrossRef]

13.	Sun, T.; Zhang, H.; Wang, Y.; Meng, X.; Wang, C. The application of environmental Gini coefficient (EGC) in allocating wastewater discharge permit: The case study of watershed total mass control in Tianjin, China. *Resour. Conserv. Recycl.* **2010**, *54*, 601–608. [CrossRef]

14.	Li, S.; Morioka, T. Optimal allocation of waste loads in a river with probabilistic tributary flow under transversmixing. *Water Environ. Res.* **1999**, *71*, 156–162. [CrossRef]

15.	Guo, H.; Ni, J.; Wang, Y. Regional waste-load allocation based on macroeconomic optimization model. *J. Basic Sci. Eng.* **2003**, *11*, 113–142. (In Chinese)

16. Qin, D.; Wei, A.; Lu, S.; Luo, Y.; Liao, Y.; Yi, M.; Song, B. Total water pollutant load allocation in Dongting lake area based on the environmental Gini coefficient method. *Res. Environ. Sci.* **2013**, *26*, 8–15. (In Chinese)

17. Qin, X.; Huang, G.; Chen, B.; Zhang, B. An interval-parameter waste-load-allocation model for river water quality management under uncertainty. *Environ. Manag.* **2009**, *43*, 999–1012. [CrossRef]

18. Liu, G.; Wang, H.; Qiu, L. Construction of a cooperation allocating initial discharge permits system for industrial source points in a lake basin. *Resour. Environ. Yangtze Basin* **2012**, *21*, 1223–1229. (In Chinese)

19. Liu, Q.; Jiang, J.; Jing, C.; Qi, J. Spatial and seasonal dynamics of water environmental capacity in mountainous rivers of the Southeastern Coast, China. *Int. J. Environ. Res. Public Health* **2018**, *15*, 99. [CrossRef]

20. Yang, Y.F.; Fu, G.W. Total pollution load distribution at national level and the regional diversity. *Acta Sci. Circum.* **2001**, *21*, 129–133. (In Chinese)

21. Anbumozhi, V.; Radhakrishnan, J.; Yamaji, E. Impact of riparian buffer zones on water quality and associated management considerations. *Ecol. Eng.* **2005**, *24*, 517–523. [CrossRef]

22. Druckman, A.; Jackson, T. Measuring resource inequalities: The concepts and methodology for an area-based Gini coefficient. *Ecol. Econ.* **2008**, *65*, 242–252. [CrossRef]

23. Ishizaka, A.; Labib, A. Review of the main developments in the analytic hierarchy process. *Expert Syst. Appl.* **2011**, *38*, 14336–14345. [CrossRef]

24. Bosi, S.; Seegmuller, T. Optimal cycles and social inequality: What do we learn from the Gini index? *Res. Econ.* **2006**, *60*, 35–46. [CrossRef]

25. Ellison, G.T. Letting the Gini out of the bottle? Challenges facing the relative income hypothesis. *Soc. Sci. Med.* **2002**, *54*, 561–576. [CrossRef]

26. Jacobson, A.; Milman, A.D.; Kammen, D.M. Letting the (energy) Gini out of the bottle: Lorenz curves of cumulative electricity consumption and Gini coefficients as metrics of energy distribution and equity. *Energy Policy* **2005**, *33*, 1825–1832. [CrossRef]

27. White, T.J. Sharing resources: The global distribution of the Ecological Footprint. *Ecol. Econ.* **2007**, *64*, 402–410. [CrossRef]

28. Wang, Y.; Niu, Z.G.; Wang, W. Application of Gini coefficient in total waste load district allocation for surface-water. *China Popul. Resour. Environ.* **2008**, *18*, 177–180. (In Chinese)

29. Rubinstein, R.Y.; Kroese, D.P. *Simulation and the Monte Carlo Method*, 3rd ed.; John Wiley & Sons: Hoboken, NJ, USA, 2016; pp. 238–245. ISBN 1118632206.

30. Yang, C.L.; Chuang, S.P.; Huang, R.H. Manufacturing evaluation system based on AHP/ANP approach for wafer fabricating industry. *Expert Syst. Appl.* **2009**, *36*, 11369–11377. [CrossRef]

31. Labib, A.W. A supplier selection model: A comparison of fuzzy logic and the analytic hierarchy process. *Int. J. Prod. Res.* **2011**, *49*, 6287–6299. [CrossRef]

32. Sipahi, S.; Timor, M. The analytic hierarchy process and analytic network process: An overview of applications. *Manag. Decis.* **2010**, *48*, 775–808. [CrossRef]

33. Chen, M.K.; Wang, S.C. The critical factors of success for information service industry in developing international market: Using analytic hierarchy process (AHP) approach. *Expert Syst. Appl.* **2010**, *37*, 694–704. [CrossRef]

34. Saaty, T.L. Decision-making with the AHP: Why is the principal eigenvector necessary. *Eur. J. Oper. Res.* **2003**, *145*, 85–91. [CrossRef]

35. Landeta, J. Current validity of the Delphi method in social sciences. *Technol. Forecast. Soc. Chang.* **2006**, *73*, 467–482. [CrossRef]

36. Okoli, C.; Pawlowski, S.D. The Delphi method as a research tool: An example, design considerations and applications. *Inf. Manag.* **2004**, *42*, 15–29. [CrossRef]

37. Ferri, C.P.; Prince, M.; Brayne, C.; Brodaty, H.; Fratiglioni, L.; Ganguli, M.; Jorm, A. Global prevalence of dementia: A Delphi consensus study. *Lancet* **2005**, *366*, 2112–2117. [CrossRef]

38. Dalkey, N.; Helmer, O. An experimental application of the Delphi method to the use of experts. *Manag. Sci.* **1963**, *9*, 458–467. [CrossRef]

39. Saaty, T.L. A scaling method for priorities in hierarchical structures. *J. Math. Psychol.* **1977**, *15*, 234–281. [CrossRef]

40. Saaty, T.L. Highlights and critical points in the theory and application of the analytic hierarchy process. *Eur. J. Oper. Res.* **1994**, *74*, 426–447. [CrossRef]

41. Gu, C.; Hu, L.; Zhang, X.; Wang, X.; Guo, J. Climate change and urbanization in the Yangtze River Delta. *Habitat Int.* **2011**, *35*, 544–552. [CrossRef]
42. Zhou, Y.; Fu, J.S.; Zhuang, G.; Levy, J.I. Risk-based prioritization among air pollution control strategies in the Yangtze River Delta, China. *Environ. Health Perspect.* **2010**, *118*, 1204. [CrossRef]
43. State Environmental Protection Administration of the China (SEPA). *Environmental Quality Standards for Surface Water (GB3838-2002)*; China Environmental Science Press: Beijing, China, 2002. (In Chinese)
44. Fetter, C.W. *Applied Hydrogeology*, 4th ed.; Prentice Hall: Upper Saddle River, NJ, USA, 2000; pp. 58–65. ISBN 978-0130882394.
45. Fu, G.W. *Mathematical Model of River Water Quality and Its Simulation*, 1st ed.; China Environmental Science Press: Beijing, China, 1987; pp. 67–146. ISBN 7-80010-002-2. (In Chinese)
46. State Statistical Bureau Ningbo Investigation Team. *2015 Ningbo Statistical YearBook*; Ningbo Municipal Statistics Bureau: Ningbo, China, 2016. (In Chinese)
47. Wang, J.; Lu, Y.; Zhou, J.; Li, Y.; Cao, D. Analysis of China resource-environment Gini coefficient based on GDP. *China Environ. Sci.* **2006**, *26*, 111–115. (In Chinese)
48. Tim, U.S.; Mostaghimi, S.; Shanholtz, V.O. Identification of critical nonpoint pollution source areas using geographic information systems and water quality modeling. *JAWRA J. Am. Water Resour. Assoc.* **1992**, *28*, 877–887. [CrossRef]
49. Young, R.A.; Onstad, C.A.; Bosch, D.D.; Anderson, W.P. AGNPS: A nonpoint-source pollution model for evaluating agricultural watersheds. *J. Soil Water Conserv.* **1989**, *44*, 168–173.
50. Pionke, H.B.; Gburek, W.J.; Sharpley, A.N. Critical source area controls on water quality in an agricultural watershed located in the Chesapeake Basin. *Ecol. Eng.* **2000**, *14*, 325–335. [CrossRef]
51. Poudel, D.D.; Lee, T.; Srinivasan, R.; Abbaspour, K.; Jeong, C.Y. Assessment of seasonal and spatial variation of surface water quality, identification of factors associated with water quality variability, and the modeling of critical nonpoint source pollution areas in an agricultural watershed. *J. Soil Water Conserv.* **2013**, *68*, 155–171. [CrossRef]
52. Zhao, J.; Luo, Q.; Deng, H.; Yan, Y. Opportunities and challenges of sustainable agricultural development in China. *Philos. Trans. R. Soc. B Biol. Sci.* **2007**, *363*, 893–904. [CrossRef]
53. Shaviv, A.; Mikkelsen, R.L. Controlled-release fertilizers to increase efficiency of nutrient use and minimize environmental degradation-A review. *Fertil. Res.* **1993**, *35*, 1–12. [CrossRef]
54. Nikoo, M.R.; Kerachian, R.; Karimi, A. A nonlinear interval model for water and waste load allocation in river basins. *Water Resour. Manag.* **2012**, *26*, 2911–2926. [CrossRef]
55. Reynard, E.; Bonriposi, M.; Graefe, O.; Homewood, C.; Huss, M.; Kauzlaric, M.; Liniger, H.; Rey, E.; Rist, S.; Schädler, B.; et al. Interdisciplinary assessment of complex regional water systems and their future evolution: How socioeconomic drivers can matter more than climate. *WIREs Water* **2014**, *1*, 413–426. [CrossRef]

 *water*

Article

Assessing the Functional Response to Streamside Fencing of Pastoral Waikato Streams, New Zealand

Katharina Doehring *, Joanne E. Clapcott and Roger G. Young

Cawthron Institute, 98 Halifax Street East, Nelson 7010, New Zealand
* Correspondence: kati.doehring@cawthron.org.nz; Tel.: +64-3-548-2319

Received: 20 May 2019; Accepted: 26 June 2019; Published: 29 June 2019

Abstract: In New Zealand, streamside fencing is a well-recognised restoration technique for pastoral waterways. However, the response of stream ecosystem function to fencing is not well quantified. We measured the response to fencing of eight variables describing ecosystem function and 11 variables describing physical habitat and water quality at 11 paired stream sites (fenced and unfenced) over a 30-year timespan. We hypothesised that (1) fencing would improve the state of stream ecosystem health as described by physical, water quality and functional indicators due to riparian re-establishment and (2) time since fencing would increase the degree of change from impacted to less-impacted as described by physical, water quality and functional indicators. We observed high site-to-site variability in both physical and functional metrics. Stream shade was the only measure that showed a significant difference between treatments with higher levels of shade at fenced than unfenced sites. Cotton tensile-strength loss was the only functional measurement that indicated a response to fencing and increased over time since treatment within fenced sites. Our results suggest that stream restoration by fencing follows a complex pathway, over a space-for-time continuum, illustrating the overarching catchment influence at a reach scale. Small-scale (less than 2% of the upstream catchment area) efforts to fence the riparian zones of streams appear to have little effect on ecosystem function. We suggest that repeated measures of structural and functional indicators of ecosystem health are needed to inform robust assessments of stream restoration.

Keywords: functional indicators; stream restoration; riparian vegetation; fencing; cotton tensile-strength loss; wood decay; ecosystem metabolism; organic matter transport; catchment restoration; structure-function relationships

---

1. Introduction

Concerns about stream degradation have led to increasing efforts worldwide over the last two decades to restore these ecosystems [1,2]. It is estimated that over US\$1 billion is spent annually on various aquatic habitat rehabilitation activities in the United States alone [1] and similar efforts are underway in Europe to rehabilitate and reconnect river habitats, such as the Skjern (€37.7 million total project cost for largest river system in Denmark) [3], Rhine (€4.4 billion total project costs; Germany) [4] and Danube River basins (€6 billion total project cost; ten European countries) [5]. A large variety of restoration techniques are applied to mitigate and reverse human impacts on rivers and streams. Fencing of waterways, for example, is a common restoration approach in New Zealand, and generally occurs in pastoral land to exclude livestock from streams, thereby reducing bank erosion and direct faecal bacteria input [6]. Once fenced, stream banks are often replanted to accelerate the re-establishment of riparian vegetation. As riparian vegetation grows over time, stream health is expected to improve due to lower water temperatures from shade, reduced nutrient, sediment and faecal bacteria input, and increased habitat provision for aquatic and riparian biota [7–9].

Understanding the effectiveness of stream restoration techniques is critical for the cost-effective design and implementation of future restoration efforts. However, the environmental outcomes of stream restoration projects are rarely evaluated. Bernhardt et al. [1] found that only 10% of ≈3700 reviewed restoration projects in the United States recorded some form of assessment or monitoring, and if restoration measures were monitored, evaluations were highly subjective, rather than based on robust scientific measures. In a more recent international review of 644 restoration projects, Palmer et al. [10] noted that when indicators of riverine attributes were measured, results were highly variable and dependent on the restoration technique and the indicators measured. For example, successful outcomes were most often recorded when riparian management was the focus of restoration and physical habitat, biophysical processes and benthic communities were the focus of assessment [10]. A lack of robust post restoration assessments, due to inconsistent or incomplete indicators for example, hinders the public and scientific community from learning from successes and failures, and thus from improving future practices [1,11,12].

Stream restoration is further complicated by temporal and spatial variation in drivers of stream health, including disturbance regimes, which can lead to hysteresis or slow recovery over time and multiple recovery pathways and endpoints [13–15]. Restoration hysteresis describes when the recovery pathway is different to the degradation pathway and may occur when feedback mechanisms that hold an ecosystem in a certain state are not fully addressed by restoration techniques (Suding & Hobbs 2009). Therefore, active stream restoration, such as large wood addition, e.g., Brooks et al. [16], in addition to passive stream restoration (such as streamside fencing) may be required to overcome restoration hysteresis. Alternative recovery trajectories include the 'rubber band' model where recovery follows closely the degradation pathway, the 'humpty-dumpty' model where recovery can result in various endpoints that are distinct from the pre-degraded condition, and the 'shifting target model' where both the recovery pathway and endpoint are unpredictable [13]. Assessing stream restoration at any single point in time without knowing the nature of the recovery pathway could provide an inaccurate assessment of restoration success. Therefore, assessing the success or failure of restoration activities becomes a question of not only what to measure, but when to measure.

In regards to what to measure, overall stream health is characterised by a combination of indicators that describe ecosystem structure and function [17,18]. Poor stream health, as a result of land use intensification, for example, is characterised by structural indicators that describe physical habitat, e.g., altered substrate composition and channel shape [19,20]; flow regime, e.g., altered velocities, [21–23]; water quality, e.g., increased water temperature and nutrient concentrations; [24,25] and biotic communities such as microorganisms and macroinvertebrates, e.g., more pollution-tolerant communities, lower numbers of taxa and higher algal biomass, [26–28]. While less commonly applied, functional indicators describe stream ecosystem processes, whereby poor stream health as a result of land use intensification, for example, is described by changes in the rates of organic matter retention and decomposition, and changes in ecosystem metabolism [29–31]. For example, Quinn et al. [32] showed that retention of coarse particulate organic matter (a major resource subsidy for invertebrate communities) is low in small pastoral streams due to a lack of in-stream structures, such as wood and roots provided by bank and riparian vegetation. Similarly, McTammany et al. [33] found that gross primary production (GPP) was positively correlated to light in agricultural streams, due to a lack of riparian shading. Reviews show that ecosystem metabolism responds to a range of environmental stressors related to a lack of established riparian vegetation, such as increased GPP and ER with increased nutrient enrichment, warmer water temperatures and increased sunlight [18,29].

To support the use of functional indicators in stream health assessments, Young, Matthaei and Townsend [18] proposed a framework to assign organic matter decomposition and ecosystem metabolism values to management bands (i.e., "healthy", "satisfactory" and "poor"). A "healthy" stream ecosystem showed characteristics close to unmodified conditions (e.g., closed canopy, lower water temperatures, moderate rates of GPP, ER and organic matter decomposition), whereas a stream ecosystem classified as "poor" showed characteristics typical of modified or impacted systems (e.g., open canopy, pasture sites, higher (or lower) rates of GPP, ER and organic matter decomposition).

To further support the use of organic matter decomposition as a functional indicator and address the inherent variability in decomposition due to organic matter type [34], Tiegs et al. [35] proposed cotton strip assays as a standardised measure of organic matter decomposition potential. Cotton strips, like leaf litter, are comprised predominantly of cellulose, but unlike leaf litter, do not contain inhibitory substances that can constrain processing by microbes or invertebrates. As such, cotton strip decomposition reflects the cellulose decomposition potential of streams and is primarily driven by the colonisation of resident microbes [36,37]. In general, cotton strip decomposition rates are higher in streams with higher temperatures, higher nutrient concentrations and high sediment input (i.e., indicative of "poor" ecosystem health), and lower in streams with a "healthy" ecosystem [35,38,39]. For example, Bierschenk, Savage, Townsend and Matthaei [39] showed that cotton strip decomposition rates increased with higher dissolved phosphorus concentrations which were linked to intensely developed catchments, and Vyšná, Dyer, Maher and Norris [38] linked increased decomposition rates to higher water temperatures due to a lack of canopy shading.

In addition to organic matter retention, organic matter processing and ecosystem metabolism, nutrient processing and nutrient assimilation into stream food-webs have been explored as functional metrics of stream health [40–42]. The δ^{15}N values of aquatic plants and animals reflect both the source of N and processes that can influence N cycling and have been shown to increase in response to land use intensification [31,43]. While technically a structural aspect of stream communities, several studies have suggested that the δ^{15}N values of aquatic biota represent an integrated signature of N cycling [44,45].

Ideally, indicators of stream ecosystem health are sensitive to change in human actions over time, such as restoration. While the primary focus of the above studies has been to quantify the response of functional indicators to land use effects, there is little if any, scientific evidence on the suitability of functional measures as indicators of restoration success [46,47]. Functional indicators are responsive to a range of reach-scale and catchment-scale drivers [48], and as such may be useful for discerning the optimum design of restoration methods to improve stream health. Further, functional indicators are predicted to respond to streamside fencing variably over time (Figure 1) following the restoration of physical metrics and other drivers discussed above. Multiple functional indicators may therefore be required to assess the restorative effect of fencing on stream functioning, considering the multidimensional character of stream function as well as its temporal character.

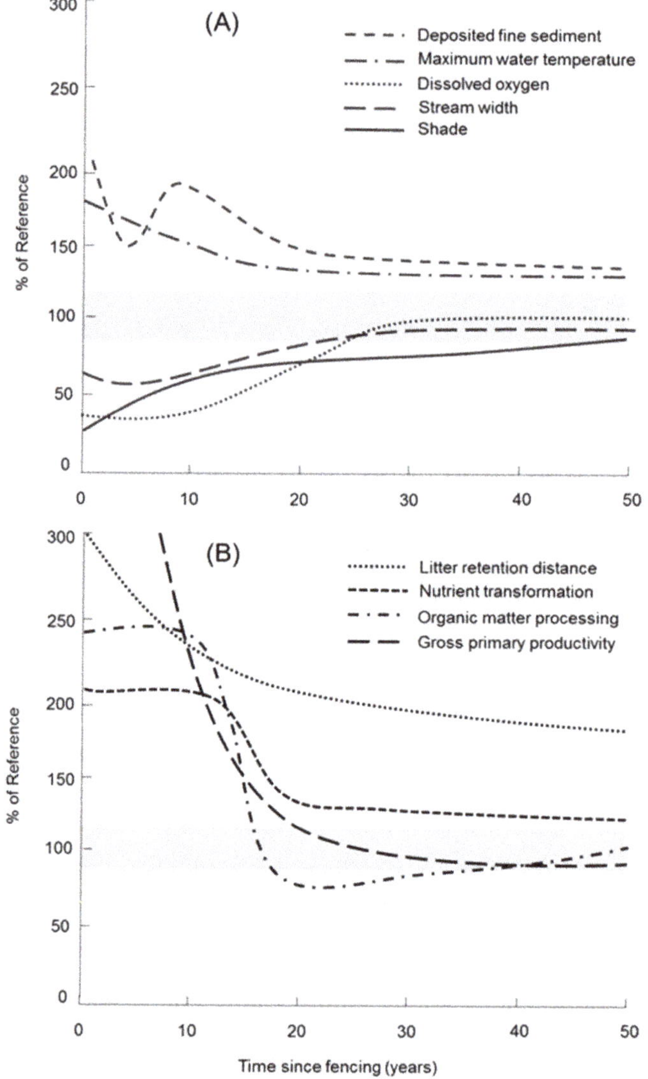

Figure 1. Hypothetical trajectory of (**A**) physical variables and (**B**) functional variables in a small (<5 m wide) pastoral stream following fencing to exclude livestock from channels and riparian zones, adapted from Parkyn et al. [49]. Similar trajectories are predicted for larger streams but the time to recover to approximately 100% of reference condition (shaded area) will take longer. Note: ecosystem respiration is not plotted because despite decreasing over time values are not predicted to reach within 300% of reference condition within 50 years.

The overall objective of this study was to assess the effectiveness of streamside fencing at a reach scale to restore stream function. We measured a range of functional metrics to indicate potential restoration recovery trajectories in the retention, transformation and absorption of nutrients and carbon into stream food-webs. Functional metrics, including GPP, ER, wood mass loss, cotton strip decomposition, organic matter retention, and δ^{15}N values of primary consumers, were measured alongside metrics describing physical stream habitat and water quality. Potential recovery trajectories were explored by surveying 11 pairs of sites, with and without riparian buffers fenced between five to 34 years previously in the central North Island, New Zealand. Overall, we expected to see a shift in stream function from impacted states for unfenced sites (i.e., "poor" ecosystem health) to less-impacted states for fenced sites (i.e., "healthy" ecosystem health). We hypothesised that

(1) fencing would improve the state of ecosystem health as described by physical and functional indicators due to riparian re-establishment, leading to a higher proportion of shade, lower average water temperature, greater hydraulic retention, decreased rates of GPP and ER, decreased organic matter processing rates, and decreased δ^{15}N values indicative of more efficient biogeochemical cycling and

(2) time since fencing would lead to a greater difference between fenced and unfenced sites as described by physical and functional indicators of ecosystem health, hence illustrating a 'rubber-band' model of recovery.

2. Materials and Methods

2.1. Study Sites and Design

Our study was conducted in the Waikato region of the central North Island, New Zealand between March 2011 and April 2012. All measurements were conducted during stable weather conditions in autumn, when streams are generally most stressed due to sustained low flows and warmer temperatures. We sampled 11 pairs of fenced and unfenced stream reaches to assess the effects of riparian fencing on stream health. Fenced reaches had been retired and fenced from adjacent grazed land between 5 and 34 years previously (Table 1), and stocking rates were consistent among unfenced sites. Fenced stream reaches were either located upstream (Raglan and Whatawhata) or downstream of unfenced stream reaches (Waitetuna, Waitomo, Mangawhara, Matarawa, Tapapakanga, Little Waipa and Waitete). At two sites, fenced and unfenced reaches were located on neighbouring, but different streams (Taupo & Kakahu), due to the difficulty of finding fenced sites equal or greater than 30 years old adjacent to grazed sites. Nine of the 11 pairs had been previously studied by Parkyn et al. [7] to test the effectiveness of fenced riparian buffers on biological and water quality indicators.

Catchment size ranged from just over 300 ha to 8600 ha (Whatawhata Stream and Little Waipa River, respectively; Table 1) as determined from the Freshwater Environments of New Zealand database, FENZ, [50]. Eight of the 11 study pairs were located in catchments with high-producing exotic grassland as the predominant land cover upstream and two study pairs with either exotic or indigenous forest, LCDB, [51]. Kakahu was the only site where predominant land use differed between Treatments, with high-producing exotic grassland at the fenced reach and indigenous forest at the unfenced reach.

The length of fenced reaches ranged from 480 m (Raglan) to 1700 m (Taupo) (Table 1). Fenced stream length as a proportion of total upstream length ranged from 0.1% (Kakahu) to 33.1% (Matarawa; Table 1). Riparian vegetation, other than grass, was present at both fenced and unfenced sites, except at unfenced Waitete, Kakahu and Taupo sites. Where riparian vegetation other than grass was present, mean vegetation buffer widths ranged from 0.6 m (Mangawhara fenced) to 75 m (Taupo fenced; Table 1) and were wider at the fenced (mean = 23.1 m) than at the unfenced sites (mean = 5.5 m). Riparian area as a proportion of total catchment area ranged from less than 0.01% (unfenced sites at Kakahu, Waitete and Mangawhara) to 1.73% (Whatawhata fenced site; Table 1).

Table 1. Site characteristics (describing treatment) listed according to number of years since fencing. SR = Study Reach; TCA = Total Catchment Area; Dominant bed substrate size range: silt (<0.063 mm), pumice/sand (0.063–2.00 mm), gravel (2.00–64.00 mm) and cobble (>64.00 mm).

Site (Site Code)	Treatment Fenced (F)/ Unfenced (U)	Years since Fencing	SR Length (m) + SR Proportion (%) of Total Upstream River Length	SR Mean Wetted Width (m)	SR Mean Depth (m)	SR Mean Riparian Width (m)	SR Mean Discharge (L/s)	TCA (ha) + (SR Riparian Area Proportion (%) of Total Catchment Area)	Dominant Bed Substrate	Slope (%)
Whatawhata (Wh)	F	5	560 (29.3)	1.5	0.21	53.5	22.6	301.6 (1.73)	Silt	0.1
	U	0	500 (20.7)	2.0	0.20	7.4	52.0	301.6 (0.24)	Gravel	1.15
Waitete (Wte)	F	8	900 (21.5)	3.4	0.34	5.0	126.9	636.1 (0.03)	Cobble	0.03
	U	0	640 (19.5)	5.2	0.33	0.0	130.3	636.1 (< 0.01)	Cobble	0.17
Raglan (Rag)	F	12	480 (8.5)	1.9	0.15	36.9	17.9	619.9 (0.12)	Silt	3.3
	U	0	470 (7.7)	2.0	0.24	4.9	14.7	619.9 (0.02)	Silt	2.5
Matarawa (Mat)	F	13	1100 (22.5)	2.2	0.35	7.7	90.0	1025.3 (0.01)	Pumice	0.10
	U	0	1250 (33.1)	1.8	0.23	3.7	69.1	850.9 (0.03)	Pumice	0.46
Little Waipa (LW)	F	14	670 (3.9)	7.1	0.34	7.8	248.0	8612.5 (0.01)	Cobble	0.27
	U	0	994 (6.1)	4.1	0.31	20.0	391.5	3467.2 (0.04)	Pumice	0.73
Waitetuna (Wai)	F	16	1600 (13.5)	6.0	0.67	8.2	783.1	5568.1 (0.02)	Silt	0.19
	U	0	920 (8.4)	4.4	0.47	1.8	604.3	5427.5 (0.01)	Silt	0.11
Mangawhara (Mg)	F	18	1200 (9.9)	6.1	0.28	14.8	188.2	3151.3 (0.01)	Gravel	0.37
	U	0	1260 (11.5)	6.7	0.27	0.6	245.0	3077.8 (<0.01)	Gravel	0.82
Tapapakanga (TP)	F	20	650 (0.1)	5.8	0.13	11.4	140.7	997.4 (0.23)	Cobble	0.87
	U	0	1000 (18.4)	5.5	0.15	9.9	173.6	997.4 (0.20)	Gravel	0.12
Kakahu (Ka)	F	30	1500 (11.1)	5.7	0.31	15.3	604.5	3715.6 (0.15)	Gravel	0.28
	U	0	715 (0.1)	2.8	0.27	0.0	151.8	1005.3 (< 0.01)	Gravel	0.73
Waitomo (Wto)	F	30	1200 (16.0)	3.9	0.49	22.7	265.7	1949.6 (0.25)	Pumice	0.15
	U	0	1200 (17.6)	2.6	0.56	3.6	287.6	1835.7 (0.04)	Pumice	0.27
Taupo (To)	F	34	1700 (16.0)	2.1	0.18	75.0	131.5	3101.9 (1.02)	Gravel	1.14
	U	0	830 (0.1)	1.1	0.21	0.0	59.6	387.7 (0.54)	Gravel	0.74

2.2. Physical Habitat

Study reach lengths were set by hydraulic travel times of at least 1 h from a defined upstream point within a fenced reach (see 'Ecosystem metabolism' below for a description of how residence time was determined). At each site, stream depth and wetted width were measured at 10 cross-sections spaced evenly along the length of the study reach to cover local variation in channel morphology. Discharge was measured on a single occasion at the downstream end of each study reach using a 2D FlowTracker Handheld Acoustic Doppler Velocimeter® (SonTek YSI, San Diego, CA, USA). Stream bed particle size distribution (including occurrence of wood) was measured using the pebble count method of Wolman [52] and the amount of in-stream macrophyte cover (emergent or submerged) was estimated in a 1-m-wide transect extending upstream of each of the 10 cross-sections. Riparian buffer width was measured on both sides of the stream at 5 cross-sections along each study reach. Stream shade was estimated visually at water level at three random points across the channel at each cross-section.

2.3. Functional Indicators

2.3.1. Organic Matter Retention

We assessed the retention capacity of stream reaches in April 2011 by using analogues of coarse particulate organic matter (CPOM) following the methodology of Quinn, Phillips and Parkyn [32]. For this, we used conditioned ginkgo (Ginkgo biloba) leaves soaked overnight to make them neutrally buoyant [53], waterproof paper triangles (4.4 cm sides) and pine dowels (30 cm long, 1 cm diameter, after Webster et al. [54] as standard CPOM types. The leaf analogues represented structurally distinct leaf types found naturally in litter fall, including freshly fallen floating leaves (triangles) and litter that had been in the stream long enough to become water saturated (ginkgo). Wood dowels were used to represent small branches.

We released 30 of each CPOM type at regular intervals across the wetted channel at the upstream end of each reach [55]. Triangles were released first, then ginkgo leaves, followed by dowels to avoid the larger dowels catching smaller analogues. Once it became clear that the analogues had been retained (minimum of 10 min), we recorded the distance travelled for each analogue. We characterised the object it was retained on including streambed type and size, leaf litter and wood, riparian and in-stream vegetation or hydraulic habitat type (pool, riffle, run).

Retention was calculated for each release of each CPOM type and averaged by type for multiple releases within a site as deposition velocity (V_{dep}, mm s^{-1}), which is a retention measure accounting for the effects of stream size [56]:

$$\text{Vdep} = \text{mean depth (mm)} \times \text{water velocity (m s}{-1})/\text{mean travel distance (Sp, m)} \quad (1)$$

Organic matter retention was measured for all analogues at all sites, except at Waitetuna where triangles could not be retrieved post release due to high water depth and turbidity.

2.3.2. Ecosystem Metabolism

We estimated the combination of gross primary production (GPP) and ecosystem respiration (ER) for each paired site using Odum's open-system two-station analysis of diel oxygen curves [57], as modified by Marzolf et al. [58]. This required measurements of the diel changes in dissolved oxygen (DO) saturation at the upstream and downstream ends of each reach [59].

At each paired site, DO saturation and water temperature were continuously recorded with optical fluorescence probes (D-Opto, Zebra-Tech Ltd, Nelson, New Zealand) during stable flows at 15-min intervals for at least a 48-h period during March and April 2012. Paired sites were recorded simultaneously. Prior to data recording, the oxygen probes were calibrated in air-saturated water at sea level. To determine the differences in the calibration resulting from instrument drift and between sondes, all loggers were deployed together for one hour at the beginning of each recording period so that any differences could be corrected before data analysis. To measure hydraulic travel times between upstream and downstream DO loggers (aimed at a minimum of 1 hour), we released 50 mL of Rhodamine® WT liquid dye across the stream width upstream of the upper DO sonde and tracked water travel time by monitoring Rhodamine® concentrations at the downstream DO sonde with a C3™ Submersible Field Fluorometer (Turner Designs, San Jose, CA, USA).

The rate of ER and the re-aeration coefficient (k) were determined following the methods from Young & Huryn [59], except k was estimated using the night-time regression method [60] instead of using tracer gases. This method assumes low surface turbulence, as was the case in this study. The ratio of GPP:ER (P/R) was also determined. P/R relates to the balance between primary production and ecosystem respiration and determines whether the reach is autotrophic (P/R > 1) or heterotrophic (P/R < 1) [57].

2.3.3. Organic Matter Processing

To provide an indication of organic matter decomposition, we evaluated organic matter processing rates by measuring wood break-down (mass loss) and cellulose decomposition potential (cotton tensile strength loss). Briefly, wood break-down rates were determined by weighing birch wood (Betula platyphylla) coffee stirrer sticks pre- and postinstalment in stream water. At each paired site, five groups were deployed in riffle habitat for 90 days from April to June 2011. Wood mass loss rates were determined following Petersen and Cummins [61] using degree days (dd) as the time variable. Wood mass loss data was collected for all 11 pairs of sites, except at Mangawhara unfenced site where the temperature logger and wood sticks were lost during deployment.

Cellulose decomposition potential was measured, following Tiegs, Clapcott, Griffiths and Boulton [35], whereby five cotton replicates (Product no. 222; EMPA, St. Gallen, Switzerland) were deployed at each site for seven days in April 2011. Cotton tensile strength loss (CTSL) was reported per degree day (dd) in the same way as the wooden stick data to allow for comparison between these two measures of organic matter processing. Water temperature was continuously recorded in 15-min intervals with a Hobo pendant logger (Onset, Bourne, MA, USA) installed at the same metal stake as the wood and cotton at each site. Mean daily temperature was used to calculate organic matter processing rates per degree day.

2.3.4. Nutrient Transformation

Stable isotopes of nitrogen (N) reflect the source and transformation of N in stream systems and provide an indicator of nutrient sources and processing. The δ^{15}N of primary consumers reflects accumulated N over weeks to months and thus provides a more holistic picture of N sources rather than a snapshot of N concentrations [61]. At each paired site, we hand-picked 10 primary consumers, either mayflies (Deleatidium sp.), shrimp (Paratya curvirostris Heller) or both, from benthic samples and preserved them in 90% isopropyl alcohol. In the laboratory, samples were rinsed with deionised water and their guts were removed and discarded prior to sample drying at 80 °C in a forced-draft oven. Ground samples were analysed in a Dumas elemental analyser interfaced to an isotope mass spectrometer (Europa Scientific Ltd, Crewe, England) to obtain δ^{15}N values to provide an indication of N sources in each stream reach.

2.4. Data Analysis

To test the hypothesis that fencing would improve the state of ecosystem health, differences between Fenced and Unfenced sites across all age groups were assessed using paired t-tests (two-sample for mean). To test the hypothesis that time since fencing would improve ecosystem health, we applied an univariate permutational analysis of variance (PERMANOVA) [62] of log (x + 1) transformed physical and functional response variables measured at a site level (n = 22). The analyses included 'Treatment' as a fixed factor with two levels (Fenced and Unfenced), 'Site' as a random factor and 'Time since fencing' as a continuous covariate. All tests were performed using 9999 permutations of residual under a reduced model and type I sum of squares calculated (sequential) to allow the inclusion of a continuous covariate. Hence, *p*-values are obtained by permutation, thus avoiding the assumption of normality [63]. Univariate PERMANOVA tests were conducted using the software PRIMER 7 [64] and the PERMANOVA add-in [65].

When there was a significant effect of Treatment on CTSL, we tested for any effects of time since fencing by fitting a second linear mixed model to the data from fenced sites only (n = 252). The model also accounted for the effects of physical variables, including slope, upstream native vegetation, catchment buffer proportion, % of fine sediment, temperature, stream depth, stream width and channel width. Initial data exploration was conducted following Zuur et al. [66]. Predictor variables were centred, scaled and Box-Cox transformed before the analyses. Because water depth was negatively correlated with slope and channel depth was positively correlated with channel width they were dropped from the model (VIF > 3). Site was included as a random effect in the model to account for the spatial correlation resulting from the paired nature of the sampling design. Additionally, 'cotton strip' was considered as a random effect nested in 'Site' to account for the repeated measure design. The variability of fixed and random effects justified the use of marginal pseudo-R^2 (accounting for fixed effects) and conditional pseudo-R^2, accounting for fixed and random effects, [67]. Model outputs were represented as partial effect plots for each predictor. Linear mixed models were fitted using the library lme4 [68] of the software R [69].

3. Results

3.1. Physical Habitat

The differences in physical stream habitat were not as prominent as expected between fenced and unfenced sites. However, fenced sites had significantly wider vegetated riparian zones and higher levels of stream shading than their paired unfenced sites. Shade estimates were significantly higher for fenced (mean = 58.1%) than unfenced (mean = 25%) stream reaches ($T(10) = 3.30, p < 0.05$; Figure 2a; Table 2). Whatawhata and Waitomo fenced sites had the highest shading (95%; Figure 2) and the Kakahu fenced site the least shading within fenced sites (10%; Figure 2a). There was no difference in canopy height between the two treatments ($T(10) = 1.34; p = 0.21$) and the level of shading did not depend on time since fencing (Table 3).

Daily dissolved oxygen (DO) saturation ranged from 76.9% (Matarawa unfenced) to 128.2% (Taupo fenced) and mean DO saturation was not significantly different between fenced and unfenced sites (Table 2). Likewise, maximum daily water temperature was not influenced by treatment nor time since fencing (Tables 2 and 3), and ranged from 10.6 °C (Taupo fenced) to 21.4 °C (Tapapakanga unfenced).

Bed substrate type was similar at fenced and unfenced sites (except at Whatawhata, Little Waipa and Tapapakanga; Table 1) with no significant differences in dominant bed substrate percentage between treatments ($T(10) = 0.32, p = 0.76$). Most sites had gravel (2–64 mm) as the dominant bed substrate, followed by silt (< 0.063 mm), pumice (0.063–2 mm) and then cobble (>64 mm; Table 1).

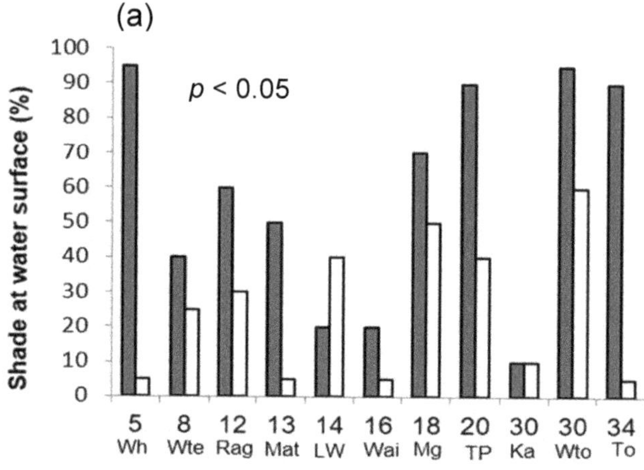

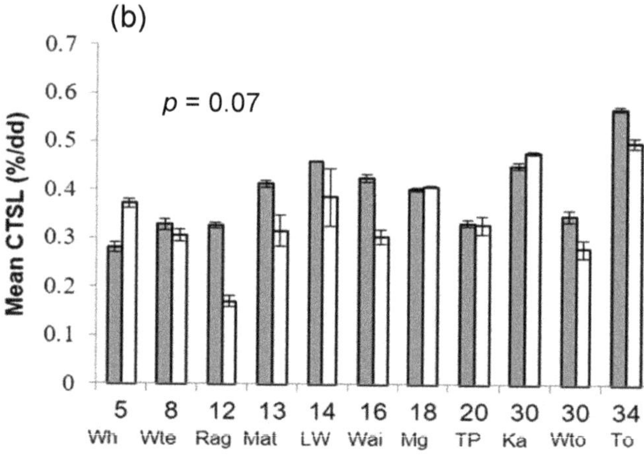

Figure 2. Bar plots showing fenced (grey) and unfenced (white) paired stream reaches for (**a**) % shade and (**b**) % cotton tensile strength loss per degree day (CTSL/dd) ordered by time since fencing.

3.2. Functional Variables

3.2.1. Organic Matter Retention

Dowel and gingko analogues were primarily retained by bank vegetation (i.e., grasses, flax and ferns), with dowels generally floating on the water surface and gingko leaves being neutrally buoyant. Triangles sank to the stream bed immediately post release and were primarily retained by stream bed vegetation and substrate. Analogues had the slowest deposition velocity at the fenced Little Waipa reach (0.27 mm/s, Gingko leaves) and fastest at the fenced Waitomo reach (18.1 m/s, Triangles). There were no significant differences in deposition velocities between treatments for all analogues (Dowel $T(10) = 1.4$, $p = 0.18$, Gingko $T(10) = 0.08$, $p = 0.93$; Triangle $T(10) = 0.43$, $p = 0.67$; Tables 2 and 3) and time since fencing did not have a significant effect on organic matter retention rates (Dowel $F(1, 10) = 0.31$, $p = 0.62$; Gingko $F(1, 10) = 0.45$, $p = 0.52$; Triangle $F(1, 10) = 1.53$, $p = 0.25$; Table 3).

3.2.2. Ecosystem Metabolism

Gross primary production ranged from 0.03 g O_2/m^2/d (Matarawa unfenced) to 10.9 g O_2/m^2/d (Little Waipa fenced; Table 2). There was no significant difference in mean GPP between the two treatments and time since fencing did not influence GPP (Tables 2 and 3). Ecosystem respiration (ER) ranged from 0.70 g O_2/m^2/d (Tapapakanga unfenced) to 17.80 g O_2/m^2/d (Little Waipa fenced). There were no significant differences between treatments and time since fencing did not affect ER (Tables 2 and 3).

3.2.3. Organic matter processing

Mean wood mass loss rates were slowest at the unfenced Raglan stream reach and fastest at the Kakahu unfenced stream reach (0.03%/dd and 0.13%/dd, respectively). There was no significant difference between treatments and time since fencing did not influence wood mass loss rates (Tables 2 and 3).

There was a strong correlation between CTSL/dd and CTSL/d (unfenced R^2= 0.95, fenced R^2 = 0.56) and so only results for CTSL/dd are presented. Across all sites, CTSL/dd ranged from 0.2% (Raglan unfenced) to 0.6% (Taupo fenced) (Table 2; Figure 2b). There was an indication of a treatment effect on CTSL/dd with slightly faster CTSL rates at fenced sites (mean = 0.40; Figure 3) compared to unfenced sites (mean = 0.34; Figure 3), but there was no significant difference between treatments (i.e., $p = 0.07$; Table 2). Slope had a positive effect on CTSL/dd, whereas both fine sediment and stream width had a negative effect on CTSL/dd (Figure 3). There was a significant positive effect of time since fencing on CTSL/dd within fenced sites (Figure 4). The fixed effect part of the model (i.e., Time since fencing) accounted for 32% of the variability in the data (marginal pseudo-R^2), whereas the random and fixed effect models combined explained 91% of the variation of the CTSL/dd data.

3.2.4. Nutrient transformation

We observed no significant difference in the δ^{15}N of primary consumers between treatments (Table 2). Further, time since fencing did not influence δ^{15}N (Table 3). Delta 15N ranged from 3.2‰ (Taupo unfenced) to 9.9‰ (Kakahu unfenced).

Table 2. Mean values and minimum/maximum ranges of variables observed (N = 11). *T* and *p*-values of paired *t*-test (two-sample for means) results and significance levels for the relationship between the change in fenced and unfenced sites for physical and functional indicators are also provided.

Metric	Variables	Fenced Sites	(Min–Max)	Unfenced Sites	(Min–Max)	*T*-value	*p*-value
Physical	Stream depth (m)	0.31	(0.13–0.67)	0.29	(0.15–0.56)	0.77	0.46
Physical	Stream width (m)	4.11	(1.50–7.10)	3.50	(1.10–6.70)	1.34	0.21
Physical	Channel width (m)	10.10	(3.10–28.70)	9.00	(4.8–15.91)	0.47	0.64
Physical	Discharge (m^3/s)	0.25	(0.02–0.78)	0.19	(0.02–0.60)	1.47	0.17
Physical	Slope (%)	0.62	(0.03–3.28)	0.80	(0.11–2.45)	1.23	0.25
Physical	Silt (%)	17.45	(0.00-89.00)	16.80	(0.00–100.00)	0.07	0.95
Physical	Pumice/Sand (%)	18.36	(0.00–77.00)	29.70	(0.00–99.00)	2.63	0.06
Physical	Fines (%)	37.36	(1.00–91.00)	43.80	(1.00–100.00)	0.63	0.55
Physical	Shade (%)	58.10	(10.00–95.00)	25.00	(5.00–60.00)	3.30	<0.05
Physical	Dissolved oxygen (% saturation)	95.80	(81.60–128.20)	93.90	(76.90–120.90)	1.20	0.25
Physical	Max daily water temperature (°C)	14.60	(10.60–20.20)	14.80	(10.70–21.40)	0.76	0.46
Functional	Retention dowel (V_{dep}; mm/s)	3.50	(0.30–14.90)	1.60	(0.60–4.10)	1.40	0.18
Functional	Retention gingko (V_{dep}; mm/s)	2.10	(0.27–5.90)	2.10	(0.50–6.00)	0.08	0.93
Functional	Retention triangle (V_{dep}; mm/s)	5.90	(0.80–18.1)	6.50	(1.70–14.4)	0.43	0.67
Functional	GPP (g O_2/m^2/day)	3.10	(0.10–10.9)	2.10	(0.03–5.20)	1.02	0.33
Functional	ER (g O_2/m^2/day)	6.00	(1.40–17.8)	6.20	(0.70–11.2)	0.13	0.89
Functional	Wood mass loss (%/degree day)	0.07	(0.04–0.10)	0.08	(0.03–0.13)	1.40	0.19
Functional	Cotton tensile strength loss (%/degree day)	0.40	(0.30–0.60)	0.30	(0.20–0.50)	2.05	0.07
Functional	Nutrient transformation (δ^{15}N; ‰)	5.60	(3.40–9.10)	5.80	(3.20–9.90)	0.28	0.78

Table 3. Univariate permutational analysis of variance testing for effects of time since fencing, fenced vs. unfenced sites (treatment) and between sites (Site) on a range of physical and functional variables tested. Significant values ($p < 0.05$) are highlighted in bold.

Physical		Stream Depth			Stream Width			Channel Width			Discharge			Slope			Shade		
	df	MS	F	p	MS	F	p	MS	F	p	MS	F	p	MS	F	p	MS	F	p
Time since fencing	1	0.00	0.11	0.74	0.00	0.00	0.99	0.35	1.32	0.29	0.03	0.66	0.43	0.00	0.00	0.97	0.00	0.00	0.96
Treatment	1	0.00	0.48	0.51	0.07	1.63	0.23	0.00	0.00	1.00	0.01	2.01	0.18	0.14	4.04	0.08	5.43	8.73	0.01
Site	9	0.02	12.92	0.00	0.35	7.92	0.00	0.26	1.62	0.22	0.05	7.69	0.01	0.28	8.33	0.00	0.97	1.57	0.26
Res	10	0.00			0.04			0.16			0.01			0.03			0.62		

		Temperature			Max. Temperature			Silt			Pumice/Sand			Fines		
	df	MS	F	p	MS	F	p	MS	F	p	MS	F	p	MS	F	p
Time since fencing	1	25.41	1.49	0.23	0.01	1.43	0.28	14.80	3.52	0.09	8.88	2.14	0.16	0.02	0.01	0.95
Treatment	1	9.06	2.46	0.07	0.00	0.69	0.43	0.04	0.02	0.85	0.89	3.19	0.11	0.08	0.18	0.65
Site	9	17.06	4.62	0.00	0.01	4.89	0.01	4.21	2.89	0.06	4.15	14.80	0.00	3.96	8.85	0.00
Res	10	3.69			0.00			1.46			0.28			0.45		

Functional		GPP			ER			Wood Mass Loss/dd			CTSL/dd			Stable Isotope			Dowel		
	df	MS	F	p	MS	F	p	MS	F	p	MS	F	p	MS	F	p	MS	F	p
Time since fencing	1	0.02	0.02	0.88	0.17	0.30	0.61	0.00	1.01	0.33	0.02	5.18	0.04	0.06	0.88	0.37	0.19	0.31	0.62
Treatment	1	0.00	0.01	0.94	0.09	0.44	0.52	0.00	3.13	0.11	0.01	4.05	0.06	0.00	0.04	0.83	0.41	1.46	0.29
Site	9	0.74	2.88	0.06	0.56	2.72	0.06	0.00	1.30	0.11	0.00	3.20	0.04	0.07	0.82	0.61	0.61	2.15	0.13
Res	10	0.26			0.21			0.00			0.00			0.08			0.28		

		Gingko			Triangle		
	df	MS	F	p	MS	F	p
Time since fencing	1	0.26	0.45	0.52	0.56	1.53	0.25
Treatment	1	0.00	0.01	0.94	0.12	0.93	0.35
Site	9	0.57	5.41	0.01	0.37	2.87	0.08
Res	10	0.11			0.13		

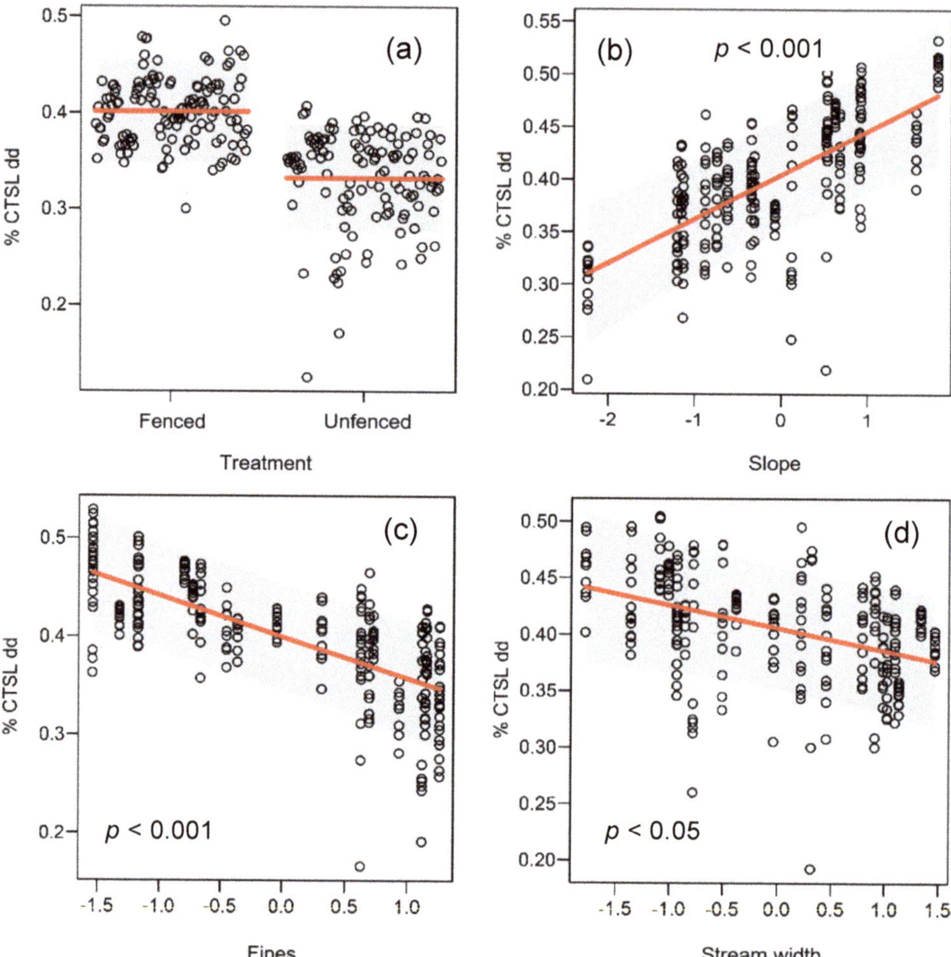

Figure 3. Partial plots of the effects of (**a**) treatment, (**b**) slope, (**c**) fines and (**d**) stream width on % cotton tensile strength loss (CTSL) per degree day (dd) fitted using a linear mixed model for fenced and unfenced sites. Note differences in scales on the y-axes.

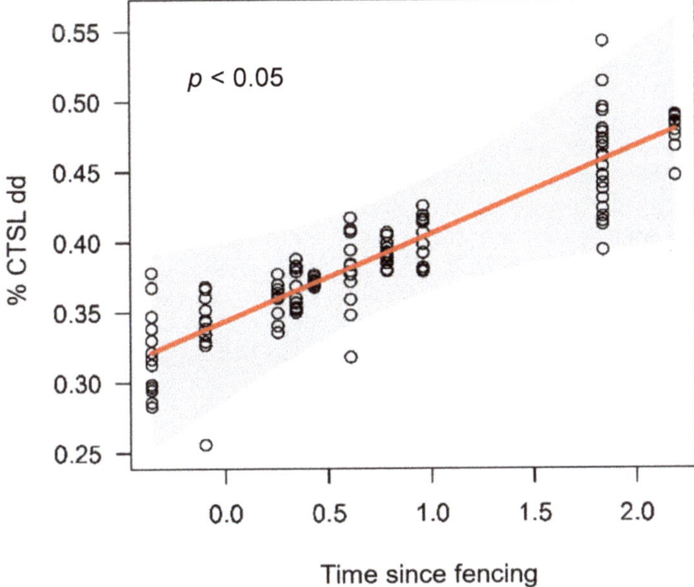

Figure 4. The effect of time since fencing on % cotton tensile strength loss (CTSL) per degree day (dd) fitted using a linear mixed model within fenced sites.

4. Discussion

In our study, we tested the effect of restoration on stream health with a focus on improved ecosystem function as an outcome. We expected to see improvements in stream ecosystem health (i.e., from modified to less modified) for several of the 19 physical and functional indicator variables measured in our study. However, only shade showed a significant difference between treatments, with higher shading observed at fenced than at unfenced sites. Our results agree with Clary [70] who found that stream shade responds reasonably quickly (i.e., 5–10 years) to livestock exclusion (for example through fencing) and the subsequent establishment of canopy cover. But because fencing is a well-recognised restoration technique for physical water quality indicators [7–9], we were surprised to see that the majority of physical measures was unresponsive to our treatment. In a previous study of nine of our 11 paired sites, positive improvements in water clarity and channel stability had been observed between five and 20 years following fencing [7]. These previous results suggest that our study design is sufficient to detect physical changes in stream health in the medium term, but does not necessarily account for temporal and spatial variation in physical and functional responses in the medium to long term.

With regards to process-based restoration, because there are few restoration activities that are explicitly designed to restore river processes [71], there is very little scientific evidence on the response of functional measures to restoration to date [30,47,72]. For example, Giling et al. [30] detected only marginal evidence of decreased GPP at replanted reaches compared to untreated reaches, despite an increase in canopy cover. In our study, both fenced and unfenced stream reaches displayed a range of stream health from 'poor' to 'healthy' condition according to Young et al. (2018). Surprisingly, a significant treatment effect on GPP was not observed in relation to the significant treatment effect on shade. Our data did not support hypothetical recovery trajectories (Figure 1) that predict diverse yet parallel recovery for physical and functional variables. Although, our analysis tested for predominantly linear responses over time and not the nonlinear trajectories predicted for deposited sediment (due to bank collapse as streams wide following fencing) or organic matter processing (due to change in organic matter input). In terms of restoration, this raises the interesting question whether stream structure needs to be restored first before improvements in stream function can be found, or if restoration of structure and function can happen simultaneously?

Of the eight functional indicators measured, cotton tensile-strength loss was the only one that responded to fencing, correlated with stream bed slope, stream width and fine sediment load. Streams with well-established bank and riparian vegetation generally have higher stream bank stability and substrate stability [73,74], which in turn reduces the physical effects of sediment flushing on microbial communities. Considering that volcanic pumice and silt are dominant substrates in our study catchments, smothering of microbes or impacting on macroinvertebrate habitat might explain the observed reduction of CTSL [75–77]. On the other hand, fine sediment can bind nutrients and consequently cotton breakdown has been shown to be faster at sites with higher sediment input due to elevated nutrients [36]. However, we did not observe any such quasi-nutrient effects, suggesting that the inhibitory physical effects of sediment were more important than sediment nutrient concentrations for CTSL in our study. In general, leaf litter breakdown rates, and in particular CTSL, are lower in streams with high levels of deposited fine sediment input [18,39,78].

Our results suggest that CTSL is also likely to increase as the streambed gets steeper and decrease as streams get wider, though it is not clear to us how these relationships relate to stream fencing. Parkyn et al. [79] demonstrated how stream widening can occur following the establishment of riparian buffers in similar agricultural streams. However, bed gradient is driven by catchment-scale rather than reach-scale processes. We did not conduct any longitudinal analysis, specifically testing for the overriding effect of the catchment on our findings. A clear avenue for future restoration studies is to view functional responses in the context of catchment-scale drivers.

Other than cotton-tensile strength loss, none of the other functional indicators measured showed a significant response to streamside fencing. This may be due to the one-off nature of sampling, with each indicator measured on a single day (or two days for ecosystem metabolism). Previous studies have shown high day-to-day variability in stream functions [80,81] that if not considered may be greater than the variation due to fencing effects. In contrast, cotton strips are deployed for seven days providing a time-integrated treatment effect on organic matter processing. However, by this logic, wooden sticks which were deployed for 90 days should also have responded significantly to the treatment, and likewise the δ^{15}N of primary consumers should provide a time-integrated indicator of nutrient transformation. Instead, our results highlight the high spatial variability in stream function and suggests that to properly test the effectiveness of streamside fencing on stream function, greater spatial and, or temporal sampling is required.

The lags and legacies constraining restoration outcomes are well recognised [14,82], so, after 30 years, we expected to see a change in stream health from impacted to less-impacted as fenced riparian buffers aged. Cotton tensile-strength loss significantly increased with time since fencing, most likely driven by increased organic matter input related to increased canopy cover up to a certain stream size (7 m in our study). Energy pathways in smaller streams, such as in our study, are supported largely by allochthonous organic matter [83], so small streams without riparian vegetation are likely to experience a lack of organic matter input and hence a lack of organic matter cycling. We, therefore, suggest that fencing is a suitable restoration technique to assist in the establishment of microbial communities through riparian leaf litter input in small streams, facilitating the recovery of stream ecosystem health. But we also highlight the phenomenon of hysteresis during stream restoration through the re-establishment of riparian vegetation post fencing and the associated long timeframes until quantifiable benefits to stream ecosystem health can be detected.

We were unable to demonstrate any effects of time since treatment for any of the physical variables tested. Shading was significantly greater at fenced sites, but time since fencing had no effect on the amount of shade on the water surface. In fact, Whatawhata (our 'youngest' fenced site) showed the highest shade records of all sites, while shading at Kakahu (one of our 'oldest' fenced sites) was the same at fenced and unfenced stream reaches. This high site-to-site variability was evident for most variables tested, and for most sites it was not possible to infer a link between restoration effort and a given response variable. This is not unusual in studies which aim to quantify restoration success on a reach-scale, while degradation occurs at the catchment-scale [71]. For example, Giling et al. [84] found that macroinvertebrates showed no response up to 22 years after replanting at the reach-scale due to the overriding effects of catchment-scale degradation.

Our study is unique in that it assessed the response of both functional and physical indicators of stream health to restoration over more than three decades. We hypothesised that a change in stream function would occur, indicative of improving stream health, over time and increasing spatial scales. However, our data did not follow a linear or 'rubber band' pattern of improvement. Tiegs et al. [35] found large variations in cotton strip tensile-strength loss among the three regions studied and among the streams within those regions, suggesting sensitivity to variation in sub-catchment and catchment-scale environmental conditions. Natural environmental variation could moderate the effects of land use and restoration efforts on stream health. In-stream habitat structure and organic matter inputs are determined primarily by local conditions such as vegetation cover at a site, whereas nutrient supply, sediment delivery, hydrology and channel characteristics are influenced by regional conditions, including landscape features and land use / cover at some distance upstream and lateral to stream sites [85]. Considering that habitat degradation often occurs at catchment scales, considerable restoration at larger scales may be required before any improving stream ecosystem responses to restoration can be seen at a reach scale [13,71]. Reach-scale restoration has been shown to have very little to no effects on biological indicators [7,12,86], however, it is well documented that width, continuous canopy and length of forested riparian zones determine the effectiveness of restoration of stream structure [87–91]. As for functional responses, Giling, Grace, Mac Nally and Thompson [30] demonstrated marginally improved ecosystem metabolism 17 years after riparian restoration in stream reaches shorter and narrower than ours and they suggested that metabolic response would take longer and be less pronounced in larger channels. Our results suggest that after 34 years metabolic recovery to a 'healthy' state may still not be achieved by passive riparian restoration via streamside fencing.

Regarding the proportion of study reaches within a catchment, our data indicated that fenced catchment proportions around 1% are likely to have some beneficiary effects on stream health. For example, shade, organic matter retention rates and water temperature for the fenced Whatawhata reach (5 years old; 1.73% of total catchment proportion) and the fenced Taupo reach (34 years old; 1.02% of total catchment proportion) pointed towards an improvement in these variables. One percent was the highest proportion within a catchment area that was fenced in our study, and we suggest that any proportion of riparian vegetation > 1% of total catchment area may have a positive effect on stream health, independent of the establishment age. In a study conducted 10 years earlier on the same sites, significant improvements in visual water clarity and channel stability were observed [7]. Similarly, Holmes [20] identified a minimum 1 km length of 5-m-wide riparian vegetation was required to improve habitat heterogeneity in small agricultural streams. We recommend further research on the effects of small-scale restoration (that is treatment size of 1%–10% of the total catchment area) on functional metrics to elucidate the optimal restoration effort to achieve improved ecosystem health in agricultural streams. Future research should include (1) increased temporal sampling to characterise restoration trajectories and improve sample size, as our study constitutes a single post restoration sample for each pair of sites; (2) concurrent measurement of riparian structure (e.g., floristic composition), channel and in-stream morphology, water quality and biological characteristics, alongside functional indicators of biophysical processes, to elucidate the relationship between structural and functional restoration; and (3) improved definition of the restoration treatment within the catchment (e.g., placement, relative scale and level of active management), to inform the optimal design of future riparian restoration. Space-for-time analysis did not yield the results we initially had hypothesised, emphasising the complex pathway of stream restoration from fencing and the ongoing need for an integrated assessment approach. This means functional as well as physical and biological structural variables need to be included in monitoring of long-term restoration efforts (>30 years) to successfully determine if an improvement in stream health has occurred.

In conclusion, our comparison of 11 paired small streams in central North Island of New Zealand showed that of the 19 physical and functional variables that were tested, only cotton tensile-strength loss and stream shade were higher within fenced sites compared to unfenced sites. This indicates that fencing is likely to influence stream ecosystem function; however, restoration needs to be at a sufficient scale, before any observable effects can be made. Small-scale (<2% of the upstream catchment area) efforts to fence the riparian zones of streams appear to have little effect on ecosystem function and we strongly emphasize the confounding effects of large site-to-site variability and the overriding effects of catchment degradation on reach-scale restoration.

Author Contributions: Conceptualization, R.G.Y.; methodology, K.D., R.G.Y, J.E.C.; formal analysis, K.D., J.E.C.; investigation, K.D.; writing, K.D., J.E.C., R.G.Y.

Funding: This research was funded by Ministry of Science and Innovation, contract number CONC-20599.

Acknowledgments: We thank all the landowners that allowed us to conduct this study on their land: Martin Koning, Sarah and Mike Moss, Colleen Hawkes, Neil Board, Jocelyn, Marc and Sarah Newton, the Van Loon Family, Matthew Wood, Phil Guest, Les and Fern Bennett, Bruce and Christine Lane, Roger and Eunice, Andrew Jenks, Auckland Council and the Department of Conservation, and apologize to those we have missed. We also thank John Quinn, David Reid, Aslan Wright-Stow, Kerry Costley, Toni Johnston and Josh de Villiers for help in the field and provision of some data. We thank Javier Atalah for assistance with the statistical analyses and three anonymous reviewers for their review of this manuscript.

References

1. Bernhardt, E.S.; Palmer, M.A.; Allan, J.D.; Alexander, G.; Barnas, K.; Brooks, S.; Carr, J.; Clayton, S.; Dahm, C.; Follstad-Shah, J.; et al. Synthesizing U.S. River restoration efforts. *Science* **2005**, *308*, 636–637. [CrossRef] [PubMed]

2. Roni, P.; Hanson, K.; Beechie, T. Global Review of the Physical and Biological Effectiveness of Stream Habitat Rehabilitation Techniques. *N. Am. J. Fish. Manag.* **2008**, *28*, 856–890. [CrossRef]

3. Pedersen, M.L.; Andersen, J.M.; Nielsen, K.; Linnemann, M. Restoration of Skjern River and its valley: Project description and general ecological changes in the project area. *Ecol. Eng.* **2007**, *30*, 131–144. [CrossRef]

4. ICPR. *Action Plan on Floods 1995–2005—Action Targets, Implementation and Results*; Brochure, Abridged Version of Technical Report No. 156; ICPR: Koblenz, Germany, 2007; p. 16. ISBN 3-935324-63-4.

5. Schwarz, U. *Assessment of the Restoration Potential along the Danube and MAIN Tributaries*; Final Draft; WWF International, Danube-Carpathian Programme: Vienna, Austria, May 2010.

6. McKergow, L.A.; Matheson, F.E.; Quinn, J.M. Riparian management: A restoration tool for New Zealand streams. *Ecol. Manag. Restor.* **2016**, *17*, 218–227. [CrossRef]

7. Parkyn, S.M.; Davies-Colley, R.J.; Halliday, N.J.; Costley, K.J.; Croker, G.F. Planted riparian buffer zones in New Zealand: Do they live up to expectations? *Restor. Ecol.* **2003**, *11*, 436–447. [CrossRef]

8. Craig, L.S.; Palmer, M.A.; Richardson, D.C.; Filoso, S.; Bernhardt, E.S.; Bledsoe, B.P.; Doyle, M.W.; Groffman, P.M.; Hassett, B.A.; Kaushal, S.S.; et al. Stream restoration strategies for reducing river nitrogen loads. *Front. Ecol. Environ.* **2008**, *6*, 529–538. [CrossRef]

9. Bragina, L.; Sherlock, O.; van Rossum, A.; Jennings, E. Cattle exclusion using fencing reduces *Escherichia coli* (*E. coli*) level in stream sediment reservoirs in northeast Ireland. *Agric. Ecosyst. Environ.* **2017**, *239*, 349–358. [CrossRef]

10. Palmer, M.; Hondula, K.; Koch, B. Ecological Restoration of Streams and Rivers: Shifting Strategies and Shifting Goals. *Annu. Rev. Ecol. Evol. Syst.* **2014**, *45*, 247. [CrossRef]

11. Follstad Shah, J.J.; Dahm, C.N.; Gloss, S.P.; Bernhardt, E.S. River and Riparian Restoration in the Southwest: Results of the National River Restoration Science Synthesis Project. *Restor. Ecol.* **2007**, *15*, 550–562. [CrossRef]

12. Collins, K.E.; Doscher, C.; Rennie, H.G.; Ross, J.G. The Effectiveness of Riparian 'Restoration' on Water Quality—A Case Study of Lowland Streams in Canterbury, New Zealand. *Restor. Ecol.* **2013**, *21*, 40–48. [CrossRef]

13. Lake, P.S.; Bond, N.; Reich, P. Linking ecological theory with stream restoration. *Freshw. Biol.* **2007**, *52*, 597–615. [CrossRef]

14. Suding, K.N.; Hobbs, R.J. Threshold models in restoration and conservation: A developing framework. *Trends Ecol. Evol.* **2009**, *24*, 271–279. [CrossRef] [PubMed]

15. Suding, K.N. Toward an Era of Restoration in Ecology: Successes, Failures, and Opportunities Ahead. *Annu. Rev. Ecol. Evol. Syst.* **2011**, *42*, 465–487. [CrossRef]

16. Brooks, A.; Howell, T.; Abbe, T.B.; Arthington, A. Confronting Hysteresis: Wood Based River Rehabilitation in Highly Altered Riverine Landscapes in South-Eastern Australia. *Geomorphology* **2006**, *79*, 395–422. [CrossRef]

17. Rapport, D.J.; Costanza, R.; McMichael, A.J. Assessing ecosystem health. *Trends Ecol. Evol.* **1998**, *13*, 397–402. [CrossRef]

18. Young, R.G.; Matthaei, C.D.; Townsend, C.R. Organic matter breakdown and ecosystem metabolism: Functional indicators for assessing river ecosystem health. *J. N. Am. Benthol. Soc.* **2008**, *27*, 605–625. [CrossRef]

19. Quinn, J. Effects of Rural Land Use (Especially Forestry) and Riparian Management on Stream Habitat. *N. Z. J. For.* **2005**, *49*, 16.

20. Holmes, R.; Hayes, J.; Matthaei, C.; Closs, G.; Williams, M.; Goodwin, E. Riparian management affects instream habitat condition in a dairy stream catchment. *N. Z. J. Mar. Freshw. Res.* **2016**, *50*, 581–599. [CrossRef]

21. Guzha, A.C.; Rufino, M.C.; Okoth, S.; Jacobs, S.; Nóbrega, R.L.B. Impacts of land use and land cover change on surface runoff, discharge and low flows: Evidence from East Africa. *J. Hydrol. Reg. Stud.* **2018**, *15*, 49–67. [CrossRef]

22. Nugroho, P.; Marsono, D.; Sudira, P.; Suryatmojo, H. Impact of Land-use Changes on Water Balance. *Procedia Environ. Sci.* **2013**, *17*, 256–262. [CrossRef]

23. Castillo, M.M.; Morales, H.; Valencia, E.; Morales, J.J.; Cruz-Motta, J.J. The effects of human land use on flow regime and water chemistry of headwater streams in the highlands of Chiapas. *Knowl. Manag. Aquat. Ecosyst.* **2013**. [CrossRef]

24. Quinn, J.M.; Stroud, M.J. Water quality and sediment and nutrient export from New Zealand hill-land catchments of contrasting land use. *N. Z. J. Mar. Freshw. Res.* **2002**, *36*, 409–429. [CrossRef]

25. Niyogi, D.K.; Simon, K.; Townsend, C.R. Breakdown of tussock grass in streams along a gradient of agricultural development in New Zealand. *Freshw. Biol.* **2003**, *48*, 1698–1708. [CrossRef]

26. Von Schiller, D.; Marti, E.; Riera, J.L.; Sabater, F. Effects of nutrients and light on periphyton biomass and nitrogen uptake in Mediterranean streams with contrasting land uses. *Freshw. Biol.* **2007**, *52*, 891–906. [CrossRef]

27. Violin, C.R.; Cada, P.; Sudduth, E.B.; Hassett, B.A.; Penrose, D.L.; Bernhardt, E.S. Effects of urbanization and urban stream restoration on the physical and biological structure of stream ecosystems. *Ecol. Appl. A Publ. Ecol. Soc. Am.* **2011**, *21*, 1932. [CrossRef]

28. Shilla, D.J.; Shilla, D.A. The effects of catchment land use on water quality and macroinvertebrate assemblages in Otara Creek, New Zealand. *Chem. Ecol.* **2011**, *27*, 445–460. [CrossRef]

29. Tank, J.L.; Rosi-Marshall, E.J.; Griffiths, N.A.; Entrekin, S.A.; Stephen, M.L. A review of allochthonous organic matter dynamics and metabolism in streams. *J. N. Am. Benthol. Soc.* **2010**, *29*, 118–146. [CrossRef]

30. Giling, D.P.; Grace, M.R.; Mac Nally, R.; Thompson, R.M. The influence of native replanting on stream ecosystem metabolism in a degraded landscape: Can a little vegetation go a long way? *Freshw. Biol.* **2013**, *58*, 2601–2613. [CrossRef]

31. Clapcott, J.E.; Young, R.G.; Goodwin, E.O.; Leathwick, J.R. Exploring the response of functional indicators of stream health to land-use gradients. *Freshw. Biol.* **2010**, *55*, 2181–2199. [CrossRef]

32. Quinn, J.M.; Phillips, N.R.; Parkyn, S.M. Factors influencing retention of coarse particulate organic matter in streams. *Earth Surf. Process. Landf.* **2007**, *32*, 1186–1203. [CrossRef]

33. McTammany, M.E.; Benfield, E.F.; Webster, J.R. Recovery of stream ecosystem metabolism from historical agriculture. *J. N. Am. Benthol. Soc.* **2007**, *26*, 532–545. [CrossRef]

34. Webster, J.R.; Benfield, E.F. Vascular plant breakdown in freshwater ecosystems. *Annu. Rev. Ecol. Syst.* **1986**, *17*, 567–594. [CrossRef]

35. Tiegs, S.D.; Clapcott, J.; Griffiths, N.; Boulton, A.J. A standardized cotton strip assay for measuring organic-matter decomposition in streams. *Ecol. Indic.* **2013**, *32*, 131–139. [CrossRef]

36. Boulton, A.; Quinn, J. A simple and versatile technique for assessing cellulose decomposition potential in floodplain and riverine sediments. *Arch. Fur Hydrobiol.* **2000**, *150*, 133–151. [CrossRef]

37. Griffiths, N.A.; Tiegs, S.D. Organic-matter decomposition along a temperature gradient in a forested headwater stream. *Freshw. Sci.* **2016**, *35*, 518–533. [CrossRef]

38. Vyšná, V.; Dyer, F.; Maher, W.; Norris, R. Cotton strip decomposition rate as a river condition indicator–Diel temperature range and deployment season and length also matter. *Ecol. Indic.* **2014**, *45*, 508–521. [CrossRef]

39. Bierschenk, A.; Savage, C.; Townsend, C.; Matthaei, C. Intensity of land use in the catchment influences ecosystem functioning along a freshwater-marine continuum. *Ecosystems* **2012**, *15*, 637–651. [CrossRef]

40. Udy, J.W.; Fellows, C.S.; Bartkow, M.E.; Bunn, S.E.; Clapcott, J.E.; Harch, B.D. Measures of nutrient processes as indicators of stream ecosystem health. *Hydrobiologia* **2006**, *572*, 89–102. [CrossRef]

41. Mulholland, P.J.; Webster, J. Nutrient dynamics in streams and the role of J-NABS. *J. N. Am. Benthol. Soc.* **2010**, *29*, 100–117. [CrossRef]

42. Von Schiller, D.; Acuña, V.; Aristi, I.; Arroita, M.; Basaguren, A.; Bellin, A.; Boyero, L.; Butturini, A.; Ginebreda, A.; Kalogianni, E.; et al. River ecosystem processes: A synthesis of approaches, criteria of use and sensitivity to environmental stressors. *Sci. Total Environ.* **2017**, *596–597*, 465–480. [CrossRef]

43. Diebel, M.W.; Zanden, M.J.V. Nitrogen stable isotopes in streams: Effects of agricultural sources and transformations. *Ecol. Appl.* **2009**, *19*, 1127–1134. [CrossRef] [PubMed]

44. Udy, J.W.; Bunn, S.E. Elevated δ^{15}N values in aquatic plants from cleared catchments: Why? *Mar. Freshw. Res.* **2001**, *52*, 347–351. [CrossRef]

45. Hamilton, S.; Tank, J.; Raikow, D.; Wollheim, W.; Peterson, B.; Webster, J. Nitrogen uptake and transformation in a midwestern U.S. stream: A stable isotope enrichment study. *Biogeochemistry* **2001**, *54*, 297–340. [CrossRef]

46. Rubin, Z.; Kondolf, G.M.; Rios-Touma, B. Evaluating Stream Restoration Projects: What Do We Learn from Monitoring? *Water* **2017**, *9*, 174. [CrossRef]

47. Sudduth, E.B.; Hassett, B.A.; Cada, P.; Bernhardt, E.S. Testing the Field of Dreams Hypothesis: Functional responses to urbanization and restoration in stream ecosystems. *Ecol. Appl.* **2011**, *21*, 1972–1988. [CrossRef] [PubMed]

48. Sheldon, F.; Peterson, E.E.; Boone, E.L.; Sippel, S.; Bunn, S.E.; Harch, B.D. Identifying the spatial scale of land use that most strongly influences overall river ecosystem health score. *Ecol. Appl.* **2012**, *22*, 2188–2203. [CrossRef] [PubMed]

49. Parkyn, S.; Collier, K.; Clapcott, J.; David, B.; Davies-Colley, R.; Matheson, F.; Quinn, J.; Shaw, W.; Storey, R. *The Restoration Indicators Toolkit: Indicators for Monitoring the Ecological Success of Stream Restoration*; National Institute of Water and Atmospheric Research: Hamilton, New Zealand, 2010; p. 134. Available online: http://www.envirolink.govt.nz/assets/Envirolink/RestorationIndicatorToolkit-stream.pdf (accessed on 20 June 2019).

50. Leathwick, J.R.; West, D.; Gerbeaux, P.; Kelly, D.; Robertson, H.; Brown, D.; Chadderton, W.L.; Ausseil, A.-G. *Freshwater Ecosystems of New Zealand (FENZ) Database*; Department of Conservation: Wellington, New Zealand, 2010.

51. Ministry for the Environment. *The New Zealand Land Cover Database (LCDB) 4*; Ministry for the Environment: Wellington, New Zealand, 2014.

52. Wolman, M.G. A method of sampling coarse river-bed material. *Trans. Am. Geophys. Union* **1954**, *35*, 951–956. [CrossRef]

53. Speaker, R.; Moore, K.; Gregory, S. Analysis of the process of retention of organic matter in stream ecosystems. *Verh. Int. Ver. Fuer Theor. Angew. Limnol.* **1984**, *22*, 1835–1841. [CrossRef]

54. Webster, J.R.; Covich, A.P.; Tank, J.L.; Crockett, T.V. Retention of coarse organic particles in streams in the southern Appalachian Mountains. *J. N. Am. Benthol. Soc.* **1994**, *13*, 140–150. [CrossRef]

55. James, A.B.W.; Henderson, I.M. Comparison of coarse particulate organic matter retention in meandering and straightened sections of a third-order New Zealand stream. *River Res. Appl.* **2005**, *21*, 641–650. [CrossRef]

56. Brookshire, E.N.J.; Dwire, K.A. Controls on patterns of coarse organic particle retention in headwater streams. *J. N. Am. Benthol. Soc.* **2003**, *22*, 17–34.

57. Odum, H.T. Primary production in flowing waters. *Limnol. Oceanogr.* **1956**, *1*, 102–117. [CrossRef]

58. Marzolf, E.R.; Mulholland, P.J.; Steinman, A.D. Improvements to the diurnal upstream-downstream dissolved oxygen change technique for determining whole-stream metabolism in small streams. *Can. J. Fish. Aquat. Sci.* **1994**, *51*, 1591–1599. [CrossRef]

59. Young, R.G.; Huryn, A.D. Effects of land use on stream metabolism and organic matter turnover. *Ecol. Appl.* **1999**, *9*, 1359–1376. [CrossRef]

60. Owens, M. Measurements on non-isolated natural communities in running waters. In *A Manual on Methods for Measuring Primary Production in Aquatic Environments*; Vollenweider, R.A., Ed.; Blackwell Scientific Publications: Oxford, UK, 1974; pp. 111–119.

61. Petersen, R.C.; Cummins, K.W. Leaf processing in a woodland stream. *Freshw. Biol.* **1974**, *4*, 343–368. [CrossRef]

62. Anderson, M.J. Permutational test for univariate or multivariate analysis of variance and regression. *Can. J. Fish. Aquat. Sci.* **2001**, *58*, 626–639. [CrossRef]

63. Anderson, M.J. Permutational Multivariate Analysis of Variance (PERMANOVA). In *Wiley StatsRef: Statistics Reference Online*; Balakrishnan, N., Colton, T., Everitt, B., Piegorsch, W., Ruggeri, F., Teugels, J.L., Eds.; Department of Statistics, University of Auckland: Auckland, New Zealand, 2017.

64. Clarke, K.R.; Gorley, R.N. *PRIMER v7: User Manual/Tutorial*; Primer-E: Plymouth, UK, 2015; p. 296.

65. Anderson, M.J.; Gorley, R.N.; Clarke, R.K. *PERMANOVA+ for Primer: Guide to Software and Statisticl Methods*; Primer-E Limited: Auckland, New Zealand, 2008.

66. Zuur, A.F.; Ieno, E.N.; Elphick, C.S. A protocol for data exploration to avoid common statistical problems. *Methods Ecol. Evol.* **2010**, *1*, 3–14. [CrossRef]

67. Nakagawa, S.; Schielzeth, H. A general and simple method for obtaining R^2 from Generalized Linear Mixed-effects Models. *Methods Ecol. Evol.* **2013**, *4*, 133–142. [CrossRef]

68. Bates, D.; Mächler, M.; Bolker, B. Fitting linear mixed-effects models using lme4. *J. Stat. Softw.* **2015**, *67*. [CrossRef]

69. R Core Team. *A Language and Environment for Statistical Computing*; R Foundation for Statistical Computing: Vienna, Austria, 2016.

70. Clary, W.P. Stream Channel and Vegetation Responses to Late Spring Cattle Grazing. *J. Range Manag.* **1999**, *52*, 218–227. [CrossRef]

71. Bernhardt, E.S.; Palmer, M.A. River restoration: The fuzzy logic of repairing reaches to reverse catchment scale degradation. *Ecol. Appl.* **2011**, *21*, 1926–1931. [CrossRef] [PubMed]

72. Lepori, F.; Palm, D.; Malmqvist, B. Effects of stream restoration on ecosystem functioning: Detritus retentiveness and decomposition. *J. Appl. Ecol.* **2005**, *42*, 228–238. [CrossRef]

73. Loades, K.W.; Bengough, A.G.; Bransby, M.F.; Hallett, P.D. Planting density influence on fibrous root reinforcement of soils. *Ecol. Eng.* **2010**, *36*, 276–284. [CrossRef]

74. Boothroyd, I.K.G.; Quinn, J.M.; Langer, E.R.; Costley, K.J.; Steward, G. Riparian buffers mitigate effects of pine plantation logging on New Zealand streams: 1. Riparian vegetation structure, stream geomorphology and periphyton. *For. Ecol. Manag.* **2004**, *194*, 199–213. [CrossRef]

75. Wagenhoff, A.; Townsend, C.R.; Matthaei, C.D. Macroinvertebrate responses along broad stressor gradients of deposited fine sediment and dissolved nutrients: A stream mesocosm experiment. *J. Appl. Ecol.* **2012**, *49*, 892–902. [CrossRef]

76. Sponseller, R.A.; Benfield, E.F. Influences of land use on leaf breakdown in southern Appalachian headwater streams: A multiple-scale analysis. *J. N. Am. Benthol. Soc.* **2001**, *20*, 44–59. [CrossRef]

77. Risse-Buhl, U.; Mendoza-Lera, C.; Norf, H.; Pérez, J.; Pozo, J.; Schlief, J. Contrasting habitats but comparable microbial decomposition in the benthic and hyporheic zone. *Sci. Total Environ.* **2017**, *605–606*, 683–691. [CrossRef]

78. Piggott, J.J.; Niyogi, D.K.; Townsend, C.R.; Matthaei, C.D. Multiple stressors and stream ecosystem functioning: Climate warming and agricultural stressors interact to affect processing of organic matter. *J. Appl. Ecol.* **2015**, *52*, 1126–1134. [CrossRef]

79. Parkyn, S.M.; Davies-Colley, R.J.; Cooper, A.B.; Stroud, M.J. Predictions of stream nutrient and sediment yield changes following restoration of forested riparian buffers. *Ecol. Eng.* **2005**, *24*, 551–558. [CrossRef]

80. Simon, K.S.; Townsend, C.R.; Biggs, B.J.F.; Bowden, W.B. Temporal variation of N and P uptake in 2 New Zealand streams. *J. N. Am. Benthol. Soc.* **2005**, *24*, 1–18. [CrossRef]

81. Clapcott, J.E.; Young, R.G.; Neale, M.W.; Doehring, K.A.M.; Barmuta, L.A. Land use affects temporal variation in stream metabolism. *Freshw. Sci.* **2016**, *35*, 1164–1175. [CrossRef]

82. Hamilton, S.K. Biogeochemical time lags may delay responses of streams to ecological restoration: Time lags in stream restoration. *Freshw. Biol.* **2012**, *57*, 43–57. [CrossRef]

83. Benfield, E.; Fritz, K.; Tiegs, S. Leaf litter Breakdown. In *Methods in Stream Ecology*; Elsevier: Amsterdam, The Netherlands, 2017; pp. 71–82.

84. Giling, D.P.; Mac Nally, R.; Thompson, R.M. How sensitive are invertebrates to riparian-zone replanting in stream ecosystems? *Mar. Freshw. Res.* **2016**, *67*, 1500–1511. [CrossRef]

85. Allan, J.D.; Erickson, D.L.; Fay, J. The influence of catchment land use on stream integrity across multiple spatial scales. *Freshw. Biol.* **1997**, *37*, 149–161. [CrossRef]

86. Ranganath, S.C.; Hession, W.; Wynn Thompson, T. Livestock Exclusion Influences on Riparian Vegetation, Channel Morphology, and Benthic Macroinvertebrate Assemblages. *J. Soil Water Conserv.* **2009**, *64*, 33–42. [CrossRef]

87. Lowrance, R. Riparian forest ecosystems as filters for non-point source pollution. In *Success, Limitations, and Frontiers in Ecosystem Science*; Pace, M.L., Groffman, P.M., Eds.; Springer: New York, NY, USA, 1998; pp. 113–141.

88. England, L.E.; Rosemond, A.D. Small reductions in forest cover weaken terrestrial-aquatic linkages in headwater streams. *Freshw. Biol.* **2004**, *49*, 721–734. [CrossRef]

89. Gore, J.A.; Shields, F.D. Can large rivers be restored? *Bioscience* **1995**, *45*, 142–152. [CrossRef]

90. Storey, R.G.; Cowley, D.R. Recovery of three New Zealand rural streams as they pass through native forest remnants. *Hydrobiologia* **1997**, *353*, 63–76. [CrossRef]

91. Scarsbrook, M.R.; Halliday, J. Transition from pasture to native forest land-use along stream continua: Effects on stream ecosystems and implications for restoration. *N. Z. J. Mar. Freshw. Res.* **1999**, *33*, 293–310. [CrossRef]

Article

A River Temperature Model to Assist Managers in Identifying Thermal Pollution Causes and Solutions

Reza Abdi * and Theodore Endreny

Department of Environmental Resources Engineering, SUNY ESF, Syracuse, NY 13210, USA; te@esf.edu
* Correspondence: reabdi@syr.edu; Tel.: +1-315-470-6565

Received: 27 April 2019; Accepted: 15 May 2019; Published: 22 May 2019

Abstract: Thermal pollution of rivers degrades water quality and ecosystem health, and cities can protect rivers by decreasing warmer impervious surface stormwater inflows and increasing cooler subsurface inflows and shading from riparian vegetation. This study develops the mechanistic i-Tree Cool River Model and tests if it can be used to identify likely causes and mitigation of thermal pollution. The model represents the impacts of external loads including solar radiation in the absence of riparian shade, multiple lateral storm sewer inflows, tributaries draining reservoirs, groundwater flow, and hyporheic exchange flow in dry weather steady flows and wet weather unsteady flows. The i-Tree Cool River Model estimates the shading effects of the riparian vegetation and other features as a function of heights and distances as well as solar geometry. The model was tested along 1500 m of a New York mountain river with a riparian forest and urban areas during 30 h with two summer storm events in 2007. The simulations were sensitive to the inflows of storm sewers, subsurface inflows, as well as riparian shading, and upstream boundary temperature inflows for steady and unsteady conditions. The model simulated hourly river temperature with an R^2 of 0.98; when shading was removed from the simulation the R^2 decreased 0.88, indicating the importance of riparian shading in river thermal modeling. When stormwater inflows were removed from the simulation, the R^2 decreased from 0.98 to 0.92, and when subsurface inflows were removed, the R^2 decreased to 0.94. The simulation of thermal loading is important to manage against pollution of rivers.

Keywords: River thermal pollution; Mechanistic model; Urban hydrology; Riparian shading; Heat balance

1. Introduction

Excessive river temperatures are detrimental to water quality and ecosystem health [1]. River warming can result from increased inflows of warm point and non-point source discharges, decreased inflows of cool sub-surface waters, removal of riparian shade, increased air temperatures, and changes in channel substrate and depth that increase absorption, conduction, and convection in heat transfer [2–4]. River thermal pollution is often associated with discharges of coolant water used by industry, but it is also associated with land-use change, including urbanization, river impoundment, channel management, and regulation [5,6]. River temperature is a critical water quality parameter for riverine systems, that affects the saturation of dissolved oxygen [7,8], kinetic reactions and resulting pollutant concentrations [9,10], and fish distribution, metabolism, growth, reproduction, and mortality [11,12]. Urbanization can elevate river temperatures through changes in riparian land cover which affects shade on the water surface, through river morphology which affects water depth, surface area, and velocity, and through flow connectivity with groundwater, stormwater, and other point and non-point source inflows [13–18]. When precipitation strikes hot impervious surfaces of urban areas, this generates warmer stormwater relative to the temperature of river water [19–21].

River temperature management and mitigation of thermal pollution are best planned with simulation models that enable scenario evaluation, to explore relationships between river temperature response and drivers that vary in space and time [22,23]. Caissie [24] provides a review of research into the spatial and temporal drivers of river temperature, and the evolution of modeling approaches, including statistical models and cause-effect deterministic models, which we call mechanistic models. An illustration of field observed spatial and temporal variation is provided by Webb et al. [25] who monitored 11 reaches in south-west England, through July 1992 to February 1993, noting a correspondence between variations in hourly river temperature and variation in discharge and drainage area (180 km^2 to 0.4 km^2), channel surface area (17 m^2 to 439 m^2), depth (0.1 m to 0.5 m), slope (0.002 to 0.05), orientation (north-south to east-west), and riparian cover (pasture to dense woodland). They found drainage basin land cover, especially riparian vegetation, overwhelmed other drivers of temperature. In two separate studies of more than16 rivers in the Washington, DC area, it was determined that runoff from impervious land uses entering rivers through urban storm sewers was the major thermal stressor, causing rapid (< 3 h) surges in temperature greater than 3 °C [1,26].

A comprehensive mechanistic model developed for river managers is Heat Source [27], which allows the user to specify local climatology, hydrology, morphology, and land use into Microsoft Excel and ESRI Arc View software, to simulate spatial and temporal variation in river temperature as a function of shortwave and longwave radiation, sensible and latent heat, riverbed conduction, and inflows from tributaries, groundwater, and hyporheic exchange. Data and algorithm limitations can constrain utilization of input-intensive mechanistic models, but these limitations can sometimes be overcome with innovations. Yearsley [28] developed a semi-Lagrangian scheme to advect river heat within a large river network when channel morphology data needed by Heat Source [27] and similar models (e.g., HSPF, CE-QUAL-W2, and QUAL2K) were not measured. To better capture abrupt changes in velocity and riparian shading, Crispell [29] created a retention time alternative to the advection-dispersion routing algorithm used in Heat Source [27], which maintains numerical stability at very fine spatial but coarse temporal discretization. For cases when observed boundary condition data are not available, Sun et al. [18] modified the DHSVM–RBM mechanistic model of river temperature to use Mohseni et al. [30] non-linear regression between weekly air temperature and river temperature to generate the upstream river temperature time series needed as a boundary condition.

The portability and accessibility of river temperature models are significant limitations for users interested in river management, pollution mitigation and restoration scenarios. The Heat Source model provides a balance between scenario simulation options and model parsimoniousness that made it our choice as the base code for developing a free, open-source, lower-complexity river temperature model useful in river pollution mitigation and restoration. The complexity of HSPF, CE-QUAL-W2, and QUAL2K is high, each containing many non-temperature routines, and CE-QUAL-W2 representing a 2-dimensional (2D) domain. These three models do not simulate ecological processes important in scenario analysis, including hyporheic exchange and temporal variation in the riparian shade; HSPF does have a pre-processor to provide temporal variation in riparian shade, which we use in our code development [14]. Glose et al. [31] noted that the major limitation of Heat Source [27] is lack of automation in making multiple simulations for parameter calibration and sensitivity analysis, given it is written in the Visual Basic for Applications language within Microsoft Excel. Glose et al. [31] addressed this limitation by using Matlab, a well-supported scientific programming language, to create the steady state model HFLUX, which represents many of the mechanistic processes in Heat Source. The HFLUX model [31] does not include the shade factor estimation and unsteady state routing algorithms of Heat Source [27], which are important in cases of temporal variation in shading and storm flow dynamics. Unfortunately, neither Heat Source [27] nor HFLUX [31] can be compiled into an executable, and therefore cannot be deployed outside of the VBA or Matlab environment.

This study created the i-Tree Cool River Model to address limitations of the Heat Source [27] and HFLUX [31] models and advance mechanistic model simulation of river management, pollution mitigation, and restoration scenarios in a parsimonious manner. The i-Tree Cool River Model is designed to allow for flexible shading factor algorithms, steady and unsteady flow, as well as other heat and mass transfer processes. The i-Tree Cool River Model is an open-source tool written in

C++, and its package contains the routines and an executable file for running the code, which can be downloaded from http://www.itreetools.org/research_suite/coolriver. The model executable is called at the command line along with a configuration extensible markup language (XML) file, which includes the required initial information. The i-Tree Cool River Model C++ algorithms can be edited and recompiled with Visual Studio 2017 Community Edition or later, which is freeware. The simulation output includes the simulated river temperature and the heat fluxes.

The objectives of this paper are to present the theory of the i-Tree Cool River Model, to apply the model in a case study with unsteady stormwater inflows, and to evaluate the importance of the heat and mass transfer processes. To that end, following the model development, the manuscript provides a model testing to address the application of the model. The science questions are: When analyzing sources of thermal pollution, and possible mitigation scenarios, what is the relative contribution of (a) storm sewer inflows, (b) subsurface inflows of groundwater and hyporheic exchange, (c) riparian shading and weather, on the accuracy of simulated river temperature?

2. Methods

2.1. Heat Flux Formulation

The i-Tree Cool River Model simulates an advection-dispersion equation with inflows and heat fluxes following Martin et al. [32]

$$\frac{\partial T_w}{\partial t} = -U\frac{\partial T_w}{\partial x} + D_L\frac{\partial^2 T_w}{\partial x^2} + R_h + R_i \tag{1}$$

where T_w is the cross-sectional averaged river temperature (°C), t is time (s), U is the reach average flow velocity (m/s), x is river distance (m), D_L is the dispersion coefficient (m^2/s), R_h is the heat flux reaction term, also known as heat transfer [3,27], and R_i is the reaction term of the external inflows. When R_i is combined with the advection and dispersion terms in Equation (1), they are collectively referred to as mass transfer [3,27]. The R_h and R_i are defined as

$$R_h = \frac{\Phi_{net}}{\rho C_p y} \tag{2}$$

$$R_i = \frac{Q_W T_W + Q_{GW} T_{GW} + Q_{Hyp} T_{Hyp} + Q_{SS} T_{SS}}{Q_i + Q_{GW} + Q_{Hyp} + Q_{SS}} - T_{W,t-1} \tag{3}$$

where Φ_{net} is the net exchange of thermal energy (W/m^2), ρ is the water density (kg/m^3), C_p is the specific heat capacity of water (J/kg °C), y is the average water column depth (m), Q is discharge (m^3/s), T is water temperature (°C), and subscripts W is the river flow, GW is groundwater flow, Hyp is hyporheic exchange, and SS is stormwater inflow, $t-1$ is prior time step. River velocities, dispersion, and inflows are calculated using standard methods, described in Supplementary Materials Section S1. The subsurface inflows distinguish between hyporheic and groundwater inflows due to their different environmental processes and use a separate mathematical formulation for each term. For surface inflows, users can include tributaries in place of storm sewers, and assign an unlimited number of surface inflows for each cross section.

The net exchange of thermal energy is defined as in Boyd et al. [27] as

$$\Phi_{net} = \Phi_{longwave} + \Phi_{shortwave} + \Phi_{latent} + \Phi_{sensible} + \Phi_{sediment} \tag{4}$$

where the Φ is the heat flux (W/m^2), and subscripts net is the net heat flux at the water surface, $longwave$ is the longwave radiation flux at the water surface, $shortwave$ is the shortwave radiation at the water surface, $latent$ is the latent heat flux from evaporation, $sensible$ is the sensible heat flux representing the convective thermal flux from the water surface, and $_{sediment}$ is the bed sediment heat flux representing conduction forcing at the water column interface.

The longwave radiation flux in Equation (4) is composed of positive downward fluxes from the atmosphere and land cover over the water surface, and a negative upward flux from the waterbody to the air, following the approach of Boyd et al. [27]

$$\Phi_{longwave} = \Phi_{longwave}^{atmospheric} + \Phi_{longwave}^{landcover} + \Phi_{longwave}^{back} \tag{5}$$

where $\Phi_{longwave}^{atmospheric}$ is the atmospheric flux (W/m^2), $\Phi_{longwave}^{landcover}$ is the land cover flux (W/m^2), and $\Phi_{longwave}^{back}$ is the back-to-air flux (W/m^2). Atmospheric longwave radiation is a function of air temperature and exposure from the river surface to the atmosphere, called the view-to-sky factor (f), calculated using Boyd et al. [27]

$$\Phi_{longwave}^{atmospheric} = 0.96\varepsilon_{atm}\sigma(T_{air} + 273.2)^4 \min(f_1,\ f_2,\ f_3) \tag{6}$$

where T_{air} is air temperature (°C), the ε_{atm} is the emissivity of the atmosphere (0 to 1), σ is the Stefan-Boltzmann constant (5.6696 $\times$ 10^{-8}, W/m^{2}K^4), and $\min(f_1, f_2, f_3)$ is the minimum of the three view-to-sky factors (0 to 1), where f_1 represents building effects, f_2 represents vegetation effects, and f_3 represents topographic effects (Figure 1). The emissivity of the atmosphere ε_{atm} is calculated using [33]

$$\varepsilon_{atm} = 1.72\left(\frac{0.1e_a}{T_{air} + 273.2}\right)^{\frac{1}{7}}(1 + 0.22C_L{}^2) \tag{7}$$

where e_a is the actual vapor pressure (mbar), and C_L is the cloudiness, which ranges from 0 for a clear sky to 1 for full cloud cover [34].

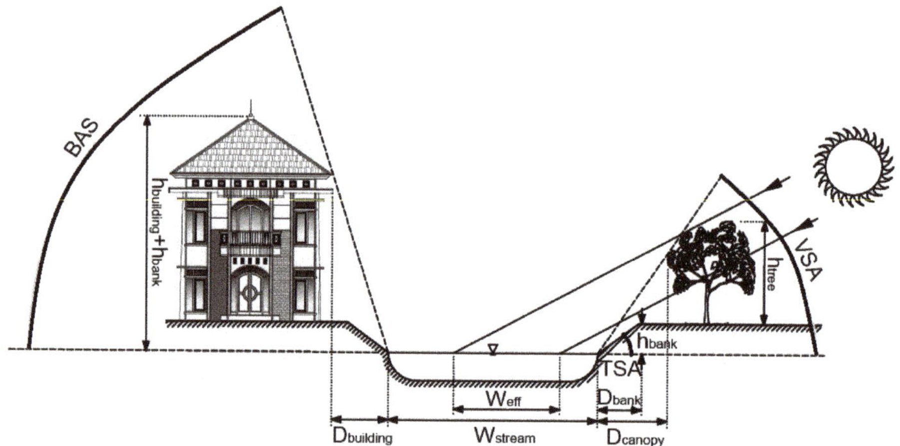

Figure 1. Shading of the river surface. A cross-sectional view, in which BSA is the building shading angle, VSA is the vegetation shading angle, and TSA is the topographic shading angle. h$_{building}$, h$_{tree}$, and h$_{bank}$ are building, vegetation, and bank heights respectively. D$_{building}$, D$_{canopy}$, and D$_{bank}$ are building to the bank, canopy to the bank, and bank.

The view-to-sky factors value of 1 indicates a full unobstructed sky view [27,35,36]. The general sky-view-factor formula for f_i is computed for each cross-section based on Chen et al. [14]

$$f_i = 1 - \frac{2}{\pi}SA_i \tag{8}$$

where i indicates the object at that cross-section, where 1 = building, 2 = vegetation, or 3 = topography; and SA is the shade angle (radians), computed as and h_c is the combined height of the objects above the water (e.g., if a tree is set on a hill, $h_c = h_{tree} + h_{bank}$), and max (D_i) is the maximum distance from all objects at that cross-section to the edge of the water.

$$SA_i = \tan^{-1}\left(\frac{h_c}{\max(D_{i,1-3})}\right) \tag{9}$$

The land cover longwave radiation in Equation (5) also uses the view-to-sky factors. The land cover radiation represents the land cover, e.g., vegetation such as trees' influence on water temperature, and the model by default sets land cover temperature equal to atmospheric temperature, following the approach of Boyd et al. [27]

$$\Phi^{landcover}_{longwave} = 0.96(1 - \min(f_1, \ f_2, \ f_3))0.96\sigma(T_{air} + 273.2)^4 \tag{10}$$

The waterbody to air radiation term in Equation (5) is a function of water temperature, representing heat flux emitted from the water surface, following the approach of Boyd et al. [27]

$$\Phi^{back}_{longwave} = -0.96\sigma(T_w + 273.2)^4 \tag{11}$$

where the T_W is the river temperature (°C).

See the Supplementary Materials, Section S2 for the methods used to find the remaining right-hand side terms in Equation (4), which are short wave radiation, latent heat flux, sensible heat flux, and bed sediment heat flux. Table S1 lists the 10 input files required by i-Tree Cool River, and names and describes the parameters in each of the files.

2.2. Study Area and Model Inputs

The i-Tree Cool River Model's accuracy in representing thermal loading was tested in unsteady state (i.e., wet weather) using unpublished data from 11 to 12 June 2007 for a 1500 m reach of Sawmill Creek, in Tannersville, New York (42.1955 N, 74.1339 W, WGS84). Sawmill Creek is a second-order mountainous river with varying watershed land use, starting in forests and transitioning to urban land. At the end of the Sawmill Creek study reach, the time of travel was approximately 30 min and the upstream watershed area is 8.16 km^2, which includes a nested urban watershed of 1.8 km^2 draining to the river in storm sewers (Figure 2a,b). Sawmill Creek flow at the upstream boundary was estimated using stage-discharge relations, monitoring stage with pressure transducers (manufactured by Global Water Instruments) at the upstream and downstream stations (0 m and 1500 m respectively) and in storm sewers. Stage was converted to discharge using the Manning equation, with stage converted to channel area and hydraulic radius using geometry relations, and the Manning roughness coefficients estimated from pebble counts Wolman [37] at each cross section by Crispell et al. [29]. We installed compound weir plates in the sewers and used the weir plate manufacturer equations to convert stage to discharge for the storm sewer inflows. Observed storm flows were corroborated with simulated flows by the i-Tree Hydro model (Yang, et al. [38]), calibrated to match the estimated baseflow.

Rainfall occurred twice during 12 June 2007 the first time with a 2 h duration totaling 3.3 mm and the second time with a 3 h duration totaling 8.4 mm. The storm sewer inflows were active during dry and wet weather, in dry weather due to illicit connections draining buildings, and in wet weather due to storm runoff. Crispell et al. [29] monitored river temperatures and storm sewer drainage temperatures at 12 river stations in the Sawmill Creek reach and the two inflow locations at 10-minute intervals using ibutton temperature data loggers, which were used to set boundary conditions. The temperature monitoring ibuttons in Sawmill Creek reach were strategically placed and considered representative of the reach temperature, capturing the influence of stormwater inflows after they had distance to mix with the channel water [29]. In the upstream section of the reach, between the cross section at station

0 m and a station at 600 m, the observed average river temperature increased by 0.03 °C per 100 m (3.3%), and between the cross section at station 900 m and a station 1500 m, temperature increased by 0.008 °C per 100 m (1%). In the middle section of the reach, between the cross section at station 600 m and station 900 m, the temperature increased by 0.1 °C per 100 m, three times the rate of the upstream reach, an increase attributed to the warming effect of the Tannersville's storm sewer inflow (Table S2).

The i-Tree Cool River was also tested in steady state mode, e.g., no rainfall events, for a 475 m reach of Meadowbrook Creek (43.0306 N, 76.0680 W, WGS84), a first order and urbanized river in the city of Syracuse, New York (Figure 3a,b). Flow at the upstream boundary, cross section survey data, and river temperature at 30 monitoring locations at 5-minute intervals were provided by Glose et al. [31], who used these data to develop HFlux. We simulated the 5-day period of 13–19 June 2012.

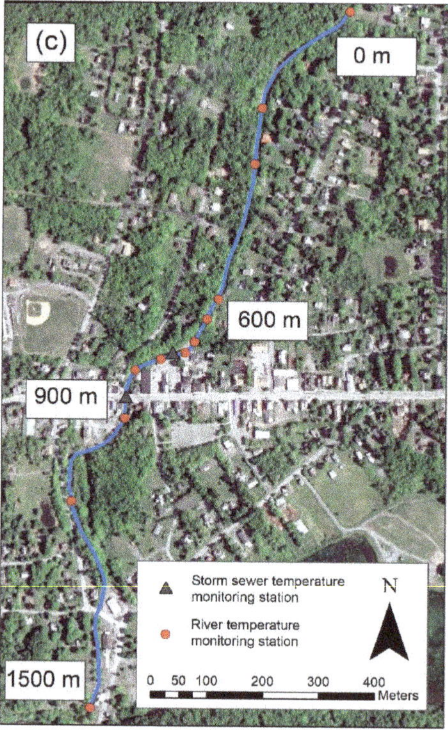

Figure 2. (**a**) New York State with the Sawmill Creek study area denoted by the star. (**b**) Monitoring stations and reach distances along Sawmill Creek.

The i-Tree Cool River Model simulated Sawmill Creek using input data from multiple sources. Specification of hourly weather data, including air temperature T_{air}, dew point temperature T_{dew}, shortwave radiation S_{in}, fraction cloudiness C, and wind speed U_{wind} were obtained from the National Solar Radiation Data Base (NSRDB) station ID#1227776, located 23 km from the study site. The NSRDB provides satellite estimated surface radiation at 30 min intervals. The shading factor, SF, in Equation (S2) and view-to-sky coefficients, f in Equation (S4) were estimated at 1 m intervals along Sawmill Creek using the TTools algorithm from observations of riparian vegetation and aerial images of the study area [29]. The river base width and bank slope were obtained from field surveys at each cross section [29], which defined the irregular pattern of river widening and narrowing. The simulated Sawmill Creek reach was delineated into 15 segments considering the locations where the temperature was observed, with segment lengths no greater than 100 m, which resulted in a simulation timestep of 0.5 seconds to satisfy the i-Tree Cool River Model stability criteria. The simulations represented the observed conditions, as well as alternative scenarios to determine model sensitivity to shading, subsurface inflow, and the calculated upstream boundary condition, which are sometimes difficult to obtain. Our calculated upstream boundary condition was derived with Mohseni et al. [30] Equation (12), a non-linear regression between air temperature and river temperature.

$$T_W = \mu + \frac{\alpha - \mu}{1 + e^{\gamma(\beta - T_{air})}} \tag{12}$$

In the equation, the coefficient α is the estimated maximum stream temperature, γ is a measure of the steepest slope of the function, and β represents the air temperature at the inflection point.

Figure 3. (**a**) New York State with the Meadowbrook Creek study area denoted by the star. (**b**) Monitoring stations and reach distances along Meadowbrook Creek.

Our observed upstream boundary condition was obtained from ibutton thermistor measurements. We analyzed the simulated and observed river temperatures for each of the cross sections averaged with respect to time to obtain a 30-hour average at each of the 12 cross sections. Simulations were written hourly for each cross section, which can be written at any timesteps for each meter of the river. We ran this simulation using Equations (5) to (11) for longwave radiation, Equation (S2) for shortwave radiation, Equation (S11) for latent heat, Equation (S12) for sensible heat, and Equation (S15) for the sediment heat.

3. Results

3.1. Model Evaluation

A scatterplot of the 30-hours of simulated and observed river temperatures for each of the 12 cross sections along the 1500 m of Sawmill Creek reach provides insights on the relative goodness of fit for each cross section and associated drivers of i-Tree Cool River Model accuracy (Figure 4). At cross section 1, along the upstream boundary, as expected, the observed boundary condition, resulted in a model fit with an R^2 of 1.0. Downstream, the fit degraded. Initial conditions of 14.7 °C for all cross sections caused the largest deviations between simulated and observed temperatures for most scatterplots. The model underestimated observed temperatures at cross sections 4 to 12 by approximately 1 °C, while upstream cross section temperatures were cooler and closer to the initial condition. The falling limb of storm event hydrograph corresponded with deviations in simulations at 22:00 hours of day 1 and 02:00 h of day 2 for cross sections 9 to 12, overestimating temperatures by approximately 0.3 °C and 0.5 °C, respectively.

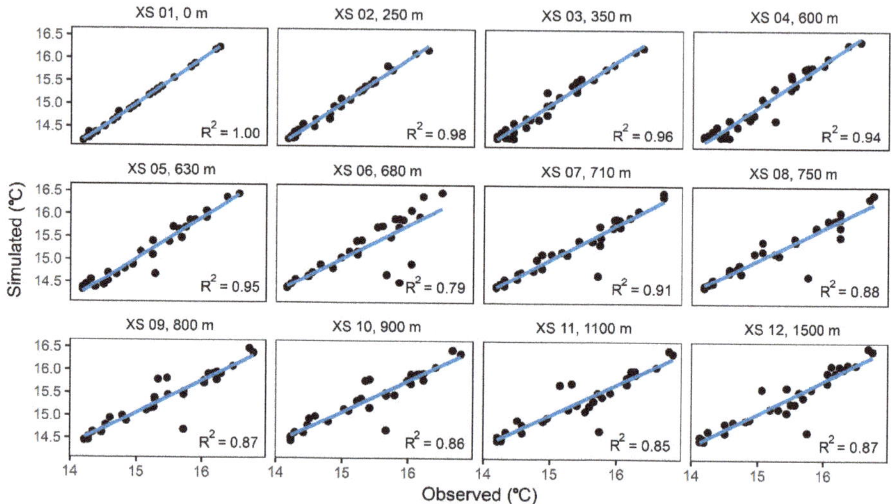

Figure 4. Observed versus simulated river temperatures for the 12 cross sections (XS) and stations from 0 m to 1500 m in Sawmill Creek. The coefficient of determination, R^2 for each cross section is shown in the plot.

The i-Tree Cool River model simulated hourly water temperatures were not significantly different than the observed, for reach averaged data, based on the p-values calculated using a paired-samples t-test and the $\alpha = 0.05$ (See Table S3 for more details). The 30-hour average simulated and observed river temperature, along the entire 1500 m Sawmill Creek reach, increased by 0.4 °C at a slope of approximately 0.02 °C per 100 m, but with longitudinal variation in that slope. For the 1500 m reach, the model had a root-mean-square error (RMSE) of 0.03 °C and a coefficient of determination (R^2) of 0.98 with a p-value of 0.87 which was greater than the α of 0.05. The model simulated the relatively rapid increase in water temperature recorded by the sensors, between cross section 600 m and 900 m, corresponding to the reach with storm sewer inflow. This relatively rapid increase in temperature leveled at station 900 m, which is the first station downstream of the last storm sewer outfall. Relatively warm water in the Tannersville storm sewer entering Sawmill Creek between cross sections at station 600 m and at station 900 m was a major driver of the i-Tree Cool River Model forecasting a rise in river temperature during both wet and dry weather conditions (Figure S1a). During the wet weather, a total of 7 hours, the rate that simulated river temperature increased from station 600 m to station 900 m at a rate of 0.32 °C per 100 m (Figure S1b), much steeper than during dry weather. During dry weather, the simulated river temperature from station 600 m to station 900 m increased at a rate of 0.04 °C per 100 m (Figure S1c), approximately 12% of the wet weather slope. The i-Tree Cool River algorithms for shading, net groundwater discharge, hyporheic exchange, and upstream boundary condition temperature influenced the simulation of longitudinal river temperature and the model goodness of fit. In all of the scenarios, the calculated paired t-test p-values were smaller than the $\alpha = 0.05$, rejecting the null hypothesis, H_0 of a significant difference between the means of the simulated and observed reach averaged river temperatures (See Table S4 for more details). When the shading algorithm was disabled, i.e., no shade was simulated, the i-Tree Cool River Model overestimated the river temperature for all cross sections by 0.34 °C, at a rate of 0.02 °C per 100 m, for the 30-hour period, 11 to 12 June 2007 (Figure S2), and the model RMSE increased to 0.36 °C and the R^2 decreased to 0.88.

Diurnal sinusoidal patterns of simulated and observed river warming and cooling were driven by the heat balance but disrupted by abrupt pulses of inflow due to warm runoff during the two storm events on 11 and 12 June 2007. The Sawmill Creek mean temperature, the average of measurements at the 12 cross sections, diurnally peaked at 15.8 °C by 15:00 h June 11, 2007 (Figure 5a), two hours after the peak in shortwave radiation (Figure 5b). By 20:00 hours, shortwave radiation has declined to 0, net radiation became negative, and river temperature has decreased from the peak of 15.8 °C to 15.1 °C. A storm event at 21:00 h 11 June 2007, and then again at 01:00 h of 12 June 2007, generates inflow of warmer water from the upstream and storm sewer, creating a temporary increase in temperature, which disrupts the sinusoidal pattern in cooling toward the diurnal minimum temperature at 05:00 h of 12 June 2007. Dry weather extends through the remainder of the simulation, and at 06:00 h of 12 June 2007, the increasing shortwave and thus net radiation reestablish heat flux as the main driver of the increasing river temperature, which peaks at 16.4 °C at 14:00 h. There was no significant difference between the simulated and observed time averaged river temperature datasets based on the paired-samples t-test and $\alpha = 0.05$ (See Table S5 for more details). The model simulations of the abrupt pulses in river temperature during the wet weather, extending from hour 20 of day 1 to hour 3 of day 2, had a Nash-Sutcliffe Efficiency (NSE) coefficient of 0.9 and a p-value of 0.80 (Figure 5a). The magnitude of the simulated river temperature changes due to the inflow of stormwater from the Tannersville storm sewer system was 0.3 °C during the first storm and 0.4 °C during the second storm. The model simulations had their poorest fit with observed river temperatures during a 6 h period on 12 June 2007, between 03:00 and 09:00 h, centered at sunrise, when it overestimated the river temperature by an average of 0.13 °C.

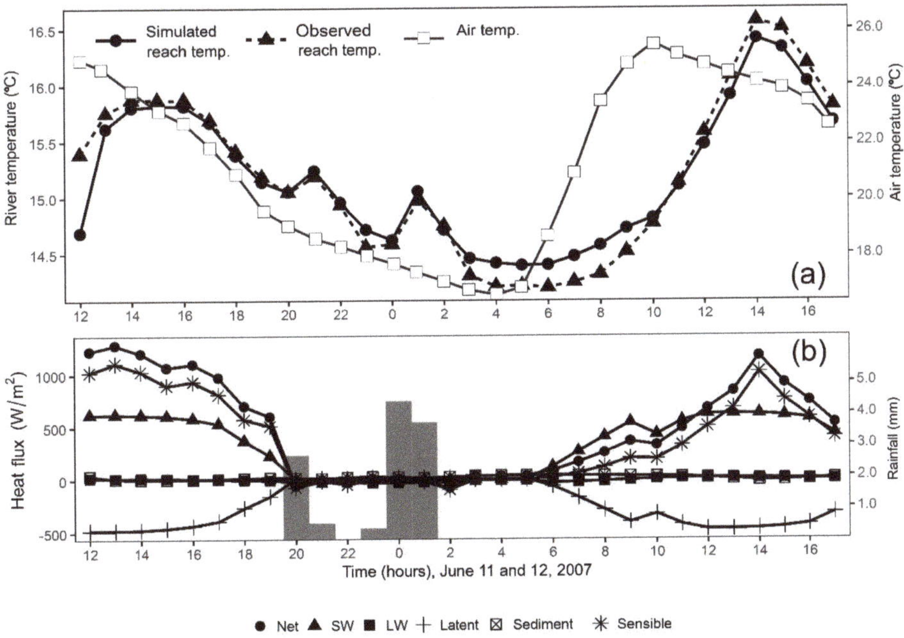

Figure 5. (**a**) Reach average air temperature and observed and simulated river temperatures in Sawmill Creek, between 12:00 h of 11 June 2007 to 17:00 h of 12 June 2007; (**b**) Simulated reach averaged heat fluxes and precipitation into Sawmill Creek between 12:00 h of 11 June 2007 to 17:00 h of 12 June 2007.

The river temperature simulated by the i-Tree Cool River Model was a function of spatially and temporally varying contributions of groundwater, hyporheic exchange, and storm sewer inflow. Analysis of these components to thermal loading can assist in developing pollution mitigation or river restoration scenarios. In cross sections without storm sewer inflow and in the absence of rain events, the river temperature was predominantly determined by river flow from the upstream reach, and groundwater only contributed approximately 1%, while hyporheic exchange contributed approximately 10% (Figure 6a; 350 m, and 1100 m). In cross sections with storm sewers, even in the absence of rain events, when the storm sewers discharged flow from illicit connections, they contributed approximately 25% of the flow influencing the cross section river temperature (Figure 6a; 800 m), which warmed the river water (see Table S2 reflecting warmer average temperature of the storm sewer in the dry weather). During wet weather, there was inflow from the storm sewer due to the flows from impervious areas, and this inflow contributed approximately 50% of the flow thereby influencing the river temperature (Figure 6b, 800 m) and provided the thermal load to the river system. Integrated along the 1500 m of Sawmill Creek reach, the contribution of groundwater summed to 15% of the total river volume, while the hyporheic exchange, which flows in and out within each sub-reach associated with a cross section, averaged approximately 10% of inflow in each sub-reach.

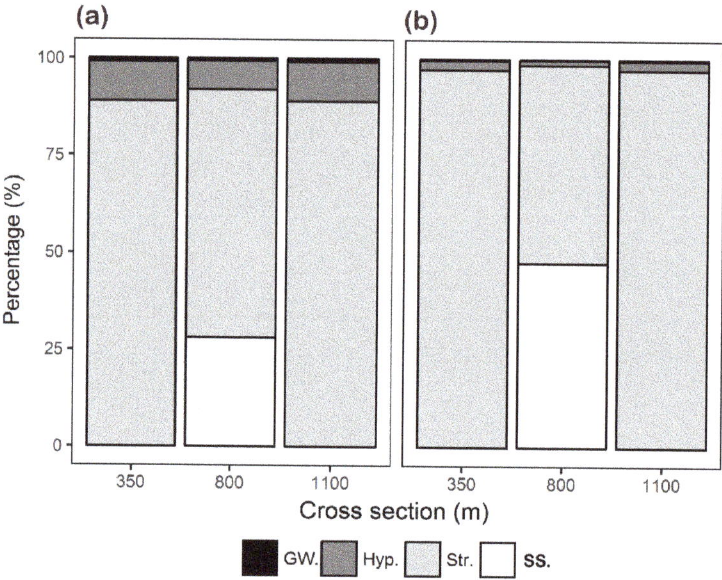

Figure 6. Simulated contribution to river temperature of the river flow (Str.), storm sewer (SS.), hyporheic exchange (Hyp.), and groundwater flow (GW.) in two timesteps including (**a**) before storm and (**b**) during the storm at three representative cross sections from the upper reaches (between cross sections at station 0 m and 600 m), middle reach (between cross sections at station 600 m and 900 m), and downstream reach (between cross sections at station 900 m and 1500 m).

In addition to the analysis of unsteady simulations in Sawmill Creek, the i-Tree Cool River Model performance was analyzed for the steady state condition in Meadowbrook Creek for 13–19 June 2012. The i-Tree Cool River Model simulated the time averaged river temperatures at 30 cross sections with an RMSE of 0.2 °C. We combined these 30 cross sections into reach averaged river temperature data to examine the diurnal pattern driven by the heat balance (Figure 7a). There was no significant difference between the simulated and observed reach averaged river temperatures based on a paired-samples t-test and $\alpha = 0.05$ (See Table S6 for more details). The model simulations of the temperature for steady state in the Meadowbrook Creek study reach for 13–19 June 2012 had a NSE coefficient of 0.9 and a p-value of 0.72. The model captures how Meadowbrook Creek cools by 0.25 °C as it flows along the 475 m reach (Figure 7b), driven by the constant inflow of cooler groundwater.

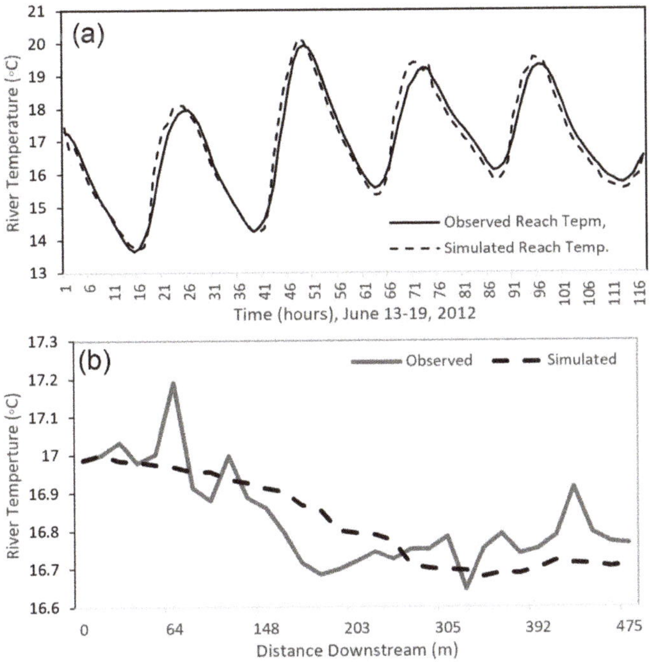

Figure 7. (**a**) Reach average observed and simulated river temperatures in Meadowbrook Creek, for 13–19 June 2012; (**b**) Time averaged observed and simulated river temperatures for 475 m reach of the Meadowbrook Creek.

3.2. Sensitivity Analysis

A sensitivity analysis of the i-Tree Cool River Model examined the fluctuation in simulated temperature with changes in input data to identify the most sensitive parameters, which is useful when considering impacts of environmental change. The sensitivity analysis was performed for steady and unsteady simulations. We used global sensitivity analysis to identify the most important parameters and coordinated this analysis with that for the Meadowbrook Creek reach in summertime, by Glose et al. [31], noting both models are based on Heat Source [27]. Glose et al. [31] used an observed boundary conditions and our Equations (1), (2), (4)–(6), (10), (11), (S2), (S11) and (S12), and varied discharge by ±10%, groundwater temperature ±15%, varied shading factor and view-to-sky factor by ±20%. They identified groundwater as the most sensitive parameter, with a ±0.2 °C change on average stream temperature. We replicated this sensitivity analysis to the 1500 m Sawmill Creek reach, confirming these sensitivities. We then extended the analysis in the 1500 m Sawmill Creek reach to consider varying parameters of storm sewer temperature, sediment temperature, and recorded boundary conditions temperature by ±15% (Figure S3) and varying parameters of substrate hydraulic conductivity (SHC), cloudiness factor (Cl), and groundwater discharge (GW) by ±20% (Figure S4). When storm sewer temperature was varied by ±15%, the reach-averaged temperature changed by 1.65% (0.27 °C). When sediment temperature was varied by ±15%, the reach-averaged temperature changed by 0.3%. When upstream boundary conditions temperature was varied by ±15%, the reach-averaged temperature changed by 9.5%. When substrate hydraulic conductivity was varied by ±20%, the reach-averaged temperature changed by 0.15%. When cloudiness factor was varied by ±20%, the reach-averaged temperature changes by 0.02%. When groundwater discharge factor was varied by ±20%, the reach-averaged temperature changes by 0.03%. Based on this analysis, the most sensitive model parameters, ranked in order of importance, are upstream boundary conditions, storm sewer temperature, sediment temperature, substrate hydraulic conductivity, groundwater discharge, and cloudiness (additional sensitivity analysis is presented in supplementary materials Figures S3–S6).

4. Discussion

The i-Tree Cool River Model simulated the warming effects of the many potential sources of thermal pollution, including radiation fluxes and urban runoff, both dry weather illicit connections and wet weather stormwater. The model also shows how groundwater and hyporheic exchange inflows can provide a cooling effect, providing a comprehensive approach to assessing and perhaps mitigating thermal loading. To determine which factors are most effective in such management, this discussion provides some perspective on the effects of each warming and cooling effect.

The impact of urban runoff on the average temperature was rapid, within 1 hour of the onset of precipitation, and caused a temperature increase of 0.3 °C for the first storm, and 0.4 °C for the second storm of 12 June 2007. The rapid and large change in river temperature can be attributed to the short duration event, which Herb et al. [1] suggest when rainfalls only last 2 to 3 h will make the largest impact on raising stormwater and water temperature. The urban storm sewer area was approximately 21% of the watershed drainage area and had 35% impervious cover, which contributed to a relatively large volume of flashy, warm, stormwater response. Relative differences between air and water temperatures contribute to the warming, as noted by Herb et al. [1]; for the Sawmill Creek reach the average air temperature was 21.2 °C and average dew point temperature was 18.0 °C, both warmer than the average river temperature which was 15.2 °C. Even though the two rainfall events occurred at night during 12 June 2007, when solar radiation was not present, the prior day averaged 20% cloud coverage, allowing 80% of summer shortwave radiation to reach the small albedo impervious surface.

Simulation of the effects of the nighttime stormwater thermal load in riverine receiving waters on Sawmill Creek contributes additional data and tools to the investigation and management of the urban heat island. A common signature of the urban heat island is elevated nighttime air temperatures in urban areas relative to rural areas, due to physical differences affecting solar heating, such as albedo and thermal capacity, and anthropogenic heat sources [39]. Daytime insulation is a common driver of thermal loading of receiving waters [40], but for 12 June in Tannersville, the daytime solar heating of impervious area did not cause thermal loading of the river until the nighttime wet weather event. The nighttime precipitation landed on warm impervious surfaces, retaining much of their daytime elevated surface temperatures due to high capacitance, and this surface warmth was conducted into the stormwater entering the relatively cool, rural origin, receiving water. The effect of urban heat islands on rivers was studied by Somers et al. [41], who noted a 1.6 °C higher warm season temperature in urban rivers than forested rivers, and 8 °C greater spatial variation in urban rivers than in rural river temperatures along a 1 km transect. During a daytime storm event affecting all rivers, the temperatures in urban rivers rose as much as 4 °C, compared with a negligible rise in temperature in the forested rivers [41]. Nelson et al. [26] forecasted the thermal impact of individual storm events and found storm-induced river temperature surged by 3.5 °C for the warm season in urban watersheds near Washington, DC, USA; with drainage areas averaging 8 km^2.

Proper simulation with the i-Tree Cool River Model of unsteady flows and their thermal pollution of receiving waters requires consideration of model goals and limitations. Typical model goals are either model inter-comparison for contrasting scenarios, such as varying impervious or tree cover, or model simulation for hindcasting or forecasting. In cases of model inter-comparison, the model has fewer limitations and the model physics will allow users to consider changes in river temperature for changes in study site conditions; model simulation has accuracy constrained by the accuracy of inputs as well as a model epistemic error [31,42]. This project attempted to improve accuracy of model simulations in Sawmill Creek by obtaining accurate data of the storm volumes and temperatures entering at the upstream and storm sewer locations along the boundary, using ibutton sensors, which are widely used for river temperature monitoring [31,43,44]. In cases where point or diffuse sources enter at multiple, unspecified locations along the river channel, such as with groundwater seeps, ibuttons may be inefficient and a better monitoring approach may involve using distributed temperature sensing system [8,35] for high-frequency time series, or from forward-looking infrared radar [45] temporally coarser data. An alternative to monitoring storm sewer inflow temperatures is estimating those temperatures using models of impervious runoff [25,29], and upstream boundary conditions can be estimated using air temperature records with the Mohseni et al. [30] non-linear regression.

Groundwater and hyporheic exchange were significant factors of temperature regulation during wet and dry weather. The section of Sawmill Creek simulated by the i-Tree Cool River Model had groundwater flow rates of approximately $0.0024\,m^3/s$ per 100 m, or 1% of flow, and riverbed longitudinal slopes that generated 10% contributions of hyporheic exchange (Figure 6). This combined subsurface flow, through the model inflow routines, contributed a cooling effect for the simulated summer period in 11–12 June 2007. Removing these inflows from the simulation caused the model to achieve RMSE of 0.18 °C and R^2 of 0.94, less important than the cooling through shading and the heat flux routines, when removed generated a RMSE of 0.36 °C and R^2 of 0.88. In winter, when river temperature is typically below subsurface water temperatures, this inflow would likely contribute a warming effect, as observed in other rivers by Risley et al. [46] and Kurylyk et al. [47]. From a survey of other studies, the relative contribution of groundwater and hyporheic exchange inflow with river water varies by site conditions and time. Poole et al. [48] working in mountain rivers, with bed slopes above 2%, also found the surface water received a larger volumetric inflow from hyporheic exchange than from groundwater, while Glose et al. [31] working in valleys with approximately 1% slopes did not identify significant hyporheic exchange and set groundwater as the only subsurface source of inflow.

Riparian shading from tree canopy, hillslope, and buildings provided the only land-based reduction in shortwave radiation and the view to the sky for the river, which influenced the longwave radiation. We used model inter-comparison simulations to contrast a scenario with and without shading and determined shading cooled river temperatures by an average of 0.34 °C during the 30 hours period. The landscape contribution to shading varied longitudinally along the reach, and at cross-section 9 was primarily from building shade, while upstream at cross sections 1 to 5 was primarily forest; hillslope topography provided minimal shading at this site. Shading is a concern in river thermal loading, and shallow and slow moving water is more vulnerable to such warming, and others have modeled this effect. Sun et al. [18] simulated 6 years along 6 separate reaches ranging in length from 85 to 1185 m of Mercer Creek in Washington State, and determined that tree and hillslope shading reduced the annual maximum temperatures by 4 °C. Roth et al. [49] simulated three cloud-free summer days in August 2007 along a 1260 m section of the Boiron de Morges River in southwest Switzerland, and determined riparian shading, by decreasing shortwave radiation, decreased daily average water temperatures by 0.7 °C. Guoyuan et al. [50] demonstrated predictions of shade from riparian vegetation (e.g., the Chen et al. [14] method used in i-Tree Cool River) were sensitive to the interaction of river azimuth and latitude, with E-W rivers in low latitudes benefiting least from riparian shade. Lee et al. [51] recommended for effective reduction in shortwave radiation, riparian areas utilize shading angles of 70° (1.22 radian) and view-to-sky factors smaller than or equal to 0.22. In the 1500 m Sawmill Creek reach, less than 10% of the view-to-sky factors were smaller or equal to 0.22 and shading angles averaged 50°, and therefore additional thermal management opportunities are present.

The i-Tree Cool River Model was designed to assist river managers assess mechanistic causes of thermal pollution using free, open source, relatively simple algorithms in order to negotiate the balance between complexity and accuracy. While the model requires several input files, many of these can be obtained from publicly available data, site surveys, or estimation approaches; for model inter-comparison studies the accuracy of input data become less critical than in forecast simulations. The number of input files required by the model is comparable to other mechanistic models simulating river temperature, such as HFlux, HSPF, and QUAL2K, which require approximately 25 to 40 parameters, spatial data of river geometry and riparian features, and time series data describing the weather and discharge. Obtaining these inputs is a potential limitation of the i-Tree Cool River Model, and methods to obtain or estimate these input files are discussed above.

5. Conclusions

In this study, we developed the one-dimensional mechanistic i-Tree Cool River Model to simulate river temperature considering a combination of advection, dispersion, heat flux, and inflow processes. The i-Tree Cool River Model has the ability to analyze the impacts of external loads including multiple lateral storm sewer inflows, groundwater flow, and hyporheic exchange flow in steady and unsteady flows. The i-Tree Cool River Model estimates the shading effects of the riparian vegetation and other features as a function of heights and distances as well as solar geometry. The model performance was tested in steady and unsteady modes for the Meadowbrook reach in Syracuse, New York and Sawmill Creek in Tannersville, New York, respectively. The i-Tree Cool River Model performed satisfactorily in both simulations. The model can be used to conduct thermal pollution analysis of urban areas and investigate land cover and hydrology-based mitigation methods. The simulated river temperature of the i-Tree Cool River Model can be used for other environmental models, such as urban development models, atmospheric models, climate change models, and hydrology models.

Supplementary Materials: The following are available online at http://www.mdpi.com/2073-4441/11/5/1060/s1, Table S1: List of the input files required for the simulation process of the i-Tree Cool River Model, Table S2: Observed water temperatures for three reaches of Sawmill Creek and for the Tannersville storm sewer, during 11 and 12 June 2007, as the average for all time steps during the dry or wet weather conditions, Table S3: Statistical analysis (paired t-test) of the reach averaged observed and simulated river temperature in Sawmill Creek for the (a) original condition including both wet and dry weather (b) wet weather, and (c) dry weather, Table S4: Statistical analysis (paired t-test) of the reach averaged observed and simulated river temperatures using the scenarios for the (a) no shading effect, (b) no groundwater and hyporheic exchange inflows, and (c) calculated boundary condition, Table S5: Statistical analysis (paired t-test) of the observed and simulated river temperatures in Sawmill Creek, between 12:00 h of 11 Jun 2007 to 17:00 h of 12 June 2007, Table S6: Statistical analysis (paired t-test) of the observed and simulated river temperatures in Meadowbrook Creek, for 13–19 June 2012, Figure S1: Time averaged observed and simulated river temperature in Sawmill Creek for the (a) original condition including both wet and dry weather (b) wet weather, and (c) dry weather, Figure S2: Time averaged observed and simulated river temperatures using the scenarios for the (a) no shading effect, (b) no groundwater and hyporheic exchange inflows, and (c) calculated boundary condition, Figure S3: Simulated time averaged river temperatures along the 1500 m Sawmill Creek reach for the original condition (Base) and for conditions with ±15% changes in (a) storm sewer temperature (T_{SS}), (b) sediment temperature, and (c) boundary conditions temperature, Figure S4: Simulated time averaged river temperature along the 1500 m Sawmill Creek reach for the original condition (Base) and for conditions with ±20% changes in (a) substrate hydraulic conductivity (SHC), (b) cloudiness factor (Cl), and (c) groundwater discharge (GW), Figure S5: Fluctuations of shading factors and daily average shortwave radiation along the 1500 m Sawmill Creek reach. The shading factors denoted by a triangle are measured at each of the 12 monitoring stations, and the minimum and maximum shading factors were selected from the 5 m interval set of shading factors measured between each station, Figure S6: Temperature differences between the observed and simulated river temperature when using Mohsni et al. [30], ΔT_{calc} versus recorded, $\Delta T_{recorded}$ boundary conditions, for (a) nighttime and (b) daytime.

Author Contributions: Conceptualization, R.A. and T.E.; methodology, T.E.; software, R.A.; validation, R.A. and T.E.; formal analysis, R.A. and T.E.; investigation, R.A. and T.E.; resources, R.A. and T.E.; data curation, R.A. and T.E.; writing—original draft preparation, R.A.; writing—review and editing, R.A. and T.E.; visualization, R.A.; supervision, T.E.; project administration, T.E.; funding acquisition, T.E.

Funding: This research was supported by the USDA Forest Service Northern Research Station i-Tree grant No. 15-JV-11242308-114.

Acknowledgments: We would like to thank Omid Mohseni, Laura Lautz, Ning Sun, and Jill Crispell for explaining their models.

Conflicts of Interest: The authors declare no conflict of interest.

References

1. Herb, W.R.; Janke, B.; Mohseni, O.; Stefan, H.G. Thermal pollution of streams by runoff from paved surfaces. *Hydrol. Process.* **2008**, *22*, 987–999. [CrossRef]
2. Parker, F.L.; Krenkel, P.A. *Thermal Pollution: Status of the Art*; Report 3; Department of Environmental and Resource Engineering, Vanderbilt University: Nashville, TN, USA, 1969.

3.   Wunderlich, T.E. *Heat and Mass Transfer between a Water Surface and the Atmosphere*; Report No. 14; Water Resources Research Laboratory, Tennessee Valley Authority: Norris Tennessee, TN, USA, 1972.

4.   Poshtiri, M.P.; Pal, I. Patterns of hydrological drought indicators in major U.S. River basins. *Clim. Chang.* **2016**, *134*, 549–563. [CrossRef]

5.   Deas, M.L.; Orlob, G.T. *Klamath River Modeling Project and Appendices, Report No. 99–04*; University of California: Davis, CA, USA, 1999; 376p.

6.   Langan, S.J.; Johnston, L.; Donaghy, M.J.; Youngson, F.; Hay, D.W.; Soulsby, C. Variation in river water temperatures in an upland stream over a 30-years period. *Sci. Total Environ.* **2001**, *265*, 195–207. [CrossRef]

7.   Abdi, R.; Yasi, M. Evaluation of environmental flow requirements using eco-hydrologic-hydraulic methods in perennial rivers. *Water Sci. Technol.* **2015**, *72*, 354–363. [CrossRef] [PubMed]

8.   Sand-Jensen, K.; Pedersen, N.L. Differences in temperature, organic carbon and oxygen consumption among lowland streams. *Freshw. Biol.* **2005**, *50*, 1927–1937. [CrossRef]

9.   Ficke, A.D.; Myrick, C.A.; Hansen, L.J. Potential impacts of global climate change on freshwater fisheries. *Rev. Fish Biol. Fish.* **2007**, *17*, 581–613. [CrossRef]

10. Segura, C.; Caldwell, P.; Sun, G.; Mcnulty, S.; Zhang, Y. A model to predict stream water temperature across the conterminous USA. *Hydrol. Process.* **2015**, *29*, 2178–2195. [CrossRef]

11. Elliott, J.M.; Hurley, M.A.; Fryer, R.J. A new, improved growth model for brown trout, Salmo trutta. *Funct. Ecol.* **1995**, *9*, 290–298. [CrossRef]

12. Ahmadi-Nedushan, B.; St-Hilaire, A.; Quarda, T.B.M.J.; Bilodeau, L.; Robichaud, E.; Thiemonge, N.; Bobee, B. Predicting river water temperatures using stochastic models: Case study of the Moisie River (Quebec, Canada). *Hydrol. Process.* **2007**, *21*, 21–34. [CrossRef]

13. LeBlanc, R.T.; Brown, R.D.; FitzGibbon, J.E. Modeling the effects of land use change on the water temperature in unregulated urban streams. *Environ. Manag.* **1997**, *49*, 445–469. [CrossRef]

14. Chen, Y.D.; Carsel, R.F.; McCutcheon, S.C.; Nutter, W.L. Stream temperature simulation of forested riparian areas: I. Watershed-Scale model development. *J. Environ. Eng.* **1998**, *124*, 304–315. [CrossRef]

15. Van Buren, M.; Watt, W.E.; Marsalek, J.; Anderson, B.C. Thermal enhancement of stormwater runoff by paved surfaces. *Water Res.* **2000**, *34*, 1359–1371. [CrossRef]

16. Sridhar, V.; Sansone, A.L.; LaMarche, J.; Dubin, T.; Lettenmaier, D.P. Prediction of stream temperature in forested watersheds. *J. Am. Water Resour. Assoc.* **2004**, *40*, 197–213. [CrossRef]

17. Herb, W.R.; Janke, B.; Mohseni, O.; Stefan, H.G. Runoff temperature model for paved surfaces. *J. Hydrol. Eng.* **2009**, *14*, 1146–1155. [CrossRef]

18. Sun, N.; Yearsley, J.; Voisin, N.; Lettenmaier, D.P. A spatially distributed model for the assessment of land use impacts on stream temperature in small urban watersheds. *Hydrol. Process.* **2015**, *29*, 2331–2345. [CrossRef]

19. Jones, M.P.; Hunt, W.F.; Winston, R.J. Effect of urban catchment composition on runoff temperature. *J. Environ. Eng.* **2012**, *138*, 1231–1236. [CrossRef]

20. Hester, E.T.; Bauman, K.S. Stream and retention pond thermal response to heated summer Runoff from urban impervious surfaces. *J. Am. Water Resour. Assoc.* **2013**, *49*, 328–342. [CrossRef]

21. Guzy, M.; Richardson, K.; Lambrinos, J.G. A tool for assisting municipalities in developing riparian shade inventories. *Urban For. Urban Green.* **2015**, *14*, 345–353. [CrossRef]

22. Edinger, J.E.; Duttweil, D.; Geyer, J.C. Response of water temperatures to meteorological conditions. *Water Resour. Res.* **1968**, *4*, 1137–1143. [CrossRef]

23. Aboelnour, M.; Engel, B.A. Application of remote sensing techniques and geographic information systems to analyze land surface temperature in response to land use/land cover change in greater Cairo region, Egypt. *J. Geogr. Inf. Syst.* **2018**, *10*, 57–88. [CrossRef]

24. Caissie, D. The thermal regime of rivers: A review. *Freshw. Biol.* **2006**, *51*, 1389–1406. [CrossRef]

25. Webb, B.; Zhang, Y. Spatial and seasonal variability in the components of the river heat budget. *Hydrol. Process.* **1997**, *11*, 79–101. [CrossRef]

26. Nelson, K.; Palmer, M.A. Stream temperature surges under urbanization and climate change: Data, models, and responses. *J. Am. Water Resour. Assoc.* **2007**, *43*, 440–452. [CrossRef]

27. Boyd, M.; Kasper, B. *Analytical Methods for Dynamic Open Channel Heat and Mass Transfer: Methodology for Heat Source Model Version 7.0*; Watershed Sciences Inc.: Portland, OR, USA, 2003.

28. Yearsley, J.R. A semi-Lagrangian water temperature model for advection-dominated river systems. *Water Resour. Res.* **2009**, *45*, W12405. [CrossRef]

29. Crispell, J.K. Hyporheic Exchange Flow around Stream Restoration Structures and the Effect of Hyporheic Exchange Flow on Stream Temperature. Master' Thesis, College of Environmental Science and Forestry, State University of New York, Syracuse, NY, USA, 2008; 63p.

30. Mohseni, O.; Stefan, H.G.; Erickson, T.R. A nonlinear regression model for weekly stream temperatures. *Water Resour. Res.* **1998**, *34*, 2685–2692. [CrossRef]

31. Glose, A.; Lautz, L.K.; Baker, E.A. Stream heat budget modeling with HFLUX: Model development, evaluation, and applications across contrasting sites and seasons. *Environ. Model. Softw.* **2017**, *92*, 213–228. [CrossRef]

32. Martin, J.L.; McCutcheon, S.C. *Hydrodynamics and Transport for Water Quality Modeling*; Lewis Publishers: New York, NY, USA, 1999.

33. Kustas, W.P.; Rango, A.; Uijlenhoet, R. A simple energy budget algorithm for the snowmelt runoff model. *Water Resour. Res.* **1994**, *30*, 1515–1527. [CrossRef]

34. Dingman, S.L. *Physical Hydrology*; Prentice-Hall Inc.: Upper Saddle River, NJ, USA, 1994.

35. Westhoff, M.C.; Savenije, H.H.G.; Luxemburg, W.M.J.; Stelling, G.S.; Van De Giesen, N.C.; Selker, J.S. Sciences A distributed stream temperature model using high-resolution temperature observations. *Hydrol. Earth Syst. Sci.* **2007**, *11*, 1469–1480. [CrossRef]

36. Benyahya, L.; Caissie, D.; El-jabi, N.; Satish, M.G. Comparison of microclimate vs. remote meteorological data and results applied to a water temperature model (Miramichi River, Canada). *J. Hydrol.* **2010**, *380*, 247–259. [CrossRef]

37. Wolman, M.G. A method of sampling coarse bed material. *Trans. Am. Geophys. Union* **1954**, *35*, 951–956. [CrossRef]

38. Yang, Y.; Endreny, T.A.; Nowak, D.J. iTree-Hydro: Snow hydrology update for the urban forest hydrology model. *J. Am. Water Resour. Assoc.* **2011**, *47*, 1211–1218. [CrossRef]

39. Memon, R.A.; Leung, D.Y.C.; Liu, C.H.; Leung, M.K.H. Urban heat island and its effect on the cooling and heating demands in urban and suburban areas of Hong Kong. *Theor. Appl. Climatol.* **2011**, *103*, 441–450. [CrossRef]

40. Hathaway, J.M.; Winston, R.J.; Brown, R.A.; Hunt, W.F.; McCarthy, D.T. Temperature dynamics of stormwater runoff in Australia and the USA. *Sci. Total Environ.* **2016**, *559*, 141–150. [CrossRef]

41. Somers, K.A.; Bernhardt, E.S.; Grace, J.B.; Hassett, B.A.; Sudduth, E.B.; Wang, S.; Urban, D.L. Streams in the urban heat island: Spatial and temporal variability in temperature. *Freshw. Sci.* **2013**, *32*, 309–326. [CrossRef]

42. Beven, K. So how much of your error is epistemic? Lessons from Japan and Italy. *Hydrol. Process.* **2013**, *27*, 1677–1680. [CrossRef]

43. Crispell, J.K.; Endreny, T.A. Hyporheic exchange flow around constructed in-channel structures and implications for restoration design. *Hydrol. Process.* **2009**, *23*, 2267–2274. [CrossRef]

44. Hester, E.T.; Doyle, M.W.; Poole, G.C. The influence of in-stream structures on summer water temperatures via induced hyporheic exchange. *Limnol. Oceanogr.* **2009**, *54*, 355–367. [CrossRef]

45. Loheide, S.P.; Gorelick, S.M. Quantifying stream–Aquifer interactions through the analysis of remotely sensed thermographic profiles and in situ temperature histories. *Environ. Sci. Technol.* **2006**, *40*, 3336–3341. [CrossRef]

46. Risley, J.C.; Constantz, J.; Essaid, H.; Rounds, S. Effects of upstream dams versus groundwater pumping on stream temperature under varying climate conditions. *Water Resour. Res.* **2010**, *46*, W06517. [CrossRef]

47. Kurylyk, B.L.; Moore, R.D.; Macquarrie, K.T.B. Scientific briefing: Quantifying streambed heat advection associated with groundwater-surface water interactions. *Hydrol. Process.* **2016**, *30*, 987–992. [CrossRef]

48. Poole, G.C.; Berman, C.H. An ecological perspective on in-stream temperature: Natural heat dynamics and mechanisms of human-caused thermal degradation. *Environ. Manag.* **2001**, *27*, 787–802. [CrossRef]

49. Roth, T.R.; Westhoff, M.C.; Huwald, H.; Huff, J.A.; Rubin, J.F.; Barrenetxea, G.; Vetterli, M.; Parriaux, A.; Selker, S.; Parlange, M.B. Stream temperature response to three riparian vegetation scenarios by use of a distributed temperature validated model. *Environ. Sci. Technol.* **2010**, *44*, 2072–2078. [CrossRef] [PubMed]

50. Guoyuan, L.; Jackson, C.; Kraseski, K. Modeled riparian stream shading: Agreement with field measurements and sensitivity to riparian conditions. *J. Hydrol.* **2012**, *428*, 142–151.

51. Lee, T.Y.; Huang, J.C.; Kao, S.J.; Liao, L.Y.; Tzeng, C.S.; Yang, C.H.; Kalita, P.K.; Tung, C.P. Modeling the effects of riparian planting strategies on stream temperature: Increasing suitable habitat for endangered Formosan Landlocked Salmon in Shei-Pa National Park, Taiwan. *Hydrol. Process.* **2012**, *26*, 3635–3644. [CrossRef]

Article

An Assessment of Self-Purification in Streams

Valentinas Šaulys, Oksana Survilė * and Rasa Stankevičienė

Department of Environmental Protection and Water Engineering, Faculty of Environmental Engineering, Vilnius Gediminas technical university; Saulėtekio al. 11, LT-10223 Vilnius, Lithuania; valentinas.saulys@vgtu.lt (V.S.); rasa.stankeviciene@vgtu.lt (R.S.)
* Correspondence: oksana.survile@vgtu.lt; Tel.: +37-069-980-153

Received: 14 November 2019; Accepted: 21 December 2019; Published: 25 December 2019

Abstract: The territory of Lithuania is characterized by a prevailing moisture excess, therefore in order to timely remove excess water from arable lands, the drainage systems have long been installed. In order to drain excess water people used to dig trenches, to regulate (deepen or straighten) natural streams. The length of regulated streams has reached 46,000 km and they are deteriorated ecosystems. Investigations showed that the self-purification of streams from nitrates and phosphates is more effective in natural stretches than in stretches regulated for drainage purposes. Decrease in the average concentration of nitrates in natural and regulated stretches are 8.8 ± 5.0 and 3.0 ± 2.9 mg NO_3^- L^{-1}, respectively. The average coefficient of nitrate self-purification, at a confidence level of 95% in natural stream stretches is 0.50 ± 0.22, and in regulated is -0.15 ± 0.21 km^{-1}, and this difference is essential. The change in the average concentration of phosphates in natural and regulated stretches is almost the same, 0.2 ± 0.1 and 0.2 ± 0.2 mg PO_4^{3-} L^{-1}, respectively. The average coefficient of phosphate self-purification, at a confidence level of 95%, in natural stream stretches is 0.28 ± 0.12, in regulated -0.14 ± 0.12 km^{-1}, and this difference is not essential. In terms of the need for the renovation of drainage systems it is suggested that soft naturalization measures are first applied in the streams of Western (Samogitian) Highlands, Coastal Lowlands, and South-Eastern Highlands to improve their self-purification processes.

Keywords: water quality in streams; self-purification; nitrates; phosphates

1. Introduction

Small-sized rivers, their headwaters, and wetlands have a significant effect on the adjustment of river flow, retention of outwash and pollutants, and preservation of biological diversity. Unfortunately, intensive urbanization affects the channels of small rivers and adjacent ecosystems and strongly influences their ecological status [1]. Intensive use of European rivers for human purposes in the last hundred years has changed natural water flows, their physical and chemical properties, channel morphology, and species composition of local flora and fauna. The former natural, curving river channels were straightened and deepened, artificial slopes were formed having nothing to do with nature. Seeking to adjust flow rate, and due to recreational purposes, people built dams, and sometimes water energy was used in hydropower engineering [2–4].

At the end of the 20th century, with a constant growth of world population the cereals production almost doubled, the use of nitrogen (N) and phosphorus (P) fertilizers, related to the increasing food demand, has grown by 6.9 and 3.5 times. From environmental point of view, a large amount of N and P fertilizers was used improperly in the lands of agricultural designation and has essentially changed biogeochemical cycles of these two main nutrients [5]. The expanded network of small rivers and trenches was filled with a large amount of pollutants, including nutrients, pesticides, heavy metals, and soil particles washed out from adjacent fields. Due to the physical and chemical impact on river channel and its water flow, part of local flora and fauna species became extinct, and part of them were replaced by invasive or introduced species [1].

Baczyk A. et al. [6] maintains that this negative anthropogenic effect has been widely described in many world-wide countries and even 96% of the articles analyzed reported a unilateral negative anthropogenic impact on water ecosystems, especially on invertebrates and fishes.

The territory of Lithuania is characterized by a prevailing moisture excess; the average annual amount of precipitation reaches 750 mm, evaporation comes to 512 mm [7,8], and therefore, the excess water is frequently accumulating in the soils of low-gradient and heavier gradation soils (clays, loam) of plains and depressions of hilly relief, due to low infiltration. To timely remove excess water from arable lands the drainage systems has long been installed, i.e., a complex of functionally related hydraulic structures located at the territory to be drained and intended for adjusting the soil moisture regime and creating favorable growing conditions for vegetation.

At the beginning of the 19th and 20th century, the largest attention when draining lands was paid to the regulation of rivers and streams and to digging of discharge trenches. Until 1958, lands were still drained by trenches, but later people started using drainage, since this was a world-recognized and more effective draining method. For the streams to be regulated, water intakes are usually curved, waterlogged, overgrown with water plants, having the already formed natural riparian lanes. Water depth and flow rate in these channels are usually low, water lies close to the surface, and floods take a long period.

Human activities, aimed to increase used agricultural and wooded areas, most of all changed small streams and the upper reaches of rivers: they were regulated to remove excess water collected by drainage, straightened to increase the flow rate and water capacity, or deepened to accommodate for at least 1.50 m depth of the river channel required for installing drainage system.

Thus, the regulated stream does not differ from a trench that was dug. The regulated streams and trenches served as a draining network.

During almost 100 years of land drainage in Lithuania, the streams were regulated, trenches were dug, and drainage systems were installed. Within this period, the area of drained land reached 3.02 million ha (47% of the total area of the country), of which 2.62 million ha were drained with the help of drainage, and 52,454 km of main drainage trenches were regulated and dug.

Land drainage has changed landscape structure, especially by a majority of newly dug trenches (Figure 1). The lowest density of trenches was found in plain regions where the most fertile clayey soils were drained. In hilly relief conditions the density of trenches remained higher.

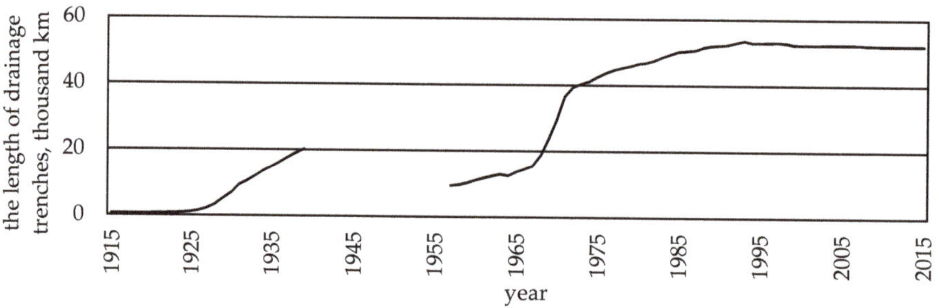

Figure 1. Dynamics in regulated stream channels and main drainage trenches in Lithuania.

The length of drainage trenches had an effect on the density of the total Lithuanian hydrographic network. What concerns the total length of trenches in the territory of Lithuania it was obviously increasing. In 1930, the density of trenches was 0.1 km/km^2, in 1945—0.34, in 1960—0.69, in 1975—0.76, and in 1999 even 0.96 km/km^2. Such density of ditches was determined by the deepening, widening and straightening of natural stream channels, i.e., after they became main drainage trenches, and later—regulated streams. In such a way the length of regulated streams reached 46,000 km [8,9].

If the concept of regulated stream can be related to the adaptation of stream to serve draining function (to timely discharge certain amount of excess water necessary to drain the area), from the ecological point of view this is a deregulation of the natural stream and should be related to the destruction of the stream's ecosystem.

When carrying out regulation of streams the fact of deepening and straightening of natural stream channels used to be emphasized, though at the same time the riparian zones of streams were destroyed. There are not many studies focused on the retention processes of biogenic materials in protective riparian zones. Results of nitrogen retention in protective riparian zones are different [10–13], however water running through the ecosystem of riparian zone is able to retain up to 74% ± 4% of nitrogen. Data shows that with the increasing width of protective riparian zone the retention efficiency also increases [14], however this relationship is statistically stronger when the zone between the cultivated land and the river border >50 m.

Some authors argue that the efficiency of protective zones in retaining phosphorus is highly dependent on parameters such as vegetation cover type and density, topography, soil, climatic conditions, however, it is of short-term, seasonal in character, as long as plant vegetation continues [11]. The largest amount of total phosphorus is washed out in early spring before the start of vegetation. At that time, the efficiency of protective zone is poor, especially in the case where riversides are overgrown mostly with grass. The outwash is slightly better retained in protective zones overgrown with trees and bushes, though due to decomposition of organic material the amount of mobile phosphorus in these zones is increased [15]. Despite this, many agricultural consultants, planners and practitioners recommend the grass-overgrown protective riparian zones as a measure to settle phosphorus compounds in the areas of intensive agriculture.

There have been various recommendations for defining the main parameter of protective zones, i.e., width [16–18]. It was proven that a long-term efficiency can be reached where the zones are wider, 30 m-wide. In agricultural landscape, as suggested by foreign authors, the absolute minimum width should be 10 m. When designing protective riparian zones, it is necessary to take into account not only width requirements but also no less important parameters, such as vegetation species and distribution, slope gradients, soil types, topography, rainfalls [19,20].

In Lithuania, the minimum required depth of stream channel for the installation of drainage systems was about 1.5 m. In the upstream, where the basin area is not large, the stream channel has excessive depth and water collected from the area to be drained does not fill the whole cross-section area with water. Thus, there is a possibility to the improve ecological situation by growing bushes and small trees on the upper part of channel slope, and to manage the lower part of channel in a way that channel capacity corresponds to the requirements for draining function.

The authors suggest an alternative way of maintaining regulated streams (trenches): to allow the overgrowing of slopes with woody vegetation, the crowns of which would shadow the channel and strengthen the slopes [8,21,22]. Depending on the tendencies of species distribution in landscape and in cross-section of the trench, the restoration of dendrology in a desirable direction can be promoted artificially by planting special species or correcting their varietal composition in slopes, also by forming protective zones. This makes it possible when operating trenches to develop their natural functions and to also preserve drainage functions. The process of overgrowing of drainage network with trees can be assessed with ambiguity: reduces hydraulic capacity but makes a positive effect on landscape structure, reduces deflation probability, accumulation of outwash, and pollution of water bodies. Once the right of land ownership was given back to people, the land use intensity started to change. In less-favored areas for developing agriculture the drainage systems is used not intensively. Still, there are locations where drainage systems are outdated, their condition is bad. More than 45% of drainage systems are older than 40 years [23].

Over the last few decades, the number of projects of river naturalization has highly increased in most developed countries in Europe and all over the world [24–27]. Implementation of river naturalization projects is usually very expensive and the real benefit is not always evident. The analysis, carried out by German specialists [28], showed that the accomplished naturalization processes made no improvements to the population of benthic invertebrates. It was also reported that restoration of this population depends on whether there are any specimens of this species in adjacent waters. This research shows that the processes are not yet fully investigated and that slightly different results may be obtained under different natural conditions.

A very important indicator, showing an overall ecological status of the river, is self-purification [29]. It occurs due to water attenuation with surface and ground waters or due to certain hydrological, biological and chemical processes, such as sedimentation, coagulation, evaporation, sedimentation of colloidal and their further consolidation on the bottom of water body or, finally, due to pollutants assimilation with the living organisms. The level of self-purification in each water body depends on certain factors, such as temperature, water level, river flow rate, hydrological regime, tidal regime, amount of inorganic compounds in water, sediment characteristics, amount of pollutants, phytoplankton, benthic invertebrate fauna, fish fauna, and algae species, and their distribution [30,31].

The aim of the research is to determine the distribution of nitrate and phosphate concentrations in the water of natural and regulated for drainage streams and the influence of regulation on the self-purification efficiency of the streams.

2. Materials and Methods

2.1. Location, Sampling Points

The research was carried out in Lithuania, in the Baltic Sea region, in the Nemunas and Venta river basins. The research included regulated and unregulated stretches of the following streams: Terpinė (T), Žalesa (Z), Kuosinė (K), Mėkla (M), Durbinis (D), and Uogis (U). Distribution of the relevant stretches in the territory of Lithuania is given in Figure 2 and Table 1. The relevant stretches were coded. The first letter indicates the relevant stream according to the first letter of its name. The second letter indicates whether the stream is regulated or natural (the sampling place in the natural stream channel is marked by N, in the regulated—by R). The beginning of the stretch is marked by s; the end—by e.

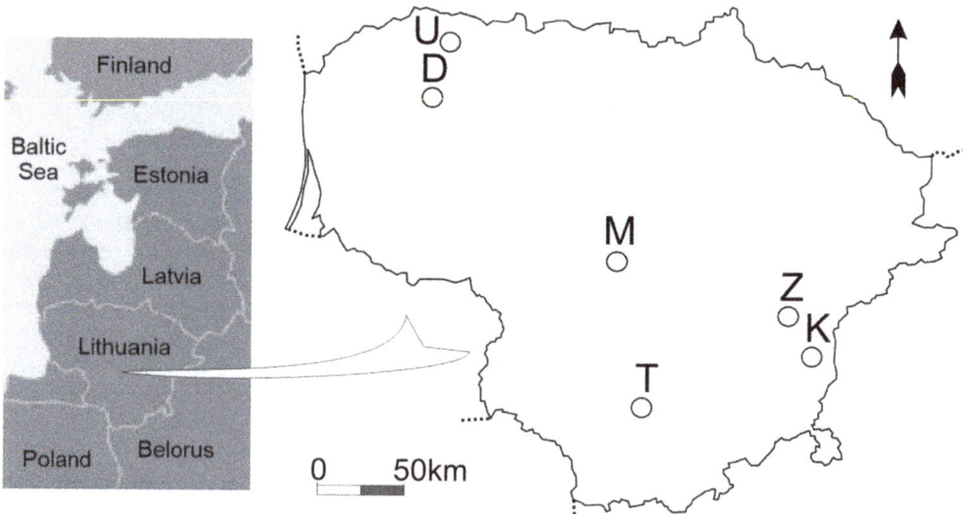

Figure 2. Distribution of the relevant stretches in the territory of Lithuania.

Table 1. Characteristics of used streams.

Streams	Hydrological Regions	It Is Regulated	Length of Stream	The Basin Area	Sampling Points
Terpinė	South-Eastern Highlands	from sources to 7.4 km; from 3.0 km to the estuary	9.0 km	12.2 km^2	TNs; TNe; TRs; TRe
Žalesa	South-Eastern Highlands	from the sources to 7.6 km.	18.8 km	97.1 km^2	ZRs; ZRe; ZNs; ZNe
Kuosinė	South-Eastern Highlands	from the sources to 16.0 km; from 11.0 to 7.0 km.	20.1 km	45.3 km^2	KRs; KRe; KNs; KNe
Mėkla	Middle Plains	from the sources to 20.0 km;	26.9 km	93.3 km^2	MRs; MRe; MNs; MNe
Durbinis	Western (Samogitian) Highlands	from the sources to 7.8 km, from 7.4 to 7.2 km and from 4.8 to 3.0 km.	9.1 km	15.5 km^2	DRs; DRe; DNs; DNe
Uogis	Western (Samogitian) Highlands	from the sources to 13.0 km.	27.6 km	68.2 km^2	URs; URe; UNs; UNe

2.2. Time of Research, Nitrogen and Phosphorus Concentrations

Water samples for water quality analysis were taken in accordance with the sampling standard [32] taking into consideration all the water sampling aspects. Water samples were taken once per month in the period from August 2013 to February 2019. The investigation in different stretches lasted 12–24 months. Concentrations of nitrates (NO_3^-) and phosphates (PO_4^{3-}) were determined. The amount of nitrates in water was measured by the photometer HANNA HI 83,205 using the cadmium reduction method and HI 93728-01 reagents. Concentrations of nitrates and phosphates in the samples taken were determined in the Laboratory of Hydraulics of Vilnius Gediminas Technical University.

According to the average annual value of each index a water body is attached to one of the five classes of ecological status. The obtained average annual values were compared to the values presented in the ecological status classification according to the physicochemical quality elements (Table 2) [33–35].

Table 2. Ecological status classification according to the physicochemical quality elements [33–35].

Index	Ecological Status of Rivers				
	High	Good	Moderate	Poor	Bad
mg NO_3^- L^{-1}	<5.75	5.75–10.18	10.19–19.92	19.93–44.27	>44.27
mg PO_4^{3-} L^{-1}	<0.15	0.15–0.28	0.28–0.55	0.55–1.23	>1.23

In the assessment many parameters (physicochemical, hydromorphological, biological) are used to determine the ecological status class. In the given article we have only rated according to two of them, NO_3^- and PO_4^{3-}.

For the assessment of river self-purification from biogenic substances, the following simplified river purification formula was used [36]:

$$\alpha = ln\left(C_0 C_L^{-1}\right) L^{-1}, \tag{1}$$

where: C_0—concentration of chemical material at the beginning of the relevant river stretch mg L^{-1}; C_L—concentration of chemical material at the end of the relevant river stretch mg L^{-1}; L—length of river stretch km; α-river purification coefficient km^{-1}.

The statistical method of one-way analysis of variance (ANOVA), at confidence level of 95%, was used to determine whether the mean values of the self-purification coefficients for natural and regulated streams differed significantly. SPSS Statistics software was used for statistical processing.

2.3. Distribution of Sediment and Used Agricultural Areas

Distribution of sediment and used agricultural areas in the relevant river basins was studied according to the information provided in the Spatial Information Portal of Lithuania [37]. The portal provides data about typological units and gradation of the analyzed soil contour of the used agricultural area, also about land irrigation status and water logging, variety of soil cover, climatic conditions, soil stoniness, agrochemical properties, base index, and soil productivity index of the analyzed soil contour. However, the determination of the agents influencing the self-purification intensity of the stretches of the studied streams is very complex, due to the abundance and diversity of the factors and will be probably addressed in the future.

3. Results

According to the Spatial Information Portal of Lithuania, the main sediment in the relevant river basins is sandy loam and light loam which are prevailing in all basins. The river basins also contain light and medium loam, adhesive sand, light and average clay, and peat. Distribution of the used agricultural areas within the river basins showed that the largest part of the river basins is taken up by permanent grasslands, arable lands, forests, urbanized territory, and pastures (Table 3).

Table 3. Soil gradation, used agricultural areas and soil productivity index of the river basins.

Terpinė (T)	Žalesa (Z)	Kuosinė (K)	Mėkla (M)	Durbinis (D)	Uogis (U)
Surface Gradation According to Fere					
80% of sandy loam; other: 1. light loam, 2. medium–heavy loam, 3. peat	40% of sandy loam; other: 1. light loam, 2. peat, 3. humus	80% of sandy loam; other: 1. sand, 2. loam, 3. peat	60% of light loam; other: 1. sandy loam, 2. peat, 3. medium–heavy loam	50% of sandy loam; other: 1. loam, 2. clay, 3. peat	60% of sandy loam; other: 1. light loam, 2. peat, 3. sand
Used Agricultural Area in a Regulated Stream Stretch (Rs–Re)					
forest, urbanized territory	grassland, forest	arable, grassland	arable, grassland, urbanized territory	arable, grassland	arable
Used Agricultural Area in a Natural Stream Stretch (Ns–Ne)					
grassland	grassland, forest	forest	arable	urbanized territory	arable, grassland
Average Soil Productivity Index in a River Basin					
32	30	30	47	34	43

The research and analysis of results showed that nitrate concentrations according to their amount and period are distributed rather differently. Many results showed that water quality in the relevant streams exceeded the limit value of good ecological status (nitrate concentrations 10.18 mg NO_3^- L^{-1}) and could be attributed to moderate and poor ecological status. The worst results were represented by Terpinė, Mėkla, and Kuosinė streams. The best ecological status was found in Žalesa stream [38].

The worst water quality was observed in winter months since at this time of the year maximum concentrations of nitrogen oxides were determined which corresponded to the poor or bad ecological status: in Terpinė stream, at point TRe (41.5 mg NO_3^- L^{-1}, in December); in Mėkla stream, at point MNs (81.4 mg NO_3^- L^{-1}, in March); MRe (76.4 mg NO_3^- L^{-1}, in February); in Kuosinė stream, at point KRs (24.5 mg NO_3^- L^{-1}, in January); in Durbinis stream, at point DRe (26.1 mg NO_3^- L^{-1}, in December). During vegetation period nitrate concentrations were lower. This was especially characteristic of the Durbinis, Mėkla (Figure 3), and Uogis streams (Figure 4). However, in the Kuosinė (Figure 5) and Terpinė streams the relatively large nitrate concentrations were measured even in vegetation period. This could be caused by a more abundant fertilization of arable lands in summer and autumn. The average annual consumption of fertilizers in Lithuanian agriculture in recent years (2016–2018) is 162,000 tons of nitrogen (N) and 52,000 t of phosphorus (P_2O_5). Winter crops were fertilized with 100–150 kg (N) ha^{-1} and 50–75 kg (P_2O_5) ha^{-1}, sugar beet 100–130 kg (N) ha^{-1}, and 80–90 kg (P_2O_5)

ha^{-1} [39]. As a result, a higher amount of nitrate pollutants is washed out into the rivers. In summer time, due to a shallow water the river water is less diluted.

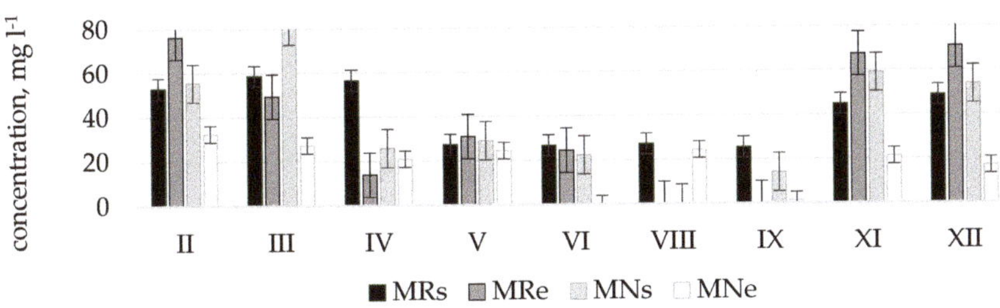

Figure 3. Nitrate concentrations (NO$_3^-$) in natural (N) and regulated (R) stretches of Mėkla stream.

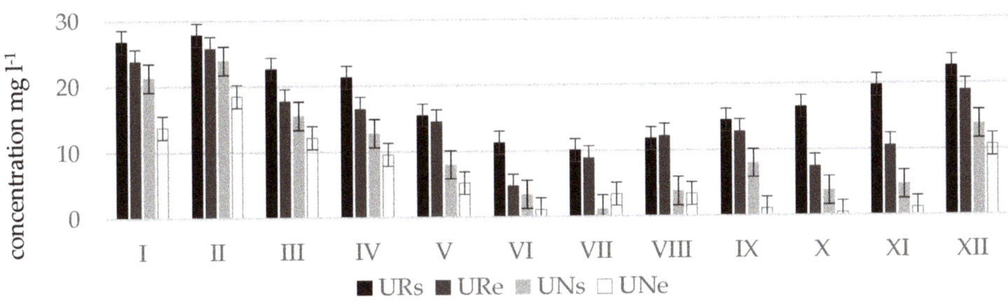

Figure 4. Nitrate concentrations (NO$_3^-$) in natural (N) and regulated (R) stretches of Uogis stream.

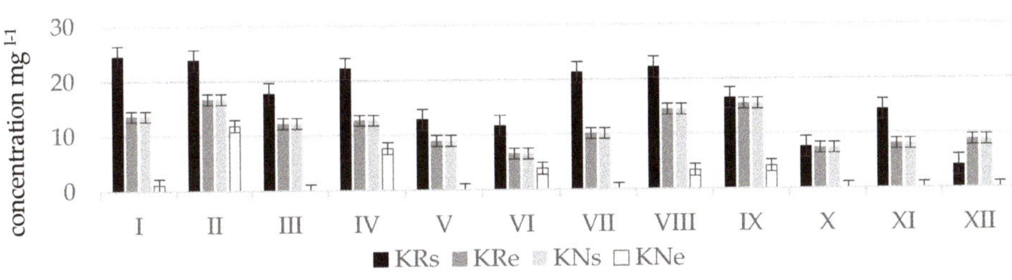

Figure 5. Nitrate concentrations (NO$_3^-$) in natural (N) and regulated (R) stretches of Kuosinė stream.

The measurement results showed that phosphate concentrations according to their amount and period, same like nitrates, are distributed rather differently and water quality in streams frequently exceeded the limit value for good ecological status (phosphate concentrations 0.28 mg PO$_4^{3-}$ L^{-1}) and corresponded to the poor or bad ecological status (Figure 6). Very high phosphate concentrations were measured in the Durbinis stream where no seasonality was observed since phosphate concentrations were very high during all months. The maximum concentration measured was 2.5 mg PO$_4^{3-}$ L^{-1}. Based on this concentration the river water corresponded to the bad ecological status (>1.23 mg PO$_4^{3-}$ L^{-1}). In Uogis stream (Figure 7), according to phosphate concentrations in water in a period of even four moths (March, May, June, and December) at the beginning of regulated stretch (point Urs) the ecological status of the stream was bad, however at the end of natural stretch (point Une) in a period of six months (March, April, June, July, November, and February) the stream had a prevailingly good ecological status, and in the remaining six months (May, August, September, October, December, and January) high ecological status. According to phosphate concentrations the Kuosinė stream (Figure 8) represented high ecological status of water quality in January, February, June, and October as well as poor ecological status in December, March, and July in KRs stretches

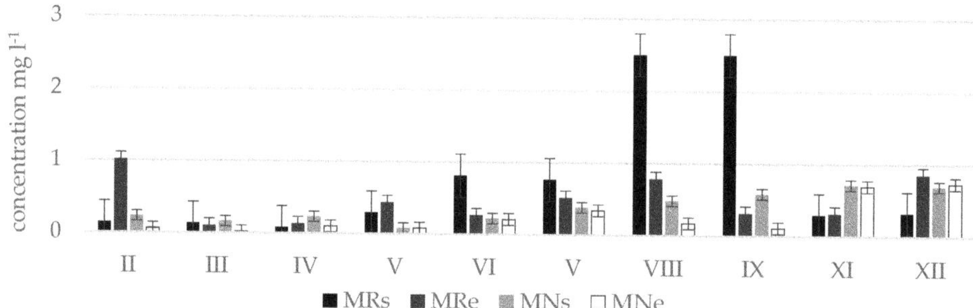

Figure 6. Phosphate concentrations (PO_4^{3-}) in natural (N) and regulated (R) stretches of Mėkla stream.

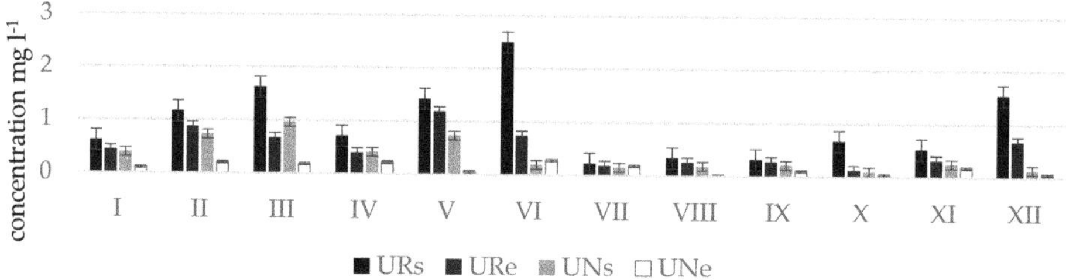

Figure 7. Phosphate concentrations (PO_4^{3-}) in natural (N) and regulated (R) stretches of Uogis stream.

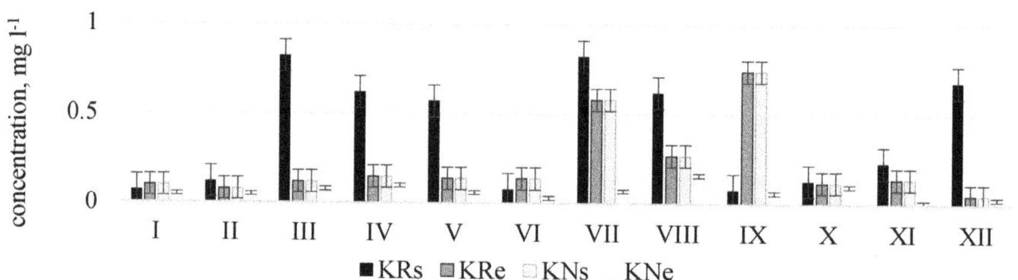

Figure 8. Phosphate concentrations (PO_4^{3-}) in natural (N) and regulated (R) stretches of Kuosinė stream.

Based on the measured concentrations of nitrates and phosphates the average values at the end of stretches and self-purification coefficients were calculated in order to find out how the river is able to purify itself from pollutants. The average concentration of nitrates in natural stretches was twice lower than that in regulated stretches, i.e., 9.3 ± 2.7 and 18.5 ± 6.1 mg NO_3^- L^{-1}, respectively. The average concentration of phosphates in natural stretches was also lower than that in regulated stretches, i.e., 0.4 ± 0.2 and 0.7 ± 0.2 mg PO_4^{3-} L^{-1}, respectively.

The average change of concentrations, determined at the beginning and at the end of river stretches, and the calculated self-purification coefficients showed that streams purify better in natural stretches, further research into organic matter could, of course, further substantiate the purification efficiency. Decrease in nitrate concentrations in natural stretches was 8.8 ± 5.0, in regulated−3.0 ± 2.9 mg NO_3^- L^{-1}, decrease in phosphate concentrations was 0.2 ± 0.1 and 0.2 ± 0.2 mg PO_4^{3-} L^{-1}, respectively (Table 4). The average nitrate self-purification coefficient of all streams was 0.50 ± 0.22 km^{-1} in unregulated stretches, and 0.15 ± 0.21 km^{-1} in regulated stretches. The average phosphate self-purification coefficient in natural stretches was 0.28 ± 0.12 km^{-1}, in regulated −0.14 ± 0.12 km^{-1}.

Table 4. Average decrease in nitrate and phosphate concentrations in natural and regulated stream stretches and self-purification coefficients (R-regulated stretch, N-natural stretch).

Stream	Decrease in Nitrate Concentrations, mg $NO_3^- L^{-1}$		Nitrate Self-Purification Coefficient α, km^{-1}		Decrease in Phosphate Concentrations, mg $PO_4^{3-} L^{-1}$		Phosphate Self-Purification Coefficient α, km^{-1}	
	N	R	N	R	N	R	N	R
Terpinė	2.7 ± 5.7	0.2 ± 4.5	0.13 ± 0.22	−0.09 ± 0.28	0.1 ± 0.8	−0.5 ± 0.9	−0.03 ± 0.18	−0.24 ± 0.53
Žalesa	3.3 ± 1.7	0.8 ± 2.8	1.01 ± 0.38	0.26 ± 0.62	0.1 ± 0.3	0.3 ± 0.2	0.32 ± 0.61	0.25 ± 0.14
Kuosinė	8.6 ± 1.7	5.3 ± 2.5	0.99 ± 0,41	0.08 ± 0.05	0.2 ± 0.1	0.2 ± 0.2	0.15 ± 0.12	0.18 ± 0.20
Mėkla	19.1 ± 14.2	3.9 ± 15.0	0.27 ± 0.90	0.39 ± 0.47	0.1 ± 0.1	0.3 ± 0.6	0.15 ± 0.10	0.04 ± 0.15
Durbinis	14.5 ± 24.5	1.2 ± 20.4	0.27 ± 0.38	0.15 ± 1.10	0.4 ± 0.5	0.4 ± 0.5	0.22 ± 0.22	0.29 ± 0.38
Uogis	3.3 ± 1.4	3.9 ± 1.7	0.38 ± 0.27	0.11 ± 0.06	0.2 ± 0.2	0.5 ± 0.3	0.74 ± 0.34	0.24 ± 0.11
Total in relevant streams [1]	8.8 ± 5.0	3.0 ± 2.9	0.50 ± 0.22	0.15 ± 0.21	0.2 ± 0.1	0,2 ± 0,2	0.28 ± 0.12	0.14 ± 0.12

[1] Average data of all relevant stream stretches.

What concerns separate streams, the best phosphate self-purification coefficient was determined in natural stretches of Uogis (0.74 ± 0.34 km^{-1}) and Žalesa (0.32 ± 0.61 km^{-1}). The nitrate self-purification coefficient in the natural stretch of Žalesa even reached 1.01 ± 0.38 km^{-1}, and in the natural stretch of Kuosinė −0.99 ± 0.41 km^{-1}. The negative values of self-purification coefficient were calculated in the regulated stretches of Terpinė stream (Table 4).

A high value of self-purification coefficient in the natural stretch of Žalesa stream could be influenced by the adjacent permanent grasslands and forests. Decrease in pollutant concentrations can also be determined by the dilution of polluted water with surface and underground waters. It should be mentioned that the right tributary flows into the Žalesa in the stretch ZNs–ZNe. The inflowing stream water could dilute the nitrate-polluted water of the Žalesa and thus contribute to the decrease in nitrate concentrations. The natural stretch of the Kuosinė stream is surrounded by forests. The self-purification was influenced not only by a natural river stretch but also by the adjacent used agricultural land, since it was likely that the access of nitrates into water was limited.

Nitrate self-purification coefficients are obviously different when comparing natural and regulated stretches of Terpinė stream (Table 4; Figure 9).

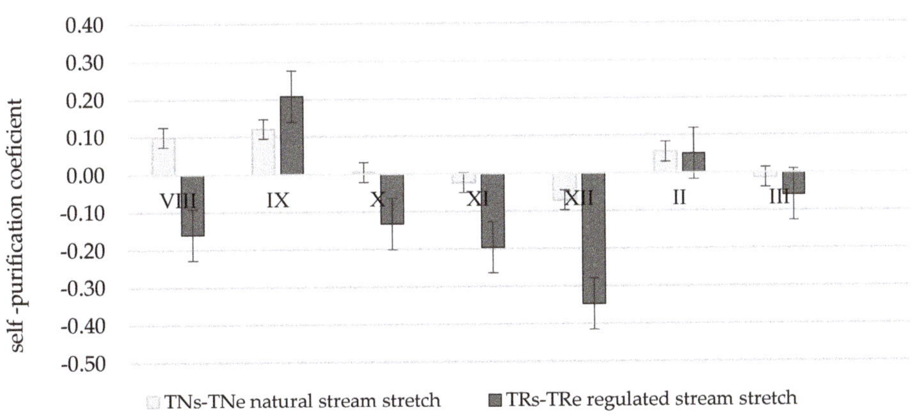

Figure 9. Nitrate (NO_3^-) self-purification coefficient in Terpinė stream.

The whole TNs-TNe stretch is a natural part of the Terpinė stream abundant in meanders thus the river flow rate is lower and detention of nitrates is higher. Nitrate (NO_3^-) self-purification coefficient α in a vegetation period (stretch TNs-TNe) is positive $\alpha = 0.10$; 0.12 km^{-1}. In winter season due to rotting vegetation and frozen land the increase in the amount of nitrates could be noticed in the stretch of natural channel $\alpha = -0.02$; −0.07 km^{-1}.

In the regulated part of stream, the amount of nitrates is frequently increasing in a flowing direction. In spite of that, in the samples of TRs the least amount of nitrates was determined in the whole period of research. This phenomenon could be explained by the fact that there is a dam before the stretch TRs–TRe and the existing pond operates as a settler, therefore retention of nitrates takes place all year round. The average self-purification coefficient of the stretch α is equal to 0.10 km^{-1}. The slopes of the stretch TRs–TRe are very high and steep, a sharply descending relief increases river flow rate thus causing intensive bank erosion and slope landslips, whereas vegetation prevails on the upper part of the slope and in a protective riparian zone [40]. The average self-purification coefficient of the stretch $\alpha = -0.09$ km^{-1}.

The one-factorial dispersion analysis showed that the average self-purification coefficients, at a confidence level of 95%, vary essentially in the Žalesa, Terpinė, and Kuosinė streams. The average self-purification coefficients, with a confidence level of 95%, do not essentially vary in the Durbinis, Mėkla, and Uogis streams. Phosphate self-purification coefficient in regulated and natural river stretches, at a confidence level of 95%, is not essential in the Kuosinė, Durbinis, Mėkla, Žalesa, and Terpinė streams, though it varies essentially in Uogis stream. The results obtained show that nitrate self-purification is more effective than phosphate self-purification. The average nitrate concentration in natural stretches is lower than that in regulated stretches, 9.3 ± 2.7 and 18.5 ± 6.1 mg NO$_3^-$ L^{-1}, respectively. Decrease in the average nitrate concentration in natural and regulated stretches is 8.8 ± 5.0 and 3.0 ± 2.9 mg NO$_3^-$ L^{-1}, respectively. The average nitrate self-purification coefficient, at a confidence level of 95%, in natural stretches is 0.50 ± 0.22 km^{-1}, in regulated stretches -0.15 ± 0.21 km^{-1}, and this difference is essential.

The average phosphate concentration in natural stretches is also lower than that in regulated stretches, 0.4 ± 0.2 and 0.7 ± 0.2 mg PO$_4^{3-}$ L^{-1}, respectively. The change in the average phosphate concentration in natural and regulated stretches is almost the same, 0.2 ± 0.1 and 0.2 ± 0.2 mg PO$_4^{3-}$ L^{-1}, respectively. The average phosphate self-purification coefficient, at a confidence level of 95%, in natural stretches is 0.28 ± 0.12 km^{-1}, in regulated stretches -0.14 ± 0.12 km^{-1}, and this difference is not essential.

The distribution of nitrate and phosphate concentrations in rivers is also influenced by the underground water, which we consider relatively small and similar, unfortunately it is not analyzed in detail in our research. However, in the future, in order to get a more detailed analysis of nitrate and phosphate changes, a more detailed assessment of groundwater (underground water fluxes, quantify the nutrient fluxes) should be carried out.

4. Discussion

The research results show that nitrate and phosphate self-purification in the regulated stretches of separate streams takes place slower or not at all since the self-purification coefficients were negative. Negative self-purification rates indicate that larger amounts of nitrate and phosphate pollutants are sometimes discharged into the river, although the inflow from the surrounding areas is similar but still variable, so regulated streams, sometimes natural also fail to treat itself. Seeking to prevent pollutants from getting into the rivers the protective riparian zones are defined along the riverbanks [41]. Industrial activities, carried out in protective riparian zones, has a direct, usually negative impact on water quality, therefore it is very important that they are observed. In order to improve water quality in rivers and streams of Lithuania it is suggested to naturalize them to the extent possible. After getting into the rivers and streams of Lithuania, nitrates and phosphates are better self-purified in natural than in regulated stretches. In order to restore a disturbed hydrological and environmental balance of regulated streams it is important to retain their drainage function without increasing maintenance costs. Therefore, seeking to improve self-purification of streams the so-called soft naturalization measures are suggested: to allow woody vegetation grow on river slopes, to form natural obstacles for water flow, and other elements in flood plains that are characteristic to wetlands [2,42]. The research evaluates the water quality of streams only in terms of nitrate and phosphate concentrations in water, but many other physicochemical, hydro morphological, and biological parameters, for instance organic fraction of nutrients that also have a significant impact on a river's ecological status should be taken into consideration to fully assess the ecological status of the river and differences in the ecological status of natural and regulated streams.

Nitrate and phosphate concentrations are greatly influenced by man-or animal-made dams. The formed ponds act as precipitators, resulting that significantly less nitrate concentrations flow out from the pond [43]. Research in the Nevėžis River Basin [22] shows that beaver ponds (comparing annual concentrations of inputs and outflows) retain relatively more phosphorus than nitrogen and more soluble mineral N and P compounds: on average 28% nitrate and ammoniacal nitrogen and 43% of orthophosphate phosphorus. But organic forms are much less retained. The retention of total N and total P is therefore 7% and 27%, respectively.

When planning renovation of drainage systems, the authors Šaulys and Barvidienė [44] suggest to take into consideration the Lithuanian hydrological regime, and natural and industrial conditions. Though the territory of Lithuania is relatively small, due to a variety of factors, which form and redistribute run-off, the feeding type and hydrological regime of surface water bodies are rather different. Lithuanian surface water bodies are divided into three hydrological regions based on relief, precipitation and soil [45,46]. Moreover, having taken into consideration the increase in soil productivity index due to drainage, abandoned agricultural land, possible breakdown of drainage systems, and dynamics in crop production, it was determined that the highest need for drainage systems renovation is in the Middle Lithuania region, the average in the Coastal Lowlands and South-Eastern Highlands, and the lowest in the Western (Samogitian) Highlands.

In terms of the need for renovation of drainage systems it is suggested that soft naturalization measures are first applied in the regulated stretches of the Durbinis and Uogis streams, and in other regulated streams of Western (Samogitian) Highlands. A more intensive application of naturalization measures is suggested for the Žalesa, Kuosinė, and Terpinė streams, and other regulated streams of South-Eastern Highlands, since the mentioned streams are crossing territories where the prevailing soil is sand and sandy loam, moreover, the soil productivity index is relatively low.

The works focused on mechanical naturalization of regulated streams in Lithuania have already started [47], though mechanical naturalization of regulated streams financially is rather expensive, the benefit for water quality and ecological diversity has not yet been comprehensively investigated and scientifically justified [24,48,49]. Therefore, no intensive mechanical naturalization works are proposed so far for the regulated streams of Lithuania. In the future, the ecological status of regulated rivers should be studied and evaluated in as many parameters as possible, at the same time contributing to the broader public awareness of the rationalization of the application of naturalization processes.

5. Conclusions

It was determined that the self-purification of streams from nitrates and phosphates is more effective in natural stretches than in stretches regulated for drainage purposes.

Decrease in the average nitrate () concentration in natural and regulated stream stretches was 8.8 ± 5.0 and 3.0 ± 2.9 mg $NO_3^- $ L^{-1}, respectively. The average nitrate self-purification coefficient, at a confidence level of 95%, in natural stretches was 0.50 ± 0.22, in regulated stretches -0.15 ± 0.21 km^{-1}, and this difference is essential.

The change in the average phosphate concentration in natural and regulated stream stretches is almost the same, 0.2 ± 0.1 and 0.2 ± 0.2 mg PO_4^{3-} L^{-1}, respectively. The average phosphate self-purification coefficient, at a confidence level of 95%, in natural stretches is 0.28 ± 0.12, in regulated stretches -0.14 ± 0.12 km^{-1}, and this difference is not essential.

During the warm vegetation period, concentrations of nitrates in water were (could be) in most cases lower than during the cold period of the year. Increased concentrations of nitrates could be caused by fertilization of arable lands. Increased nitrate concentrations could also be affected by the adjacent communities of garden-plots, this is a possible unevenness of inflow from surrounding areas.

In terms of the need for renovation of drainage systems it is suggested that soft naturalization measures are first applied in the streams of Western (Samogitian) Highlands, Coastal Lowlands, and South-Eastern Highlands to improve their self-purification processes and where the need for renovation of drainage systems is low.

Author Contributions: V.Š. conceived the study; O.S. and R.S. procured and analyzed the data, prepared the illustrations and wrote the manuscript under the supervision of V.Š. All authors contributed with ideas to the analysis, discussed the results and commented on the manuscript. All authors have read and agreed to the published version of the manuscript.

Funding: This research received no external funding.

Conflicts of Interest: The authors declare no conflict of interest.

References

1. Riley, W.D.; Potter, E.C.E.; Biggs, J.; Collins, A.L.; Jarvie, H.P.; Jones, J.I.; Kelly-Quinn, M.; Ormerod, S.J.; Sear, D.A.; Wilby, R.L.; et al. Small water bodies in Great Britain and Ireland: Ecosystem function, human-generated degradation, and options for restorative action. *Sci. Total Environ.* **2018**, *645*, 1598–1616. [CrossRef] [PubMed]

2. Groll, M. The passive river restoration approach as an efficient tool to improve the hydromorphological diversity of rivers—Case study from two river restoration projects in the German lower mountain range. *Geomorphology* **2017**, *293*, 69–83. [CrossRef]

3. Campana, D.; Marchese, E.; Theule, J.I.; Comiti, F. Channel degradation and restoration of an Alpine river and related morphological changes. *Geomorphology* **2014**, *221*, 230–241. [CrossRef]

4. Gregory, K.J. The human role in changing river channels. *Geomorphology* **2006**, *79*, 172–191. [CrossRef]

5. Isaac, M.E.; Hinsingerc, P.; Harmanda, J.M. Nitrogen and phosphorus economy of a legume tree-cereal intercropping system under controlled conditions. *Sci. Total Environ.* **2012**, *434*, 71–78. [CrossRef]

6. Bączyk, A.; Wagner, M.; Okruszko, T.; Grygoruk, M. Influence of technical maintenance measures on ecological status of agricultural lowland rivers—Systematic review and implications for river management. *Sci. Total Environ.* **2018**, *627*, 189–199. [CrossRef]

7. Lithuanian Hydrometeorological Service. Available online: http://www.meteo.lt/en/precipitation (accessed on 12 April 2019).

8. Šaulys, V. *Open Channel Hydraulics*, 2nd ed.; Technika: Vilnius, Lithuania, 2016; p. 272. [CrossRef]

9. Povilaitis, A.; Taminskas, J.; Gulbinas, Z.; Linkevičienė, R.; Pileckas, M. *Lithuanian Wetlands and Their Water Protective Importance*, 4th ed.; Apyaušris: Vilnius, Lithuania, 2011; p. 368.

10. Hickey, M.B.C.; Bruce, D. A review of the efficiency of buffer strips for the maintenance and enhancement of riparian ecosystems. *Water Qual. Res. J. Can.* **2004**, *39*, 311–317. [CrossRef]

11. Søvik, A.K.; Syversen, N. Retention of particles and nutrients in the root zone of a vegetative buffer zone—Effect of vegetation and season. *Boreal Environ. Res.* **2008**, *13*, 223–230.

12. Dosskey, M.G.; Vidon, P.; Gurwick, N.P.; Allan, C.J.; Duval, T.P.; Lowrance, R. The role of riparian vegetation in protecting and improving chemical water quality in streams. *J. Am. Water Resour. Assoc.* **2010**, *46*, 261–277. [CrossRef]

13. Rosa, D.J.; Clausen, J.C.; Kuzovkina, Y. Water quality changes in a short-rotation woody crop riparian buffer. *Biomass Bioenergy* **2017**, *107*, 370–375. [CrossRef]

14. Mayer, P.M.; Reynolds, S.K.; McCutchen, M.D.; Canfield, T.J. Meta-Analysis of Nitrogen Removal in Riparian Buffers. *J. Environ. Qual.* **2007**, *36*, 1172–1180. [CrossRef] [PubMed]

15. Dorioz, J.M.; Wang, D.; Poulenard, J.; Trévisan, D. The effect of grass buffer strips on phosphorus dynamics—A critical review and synthesis as a basis for application in agricultural landscapes in France. *Agric. Ecosyst. Environ.* **2006**, *117*, 4–21. [CrossRef]

16. Aguiar, T.R.; Rasera, K.; Parron, L.M.; Brito, A.G.; Ferreira, M.T. Nutrient removal effectiveness by riparian buffer zones in rural temperate watersheds: The impact of no-till crops practices. *Agric. Water Manag.* **2015**, *149*, 74–80. [CrossRef]

17. Elliott, K.J.; Vose, J.M. Effects of riparian zone buffer widths on vegetation diversity in southern Appalachian headwater catchments. *For. Ecol. Manag.* **2016**, *376*, 9–23. [CrossRef]

18. Schilling, K.E.; Jacobson, P. Effectiveness of natural riparian buffers to reduce subsurface nutrient losses to incised streams. *Catena* **2014**, *114*, 140–148. [CrossRef]

19. O'Toole, P.; Chambers, J.M.; Bell, R.W. Understanding the characteristics of riparian zones in low relief, sandy catchments that affect their nutrient removal potential. *Agric. Ecosyst. Environ.* **2018**, *258*, 182–196. [CrossRef]

20. Neilen, A.D.; Chen, C.R.; Parker, B.M.; Faggotter, S.J.; Burford, M.A. Differences in nitrate and phosphorus export between wooded and grassed riparian zones from farmland to receiving waterways under varying rainfall conditions. *Sci. Total Environ.* **2017**, *598*, 188–197. [CrossRef]

21. Bastienė, N.; Kirstukas, J. Principles and priorities of riparian strips rehabilitation. *Water Environ. Eng.* **2010**, *37*, 71–83.

22. Lamsodis, R.; Morkūnas, V.; Poškus, V.; Povilaitis, A. Ecological approach to management of open drains. *Irrig. Drain.* **2006**, *55*, 479–490. [CrossRef]

23. Maziliauskas, A.; Morkūnas, V.; Rimkus, Z.; Šaulys, V. Economic incentives in land reclamation sector in Lithuania. *J. Water Land Dev.* **2007**, *11*, 17–30. [CrossRef]

24. England, J.; Wilkes, M.A. Does river restoration work? Taxonomic and functional trajectories at two restoration schemes. *Sci. Total Environ.* **2018**, *618*, 961–970. [CrossRef] [PubMed]

25. Paillex, A.; Schuwirth, N.; Lorenz, A.W.; Januschke, K.; Peter, A.; Reichert, P. Integrating and extending ecological river assessment: Concept and test with two restoration projects. *Ecol. Indic.* **2017**, *72*, 131–141. [CrossRef]

26. Kail, J.; Brabec, K.; Poppe, M.; Januschke, K. The effect of river restoration on fish, macroinvertebrates and aquatic macrophytes: A meta-analysis. *Ecol. Indic.* **2015**, *58*, 311–321. [CrossRef]

27. Kim, J.J.; Atique, U.; An, K.G. Long-term ecological health assessment of a restored urban stream based on chemical water quality, physical habitat conditions and biological integrity. *Water* **2019**, *11*, 114. [CrossRef]

28. Sundemann, A.; Stoll, S.; Haase, P. River restoration success depends on the species pool of the immediate surroundings. *Ecol. Appl.* **2011**, *21*, 1962–1971. [CrossRef]

29. Tian, S.; Wang, Z.; Shang, H. Study on the Self-purification of Juma River. *Procedia Environ. Sci.* **2011**, *11*, 1328–1333. [CrossRef]

30. Ifabiyi, I.P. Self-purification of a freshwater stream in Ile-Ife: Lessons for water management. *J. Hum. Ecol.* **2008**, *24*, 131–137. [CrossRef]

31. Bakar, A.A.A.; Khalil, M.K. Study on stream ability for self-purification process in receiving domestic wastewater. *Adv. Sci. Lett.* **2016**, *22*, 1252–1255. [CrossRef]

32. *Water Quality—Sampling—Part 1: Guidance on the Design of Sampling Programmes and Sampling Techniques*; International Standardisation Organisation: Geneva, Switzerland, 2007; LST EN ISO 5667-1:2007/AC:2007.

33. Directive 2000/60/EC of the European Parliament and of the Council of 23 October 2000 establishing a framework for Community action in the field of water policy. *Off. J. Eur. Communities* **2000**, *327*, 1–73. Available online: https://www.eea.europa.eu/policy-documents/directive-2000-60-ec-of (accessed on 5 April 2019).

34. Grizzetti, B.; Pistocchi, A.; Liquete, C.; Udias, A.; Bouraoui, F.; van de Bund, W. Human pressures and ecological status of European rivers. *Sci. Rep.* **2017**, 6941. [CrossRef]

35. Environmental Protection Department under the Ministry of Environment. Available online: https://aad.lrv.lt/en/ (accessed on 25 May 2019).

36. Tumas, R. *Water Ecology*; Naujasis Lankas: Kaunas, Lithuania, 2003; p. 352.

37. The Manager of the Spatial Information Portal of Lithuania. Available online: https://www.geoportal.lt/geoportal/en/web/en (accessed on 10 April 2019).

38. Survilė, O.; Šaulys, V.; Stanionytė, A. An assessment of self-purification of regulated and natural streams. In Proceedings of the 10th International Conference Environmental Engineering, Vilnius, Lithuania, 27–28 April 2017.

39. Lithuanian Official Statistics Portal. Available online: https://osp.stat.gov.lt/statistiniu-rodikliu-analize?theme=all#/ (accessed on 12 December 2019).

40. Burneika, A.; Barvidiene, O. The influence of naturalization processes on water quality in regulated part of Terpine stream. In Proceedings of the 17th Conference for Junior Researchers, Vilnius, Lithuania, 10 April 2014.

41. Montreuil, O.; Merot, P.; Marmonier, P. Estimation of nitrate removal by riparian wetlands and streams in agricultural catchments: Effect of discharge and stream order. *Freshw. Biol.* **2010**, *55*, 2305–2318. [CrossRef]
42. Survilė, O.; Šaulys, V.; Bagdžiūnaitė-Litvinaitienė, L.; Stankevičienė, R.; Litvinaitis, A.; Stankevičius, M. Assessment and research of river restoration processes in regulated streams of Southeastern Lithuania. *Pol. J. Environ. Stud.* **2016**, *25*, 1245–1251. [CrossRef]
43. Winton, R.S.; Calamita, E.; Wehrli, B. Reviews and syntheses: Dams, water quality and tropical reservoir stratification. *Biogeosciences* **2019**, *16*, 1657–1671. [CrossRef]
44. Šaulys, V.; Barvidienė, O. Substantiation of the expediency of drainage systems renovation in Lithuania. In Proceedings of the 9th International Conference Environmental Engineering, Vilnius, Lithuania, 22–23 May 2014.
45. *Soils of Lithuania*, 2nd ed.; Lithuanian Science: Vilnius, Lithuania, 2001; Book 32; p. 1244.
46. Gailiušis, B.; Jablonskis, J.; Kovalenkovienė, M. *Lithuanian Rivers: Hydrography and Runoff*, 2nd ed.; Lithuanian Energy Institute: Kaunas, Lithuania, 2001; p. 792.
47. The Environmental Protection Agency. Available online: http://vanduo.gamta.lt. (accessed on 22 May 2019).
48. Friberg, N.; Bonada, N.; Bradley, D.C.; Dunbar, M.J.; Edwards, F.K.; Grey, J.; Hayes, R.B.; Hildrew, A.G.; Lamouroux, N.; Trimmer, M.; et al. Biomonitoring of human impacts in natural ecosystems: The good, the bad and the ugly. *Adv. Ecol. Res.* **2011**, *44*, 1–68. [CrossRef]
49. Logar, I.; Brouwer, R.; Paillex, A. Do the societal benefits of river restoration outweigh their costs? A cost-benefit analysis. *J. Environ. Manag.* **2019**, *323*, 1075–1085. [CrossRef]

Article

The Straightening of a River Meander Leads to Extensive Losses in Flow Complexity and Ecosystem Services

Tian Zhou [1] and Theodore Endreny [2],*

[1] Atmospheric Sciences and Global Change Division, Pacific Northwest National Laboratory, Richland, WA 99352, USA; tian.zhou@pnnl.gov

[2] Department of Environmental Resources Engineering, College of Environmental Science and Forestry, State University of New York, Syracuse, NY 13210, USA

* Correspondence: te@esf.edu

Received: 13 April 2020; Accepted: 8 June 2020; Published: 11 June 2020

Abstract: To assist river restoration efforts we need to slow the rate of river degradation. This study provides a detailed explanation of the hydraulic complexity loss when a meandering river is straightened in order to motivate the protection of river channel curvature. We used computational fluid dynamics (CFD) modeling to document the difference in flow dynamics in nine simulations with channel curvature (C) degrading from a well-established tight meander bend ($C = 0.77$) to a straight channel without curvature ($C = 0$). To control for covariates and slow the rate of loss to hydraulic complexity, each of the nine-channel realizations had equivalent bedform topography. The analyzed hydraulic variables included the flow surface elevation, streamwise and transverse unit discharge, flow velocity at streamwise, transverse, and vertical directions, bed shear stress, stream function, and the vertical hyporheic flux rates at the channel bed. The loss of hydraulic complexity occurred gradually when initially straightening the channel from $C = 0.77$ to $C = 0.33$ (i.e., the radius of the channel is three-times the channel width), and additional straightening incurred rapid losses to hydraulic complexity. Other studies have shown hydraulic complexity provides important riverine habitat and is positively correlated with biodiversity. This study demonstrates how hydraulic complexity can be gradually and then rapidly lost when unwinding a river, and hopefully will serve as a cautionary tale.

Keywords: river engineering; meander bend; CFD simulation; hydraulic complexity

---

1. Introduction

River meanders are a common landform in alluvial systems, characterized by sinuous patterns of planform curvature of channel banks and bedform topography within those banks. Classic bedform features include point bars on the inner bank and pools along the outer bank centered at the meander bend apex, and riffles upstream and downstream of the bend [1–3]. The meander planform curvature generates a centrifugal force on river water and creates non-uniform depths and velocities, which shapes the bedform topography in alluvial rivers. As one travels streamwise downriver, the rising and falling of bedform topography generates drag and steering on river flows, creating a complex feedbacks between topography and flow [4]. The hydraulic features of meanders include streamwise velocity separating from the inner bank about the apex and attaching to the outer bank, a superelevated water surface forming along the outer bank, eddies forming downstream of the point bar, and helical corkscrew patterns of cross-channel flow. This hydraulic complexity is known to control the pressure distributions along the riverbed and therefore induce patterns of surface water downwelling into the streambed balanced by groundwater upwelling into the channel, a mixing process important for species and water quality called hyporheic exchange [5].

In a natural alluvial system, the meander bend planform will evolve through a repeating cycle of straight to curved channels with coordinated bedform adjustments. Initially, straight channels the viscosity of water and roughness of channel banks leads to depositional bars opposite erosional pools, which then experience meander elongation and a decreasing radius of curvature. Ultimately, the evolution culminates in a meander neck cutoff, at which point an oxbow lake may occupy the abandoned meander bend and the upstream and downstream sections of channel are joined by a straight channel, with a radius of curvature reset to infinity [6,7]. This repeating cycle of meander evolution is at different stages, initiation to completion, along the streamwise path of a river. Flows within meander bends generate hydraulic complexity [8–10] . The hydraulic complexity at a meander bend further influences river ecosystems. Gualtieri et al. [11,12] demonstrated how hydraulic complexity metrics succinctly represent flow patterns and structures utilized by riverine biota, and greater complexity is correlated with more habitat types and greater biologic diversity. Common hydraulic complexity metrics include river velocity and depth gradients, which are known to support species life cycles, such as spawning, feeding, and resting [11]. The importance of hydraulic complexity to riverine species richness [13] is in keeping with other ecological systems, where a positive correlation exists between biodiversity and habitat heterogeneity [14].

Riverine meander evolution has too often been cut short by river straightening, and as communities realized the associated loss of ecosystem services they are eager to restore these meanders as well as reduce their future loss [15]. River meanders have been removed to repurpose the floodplain for development or provide local flood water relief (while exacerbating flood levels downstream). However, human modification of rivers too often leads to unintended consequences on river flows, sediment transport, riverbank stability, and aquatic suitability, making river meander loss a significant public and private concern. A successful river management target is a self-sustaining system with dynamic equilibrium for hydrological, geomorphological, and ecological processes [16]. In an analysis of how to meander bend curvature affects hydraulics and hence ecosystems, it is best to keep constant variables of flow stage, i.e., depth, and bedform topography within the channel banks. Bankfull stage, when water is at the threshold of leaving the channel, is a standard depth for such analysis.

The goal of this research is to quantify the extent that hydraulic complexity degrades when a highly evolved meander bend at the bankfull stage is straightened, while maintaining equivalent bedform topography associated with a highly sinuous meander. The research is approached from the perspective of valuing river flow hydrodynamics and surface water–groundwater exchange patterns associated with high sinuosity meanders, and views the straightening of the channel as a departure from, and the restoration of the sinuosity as a return to, a target value. In order to examine the independent adjustment of planform curvature and bedform topography, we used computational fluid dynamics (CFD) models to simulate flow dynamics in the channel water column and at or in the channel bed, including hyporheic fluxes across the channel bed [17–21]. The desired outcome of this study is the knowledge to help river managers and engineers better understand the hydraulic complexity associated with their river meanders and to guide their channel designs which increasingly rely upon river steering structures to create curvature or bedform.

2. Methods

To systematically change planform curvature while keeping equivalent bedform topography, we used a commercial CFD model (Flow3D®) as a virtual river system that coupled the surface water flow and groundwater movement in one domain. We describe the range of the channel curvature with variable C, defined as the ratio of the channel width, B to the centerline radius of channel curvature, R. A straight channel has a $C = 0$, while a high sinuosity meander bend might have a $C > 0.5$. Nine experiments with gradually changed curvature from a tight bend to a straight channel were simulated by maintaining equivalent cross-section profiles. The output of these simulations were then used to investigate the hydraulic complexity response to the channel curvature changes. Note that the purpose of this study is to investigate the flow responses to meander curvature changes solely, not to bedform topography. There was no sediment transport and associated self-adjusted river bedform evolution included in the experiment setup.

CFD Model

The CFD model used finite-element methods to solve the velocity and pressure variables using the fully 3D transient conservation of mass and momentum equations. Flow depth was controlled by the pressure boundary conditions at both upstream and downstream boundaries. Flow surface was treated as a free surface boundary by applying the volume-of-fluid (VOF) technique [22] which computed fluid volume fraction as one variable in each grid cell. The mass continuity was represented as

$$\frac{V_F}{\rho c^2}\frac{\partial p}{\partial t} + \frac{\partial u A_x}{\partial x} + R\frac{\partial v A_y}{\partial y} + \frac{\partial w A_z}{\partial z} + \xi\frac{u A_x}{x} = 0. \tag{1}$$

The momentum conservation was represented as

$$\frac{\partial u}{\partial t} + \frac{1}{V_F}\left(u A_x\frac{\partial u}{\partial x} + v A_y R\frac{\partial u}{\partial y} + w A_z\frac{\partial u}{\partial z}\right) - \xi\frac{A_y v^2}{x V_F} = \frac{-1}{\rho}\frac{\partial p}{\partial x} + f_x \tag{2}$$

$$\frac{\partial v}{\partial t} + \frac{1}{V_F}\left(u A_x\frac{\partial v}{\partial x} + v A_y R\frac{\partial v}{\partial y} + w A_z\frac{\partial v}{\partial z}\right) + \xi\frac{A_y u v}{x V_F} = \frac{-1}{\rho}\left(R\frac{\partial p}{\partial x}\right) + f_y \tag{3}$$

$$\frac{\partial w}{\partial t} + \frac{1}{V_F}\left(u A_x\frac{\partial w}{\partial x} + v A_y R\frac{\partial w}{\partial y} + w A_z\frac{\partial w}{\partial z}\right) = \frac{-1}{\rho}\frac{\partial p}{\partial x} + f_z. \tag{4}$$

In these equations, V_F is the fraction volume open to flow, ρ is the fluid density, c is a constant (approximately the sound speed), p is pressure, t is time, A_i is the area fractions for flow in the $i(x, y, z)$ direction, and $u_i(u, v, w)$ denote the velocities in i directions, f_i is the viscous acceleration terms. R and ξ are transformation coefficients depend on the choice of the coordinate system is in use. When Cartesian coordinates (x, y, z) are used, $\xi = 0$ and $R = 1$. When cylindrical coordinates (r, θ, z) are used, y derivatives in the equation will be converted to azimuthal derivatives:

$$\frac{\partial}{\partial y} \rightarrow \frac{1}{r}\frac{\partial}{\partial \theta}, \tag{5}$$

which is accomplished by using the equivalent form:

$$\frac{1}{r}\frac{\partial}{\partial \theta} = R\frac{\partial}{\partial y}, \tag{6}$$

where $R = r_m/r$, $y = r_m\theta$, and r_m is a fixed channel radius. In the meantime, $\xi = 1$. Readers are referred to the FLOW-3D users manual (https://www.flow3d.com) for more information.

Turbulence was simulated using the filtered Large Eddy Simulation (LES) model. The LES model solved the instantaneous flow fluctuations at the grid-scale and estimated the sub-grid turbulence variables using eddy viscosity terms, which provided better near boundary turbulence predictions than the classic two-equation $k - \epsilon$ models [17,23]. The sediment bed was treated as a homogenous porous media with uniform porosity. To allow for the coupled surface water–groundwater mixing, a porous riverbed was simulated within the same domain by applying a drag coefficient (DRG) on the momentum equations, determined using the Kozeny–Carman relation based on the sediment propoerties such as grain size and porosity:

$$DRG = 180\nu\left(\frac{1-n}{n}\right)^2\frac{1}{d_{50}^2}, \tag{7}$$

where v is the kinematic viscosity of water, n is the sediment porosity. Note that DRG only applies to the porous media to slow down the flow velocity and mimic the Darcy flow behavior. For the surface flow part, DRG was not activated.

The CFD model simulated river flows were verified in our previous study [21] using Ecole Polytechnique Federale de Lausanne (EPFL) curved flume experiment observations reported by Blanckaert [24]. This flume had a width $B = 1.3$ m, a 9 m long straight inflow channel with a slope of 0.002, a 193° meander bend with centerline radius of curvature $R = 1.7$ m, which results in a $C = 0.77$, and a straight outflow channel. Flow discharge was 89 l/s, with a mean flow depth $H = 0.14$ m, Reynolds number $Re = 68 \times 10^3$, and Froude number $Fr = 0.41$. A flume substrate particle diameter of 1.6 mm to 2.2 mm ($d_{50} = 2$ mm) was used to obtain an equilibrium bed, which was defined as with sediment inflow equaling outflow [24]. Bedform topography of the bend was mapped using acoustic limnimeter to a 450 × 50 grid and made digitally available to the authors. Identical flume geometry settings and boundary conditions were imported into the CFD model for river flow simulations. The computational domain was constructed based on a cylindrical coordination system with 130 grids along the radial or transverse axis n (x in Cartesian coordinate), from 0 m to 1.3 m, 300 grids along the azimuthal or streamwise axis s (y in Cartesian coordinate), from 0° to 193°, and 50 grids along the vertical axis z, from -0.3 m to 0.2 m. Bedform roughness was represented using a recommended roughness height of 0.037 m for this very experiment [25]. The goodness of fit (R^2) between observed and simulated velocity profiles were greater than 0.75 at most cross-sections, indicating the CFD model captured the major flow dynamic patterns in both streamwise and transverse directions and was adequately simulated the meander bend 3D river flows. The readers are referred to Figure 2 in Zhou and Endreny [21] for detailed model verification processes.

Given that the EPFL flume experiment did not include hyporheic exchange processes, we verified the capability of estimating hyporheic exchange rates for the CFD model using a 30 cm wide, 700 cm long, 50 cm deep Armfield® S series recirculating flume filled with 0.3 m depth sediment. The sediment bed had an average depth of 30 cm featuring a pool-riffle sequence with a wavelength of 0.5 m. Dye was released at multiple points along the surface of the sediment bed and was tracked through the transparent wall of the flume. The hyporheic flow vectors simulated by the CFD model agreed well with observations. We also successfully modeled and verified the penetration front movements with the results observed in Elliott and Brooks [26] Run9 experiment. Both model verification campaigns suggest that the CFD model adequately captured the hyporheic flow directions and velocity, as well as the depth of the active hyporheic zone in the bedform. The readers are referred to Figures 2, 4 and 6 in Zhou and Endreny [27] for more details about the hyporheic process verifications. After being verified by the observations, the CFD models have the potential to explore the flow behavior under different boundary conditions [28].

3. Hydraulic Complexity Variables

Nine planform curvatures with $C = 0.77$ (EPFL flume), 0.5, 0.4, 0.33, 0.25, 0.14, 0.08, 0.05, and 0 (straight channel) were simulated in this study (Figure 1). Each experiment used the equivalent $C = 0.77$ bedform topography by stretching the 3D bedform topography around the channel centerline to maintain uniform cross-section profiles between curvatures. Each CFD simulation used the same cylindrical computational gridding and numerical methods established for the $C = 0.77$ simulation.

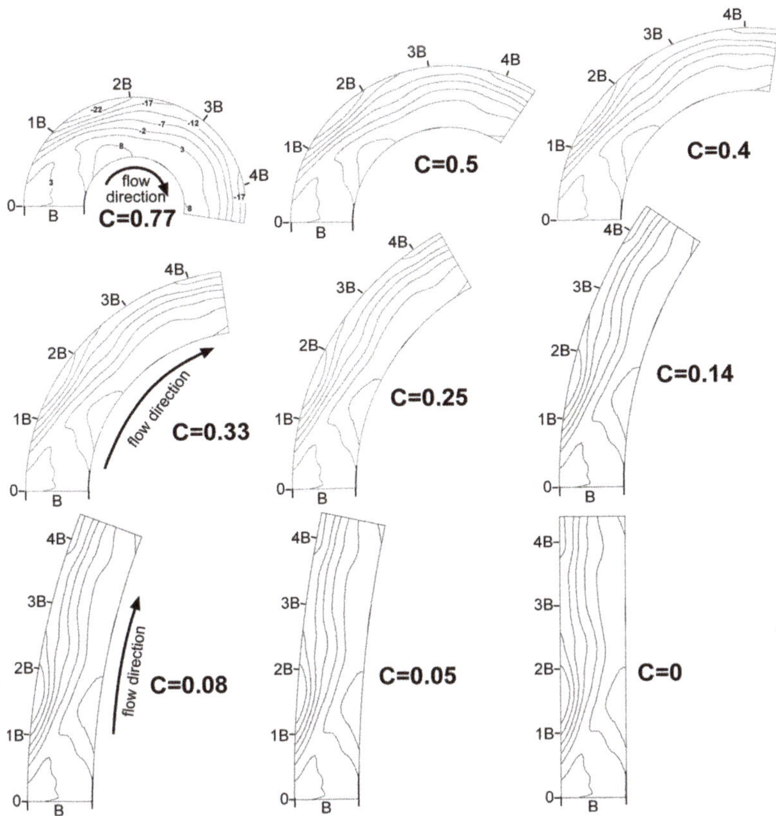

Figure 1. Geometries and bed topography settings of the nine computational fluid dynamics (CFD) simulations with channel curvature (C) changed from 0.77 to 0.

Meander bend hydraulic complexity for these simulations were characterized using CFD output variables based on the flow surface elevation and flow velocity, with v_s, v_n, and v_z velocity vectors representing flow in the s (streamwise), n (transverse), and z (vertical) directions, respectively. The hydraulic complexity variables were normalized when possible, and included (1) normalized flow surface elevation (h/H), where h is the local flow surface elevation, H is the average flow surface elevation; (2) normalized depth averaged streamwise and transverse unit discharge (q_s/UH and q_n/UH)., where q_s and q_n are the unit discharges at s and n directions, equal to $<v_s> \times h$ and $<v_n> \times h$; U is the bulk velocity; "$< >$" denotes depth averaging; (3) normalized depth averaged streamwise velocity at bankfull depth ($<v_s>/U$); (4) bed shear stresses distrubution, represented by the bed friction coefficient, $C_f = \frac{u^{*2}}{U^2}$, where u^* is the shear velocity at the boundary, $u^* = \frac{u_b}{u^+}$, u_b is velocity magnitude at the bottom grid, u^+ is dimensionless velocity for a hydraulically rough wall computed with the modified wall function [29], $u^+ = 2.5ln\frac{z_b}{k_s} + 8.5$ where z_b is the grid size and k_s is the roughness height which was 6 mm; (5) vertical hyporheic flow, represented by the vertical flux 2 cm below the water-sediment interface (v_{zbed}/U); (6) secondary circulation represented by depth averaged normalized vertical velocity ($<v_x>/U$) for the size; and normalized transverse stream function, ψ/UBH for the direction and strength (positive values indicate clockwise circulation when facing upstream). The ψ/UBH was computed as

$$\psi = \frac{1}{2}\left(\int\limits_{n_i}^{n} v_z n\, dn - \int\limits_{z_i}^{z} v_n n\, dz\right), \tag{8}$$

in which the n_i and z_i represent the cross-sectional and vertical coordinates.

The CFD simulated output of the river flows and associated hydraulic complexity for all nine channel curvatures with $C = 0.77$ topographic bedform was projected into a straight channel ($C = 0$) to facilitate inter-comparison. Distance along the streamwise axis s is reported relative to channel width B as s/B. Each hydraulic complexity variable is initially presented as the simulated values at the $C = 0.77$, $C = 0.33$, and $C = 0$ planform curvature to illustrate spatial patterns for each variable.

4. Results

4.1. Flow Surface Elevation

The normalized flow surface elevation (h/H) patterns were very sensitive to planform curvature (Figure 2), with large differences in the inlet section ($<1.5\ s/B$ longitudinal position) emerging between $C = 0.77$ and $C = 0.33$, and between $C = 0.33$ and $C = 0$. Similarities between the $C = 0.33$ and $C = 0$ did exist, and included a semicircle of maximum flow surface values over the pool (left bank centered at $1.5\ s/B$) and a minimum flow surface value along the right bank over the point bar apex. These areas of similarity in h/H were regulated by bedform topography, but curvature and superelevation effects suppressed bedform topographic control for $C > 0.33$. At the tighter curvature of $C = 0.77$ the centrifugal forces elevated the water surface to $h/H \geq 1.08$ and the zone of maximum values stretched from the inlet to $2\ s/B$ and pinched into a narrower lateral extent, while the water surface at the opposite bank dropped to $h/H \leq 0.92$ from 0 to $1.4\ s/B$, followed by other zones of lowered surface values ($h/H < 1$) at $2\ s/B$ and from $3\ s/B$ to $4\ s/B$. The complete absence of planform curvature in the $C = 0$ system did not prohibit bedform topography from creating a backwater effect behind the point bar, noted by the zone of maximum h/H along the right bank from $0\ s/B$ to $1\ s/B$.

The impact of planform curvature was illustrated with transverse plots of h/H at $1.5\ s/B$, the point bar apex (Figure 3). Considering the normalized flow surface elevation as a function of normalized transverse distance, $(h/H) = Fh(n/B)$, the second order derivative of the function $Fh''(n/B)$ was used to identify the convexity of the flow surface curves with positive and negative values indicating convex and concave surfaces, respectively. We find that $C = 0.33$ was the divide that separated the h/H patterns into two types. In tight meander bend with $C > 0.33$ the transverse pattern of h/H was convex, and for mildly curved channels with $C < 0.33$ it had a concave pattern. At a curvature of $C = 0.33$ the h/H profile $Fh''(n/B)$ was close to zero and the h/H profile was flat. The mechanism might be that the centrifugal effects piled up the water in high curvature channels and created convex h/H superelevation transverse profiles at the meander apex, while open meanders the bedform topography created backwater effects and concave h/H transverse profiles.

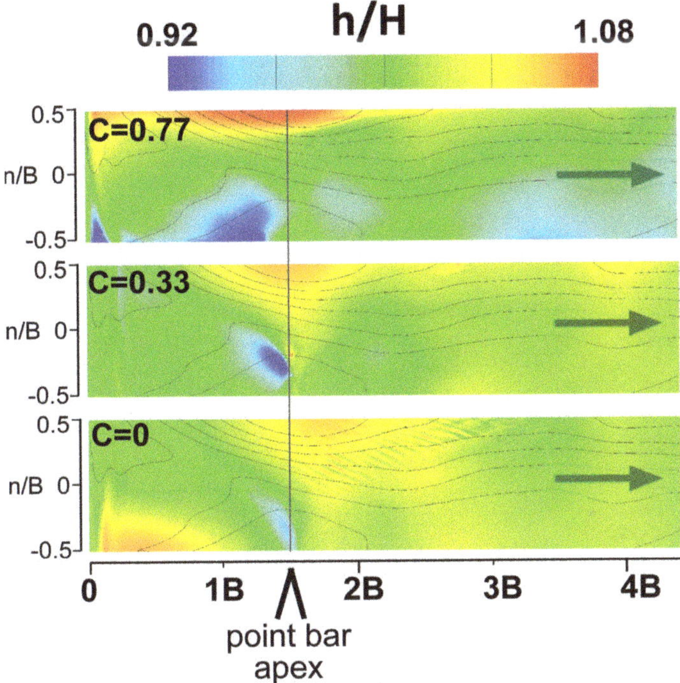

Figure 2. Flow surface elevation (h) normalized by H at $C = 0.77$, $C = 0.33$, and $C = 0$ conditions. n denotes the lateral coordination with $n = 0$ at channel center and B denotes the channel width.

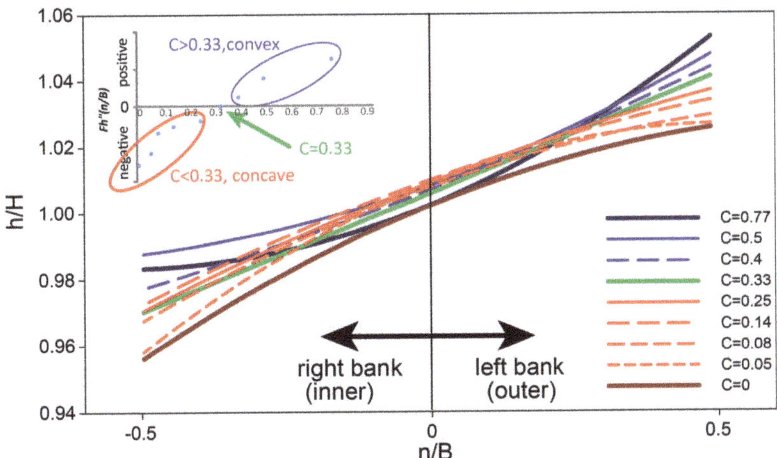

Figure 3. Normalized flow surface profiles for the nine simulations at the point bar apex 1.5 s/B. The insert plot shows the second order derivative of normalized flow surface elevation in the transverse direction, $Fh''(n/B)$, which gives the convexity or concavity of the surface profile curves.

4.2. Streamwise and Transverse Unit Discharge

The unit discharge, which represents a mass redistribution of water within the channel, was less sensitive to channel curvature. The unit discharge in the streamwise (q_s/UH) (Figure 4) direction had relatively stable spatial patterns of maximum and minimum values with changing curvature.

This suggests their spatial patterns were regulated by bed topography and the larger conveyance regions opened by pools and smaller conveyance regions constrained by point bars. This topographic control was illustrated by comparing streamwise unit discharge for $C = 0.77$, 0.33, and 0 at the left bank pool and the right bank point bar conveying. Nearly 90% of the downstream mass was conveyed along the left half (i.e., the pool side) of the channel for all curvatures, while along the right bank at station $2\,s/B$ there was upstream mass transport, indicating the existence of backwater eddy and flow separation for all curvatures. The maximum streamwise unit discharge occurred at about $1.5\,s/B$, right above the deepest part of the pool. Differences in streamwise unit discharge patterns between $C = 0.77$ and $C = 0$ emerged along the left bank stations $2.5\,s/B$ to $3.5\,s/B$, where mass transport increased by 50% as curvature increased. In this post-apex section of the meander bend there was greater sensitivity of streamwise mass flux to curvature.

The distribution of transverse unit discharge, q_n/UH, were also largely insensitive to curvature, where $C = 0.77$, 0.33, and 0 each experienced a pronounced transport to the outer bank (positive values) upstream of the apex and transport to the inner bank further downstream (Figure 5). The impact of bed topography was noted in the damped oscillation patterns of q_n/UH with the maximum leftward mass transport at $1\,s/B$, just upstream of the point bar, and rightward transport at about $2.3\,s/B$, downstream of the pool. As curvature increased, the single ovate shaped hot-spot of outer-bank directed mass transport that extended beyond the apex for $C = 0.77$, became split at the tip for $C = 0.33$, and then reduced to a smaller region for $C = 0$ (Figure 5). A parallel set of changes occurred in inner-bank-ward mass transport, where the initiation of this inner bank transport region moved further downstream as curvature reduced from $C = 0.77$ to $C = 0$. The differences in mass transport between the $C = 0.33$ and $C = 0$ simulations were only about 5% based on transversely averaging ($< q_n/UH >$) (Figure 6). The inset of Figure 6 shows the likelihood (here we use coefficient of determination, R^2) of $< q_n/UH >$ between each curvature and the straight channel condition ($C = 0$, $R^2 = 1$). The lower the R^2, the transverse unit discharge is more complex than in the straight channel.

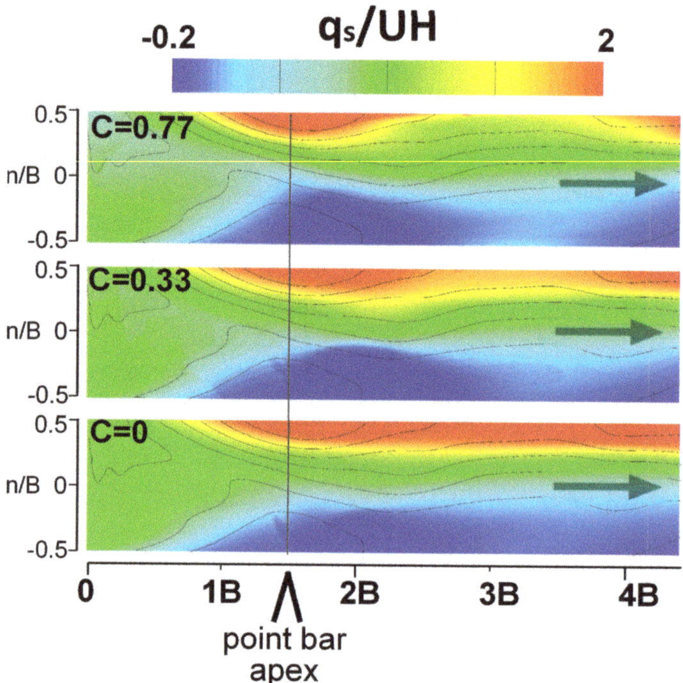

Figure 4. Streamwise unit discharge q_s/UH for channel curvature $C = 0.77$, 0.33, and 0 conditions.

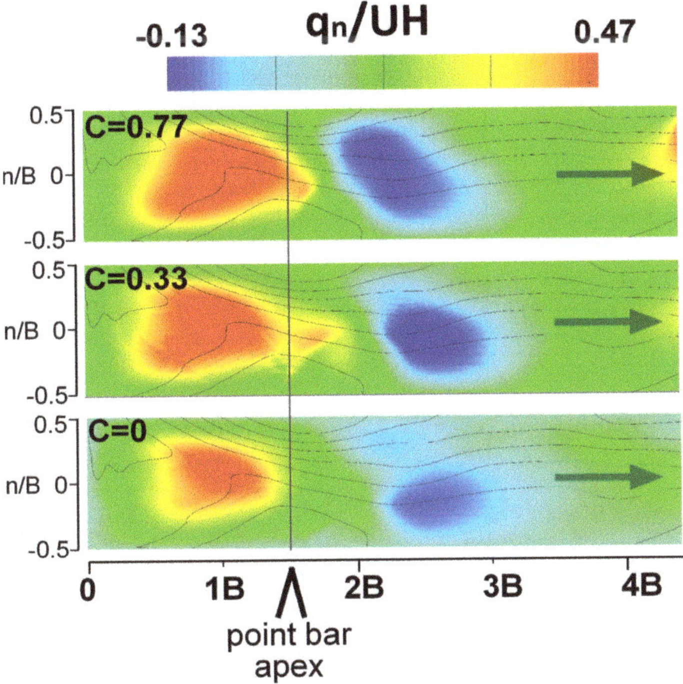

Figure 5. Transverse unit discharge q_n/UH for channel curvature $C = 0.77, 0.33$, and 0 conditions.

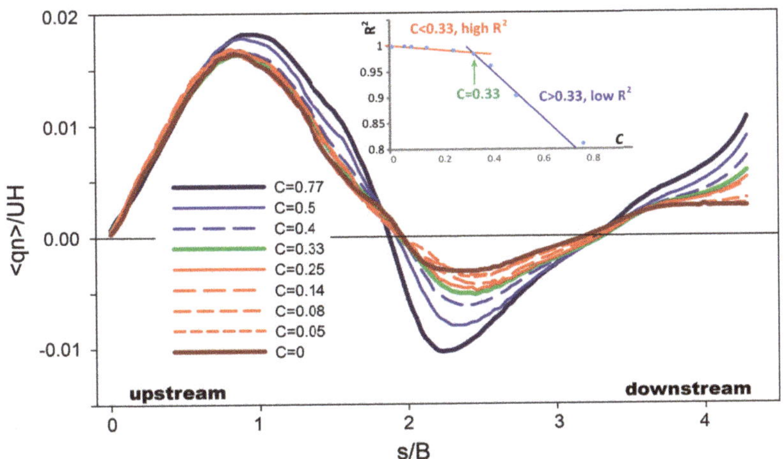

Figure 6. Transverse unit discharge averaged over the transverse direction. The inset shows the R^2 of transverse unit discharge $< q_n/UH >$ between each curvature, C, and the straight channel condition ($C = 0$, $R^2 = 1$); a lower R^2 suggests greater hydraulic complexity for transverse unit discharge.

The distribution of transverse unit discharge, q_n/UH, were also largely insensitive to curvature, where $C = 0.77, 0.33$, and 0 each experienced a pronounced transport to the outer bank (positive values) upstream of the apex and transport to the inner bank further downstream (Figure 5). The impact of bed topography was noted in the damped oscillation patterns of q_n/UH with the maximum leftward mass transport at $1\ s/B$, just upstream of the point bar, and rightward transport at about $2.3\ s/B$, downstream of the pool. As curvature increased the single ovate shaped hot-spot of outer-bank directed mass transport that extended beyond the apex for $C = 0.77$, became split at the tip for $C = 0.33$, and then reduced to a smaller region for $C = 0$ (Figure 5). A parallel set of changes occurred in inner-bank-ward mass transport, where the initiation of this inner bank transport region moved further downstream as curvature reduced from $C = 0.77$ to $C = 0$. The differences in mass transport between the $C = 0.33$ and $C = 0$ simulations were only about 5% based on transversely averaging ($< q_n/UH >$) (Figure 6). The inset of Figure 6 shows the likelihood (here we use coefficient of determination, R^2) of $< q_n/UH >$ between each curvature and the straight channel condition ($C = 0$, $R^2 = 1$). The lower the R^2, the transverse unit discharge is more complex than in the straight channel.

The results showed that the R^2 was greater than 0.8 in all channels and greater than 0.98 in $C < 0.33$ channels. With increasing channel curvature, the R^2 remained constant until it started to drop rapidly from 0.95 at $C = 0.33$ to 0.8 at $C = 0.77$. This analysis suggested that the transverse unit discharge is generally insensitive to curvature, especially in moderately curved channels ($C < 0.33$).

4.3. Streamwise Velocity Distribution

The streamwise velocity distributions, quantified by the first moment of depth-averaged streamwise velocity $< v_s > /U$, had spatially varying sensitivity to curvature, which was most pronounced at the channel inlet (Figure 7). In the inlet the maximum was along the right bank in $C = 0.77$ but was along the left bank in $C = 0$. These differences in streamwise velocity are explained by curvature and its role in transverse velocity gradients. In $C = 0.77$ the shorter streamwise distance along the inner bank caused an increase in $< v_s > /U$ from the inner bank to the outer bank [30]. Curvature also caused differences in magnitude of $< v_s > /U$ between the $C = 0.77$ and $C = 0$ channels from 2.5 s/B to 4 s/B, with the maximum $< v_s > /U$ increased and concentrated near the outer bank area.

The first moment of depth averaged streamwise velocity $< v_s > /U$, which is the center of gravity of the streamwise flow distribution, has an exact transverse location denoted by n/B, where $n = 0$ at the channel center and $n/B = 0.5$ at the left bank (Figure 8). For all planform curvatures the first moment of $< v_s > /U$ oscillated, originating near $n/B = 0$ at the channel inlet, approaching the left bank distance $n/B > 0.2$ at $s/B = 1.5$, and then returning toward the channel center $n/B = 0.1$ at $s/B = 2$. This oscillation was explained by flow separation upstream of the point bar, the subsequent relocation of maximum streamwise velocities near the outer bank, and the return of those maximum velocities toward the channel center at the meander outlet. Sensitivity to curvature was observed from 2 s/B to 4 s/B, where the amplitude of oscillation in the first moment of $< v_s > /U$ diminished with decreasing curvature; oscillating by 0.35 n/B for $C = 0.77$ and oscillating by 0.15 n/B for $C = 0$. For high curvature channels ($C > 0.33$) with tight meanders, flow entered the meander bend along the right half of channel due to a curvature induced potential vortex, while in moderately curved $C < 0.33$ channels flow entered the meander along the left half of the channel due to a point bar backwater effect. At channel location 2 s/B the maximum leftward amplitude was $n/B = 0.4$ from the channel center for the $C = 0.77$ simulation and $n/B = 0.25$ from the center for the $C < 0.33$ simulations. The coefficient of determination for the first moment of $< v_s > /U$ between $C = 0$ and other simulations resulted in $R^2 > 0.95$ for $C < 0.33$ channels, but ranged from 0.6 to 0.88 for the $C > 0.33$ channels (the insert in Figure 8). As the meander bend tightened and C approached 1.3 the R^2 varied greatly with curvature, while in channels with $C < 0.33$ the R^2 was relatively insensitive to changes in curvature, suggesting bedform topography was a stronger influence than planform curvature on the streamwise velocity distribution.

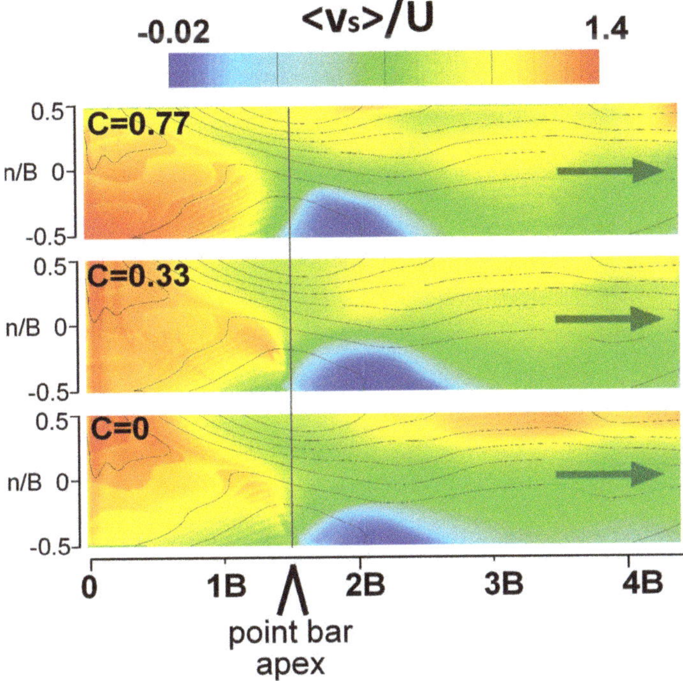

Figure 7. Normalized depth averaged streamwise velocity $< v_s > /U$ for channel curvature $C = 0.77, 0.33,$ and 0 conditions.

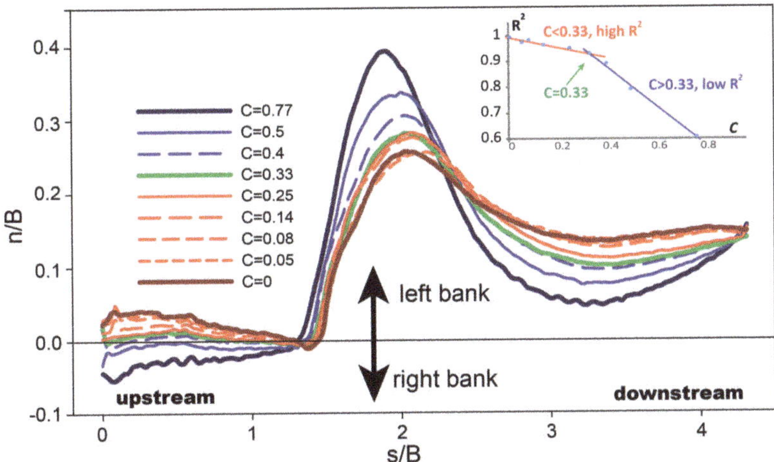

Figure 8. The first moment of normalized depth averaged streamwise velocity $< v_s > /U$, which represents center of gravity of the streamwise flow distribution, along the channel. The inset shows the R^2 of the first moment of $< v_s > /U$ between each curvature and the straight channel condition ($C = 0$, $R^2 = 1$); a lower R^2 suggests greater hydraulic complexity for the first moment of depth averaged streamwise velocity.

4.4. Bed Shear Distribution

The bed shear distribution, Cf, is determined in part by near-bed velocities (Figure 9). Along the right bank side at the channel inlet, the bed shear was evidently controlled by the bed topography, with maximum values for all planform curvatures extending up to the point bar. At the point bar cross-section the bed shear was controlled by bedform topography over the point bar, with a minimum Cf values for $C \leq 0.33$ curvatures, suggesting conditions for sediment deposition. However, opposite the point bar the bed shear was controlled by planform curvature, with maximum Cf values in the $C = 0.77$ simulation and minimum Cf values in the $C = 0$ simulation. These maximum Cf values in the $C = 0.77$ were over the pool area from $1\ s/B$ to $2.5\ s/B$, where sediment erosion was most active. Note that the artificial strips observed in bed shear Cf distribution are caused by the structured meshing approach used to represent the bedform topography. This factor could be reduced/smoothed if a finer mesh is applied.

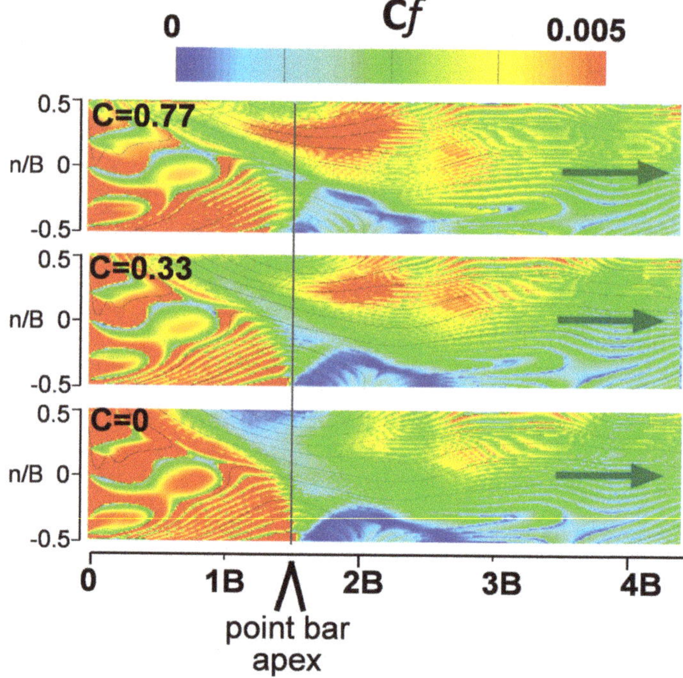

Figure 9. Distribution of river channel bed shear Cf for channel curvature $C = 0.77, 0.33,$ and 0 conditions.

4.5. Vertical Hyporheic Exchange

The vertical hyporheic exchange patterns were sensitive to planform curvature, but more than half of the bed was controlled by bedform topography (Figure 10). Since the magnitude of vertical flux was very small (about three orders of magnitude lower) compared with the bulk velocity (U), here we only analyzed the direction of the hyporheic flux direction (i.e., upwelling, downwelling, or neutral), among the simulations. In $C = 0.77$ the upwelling area occupied a dominant amount of cross-sectional bed area from the inlet to $2\ s/B$, followed by another large swath of upwelling between 2.5–$3.5\ s/B$, while the downwelling area occupied about 20% of the total area and was primarily located to the left bank from $0.5\ s/B$ to the outlet due to the strong downward flow along the pool area. By contrast, the largest patch of upwelling in the $C = 0$ simulation extended from the inlet to the point bar, with another patch connected to the left bank centered at $3\ s/B$. Bedform topographic controls on hyporheic flux were identified by similar patterns of flux direction for $C = 0.77, 0.33$, and 0 and were located in the pool opposite the point bar, on the lee side of the point bar at 2 s/B, and in a patch attached to the left bank at 2.5 s/B.

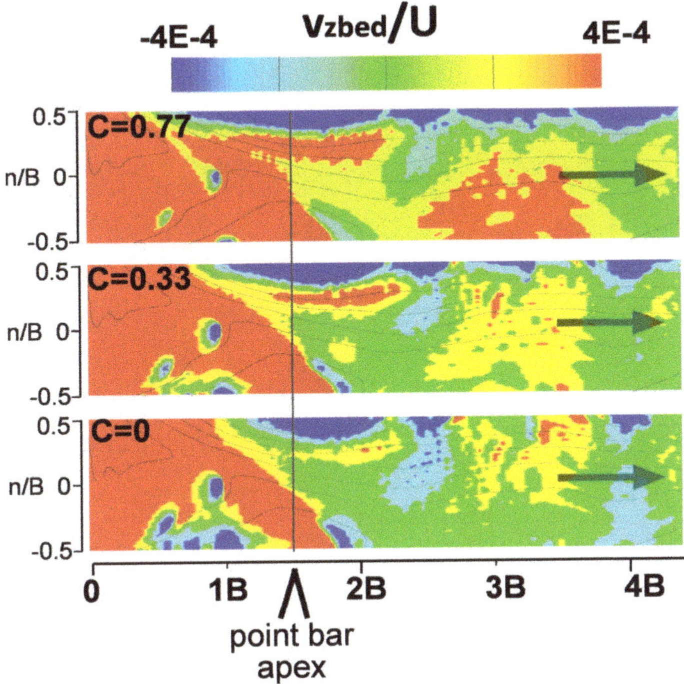

Figure 10. Normalized vertical hyporheic flux v_{zbed}/U at 2 mm below sediment surface for channel curvature $C = 0.77, 0.33$, and 0 conditions. Positive indicates upwelling of groundwater into the river channel.

4.6. Secondary Circulation Patterns

Vertical velocity distributions and transverse stream functions are highly related to the secondary circulation patterns. The normalized depth-averaged vertical velocity distribution had areas sensitive to planform curvature and other areas more sensitive to bedform topographic control (Figure 11). All three curvatures had strong negative vertical velocities at the left bank from 0.5 s/B to 1.5 s/B, with flow downwelling into the pool, and all curvatures had positive vertical velocities centered at 2.5 s/B where the rise in bed caused upwelling currents. However, only the tight meanders ($C > 0.33$) had maximum vertical velocities in the pool, initiating at 1.5 s/B, while the straighter channels had a small region of maximum vertical velocities over the right bank point bar apex. Only in the $C > 0.33$ channels was the presence of a strong secondary circulation cell present, with maximum and downwelling and upwelling collocated over the pool.

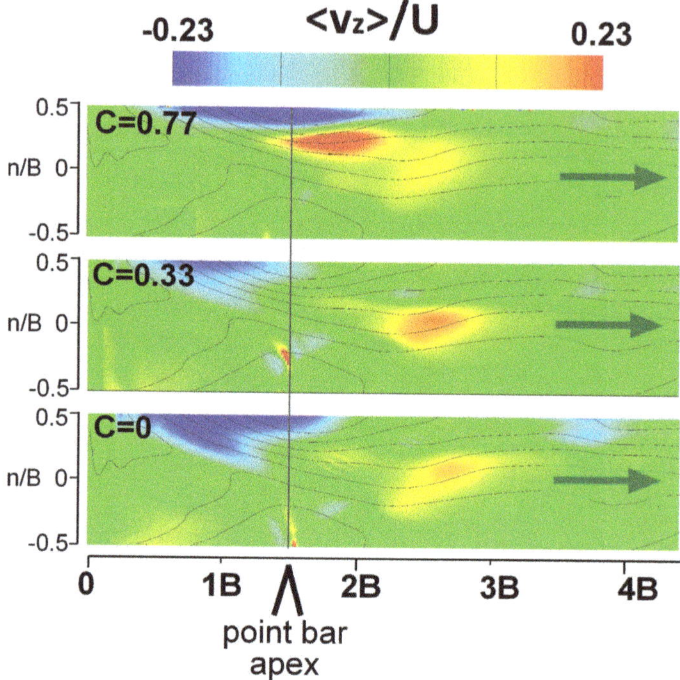

Figure 11. Normalized vertical velocity $<v_z>/U$ for channel curvature $C = 0.77, 0.33$, and 0 conditions, with positive values upward flows, negative values downward flows.

The transverse stream function distributions (ψ/UBH) at the meander apex revealed the presence of flow cells rotating around the streamwise axis for the three curvatures ($C = 0.77, 0.33$, and 0). A cross-sectional view of the stream function and transverse flow cell vectors for $1.8\ s/B$ (Figure 12) showed a clockwise (looking upstream) secondary circulation present in all curvatures, centered in the pool. A patch of negative stream function values was also present for all curvatures, located against the top half of the outer bank. The intensity of rotation increased with decreasing curvature, where $C = 0.77$ had maximum stream function values of 0.3 and $C = 0$ had maximum values of 0.1. The centroid of the rotation lifted higher into the water column with increasing curvature, and the lateral extent of rotation was reduced with decreasing curvature. This was seen by the stronger $C = 0.77$ circulation cell tightly centered in the pool z and n domain, while the weaker $C = 0$ circulation cell was lower in the water column and stretched into the point bar bench and outer bank. The pattern and strength of vertical and rotational flow patterns were strongly influenced by curvature and its centrifugal effects. For mild curvatures of $C < 0.33$ there were similarities in the pattern and strength of the secondary circulation, indicating the strong influence of bedform topography in these more open curvature channels.

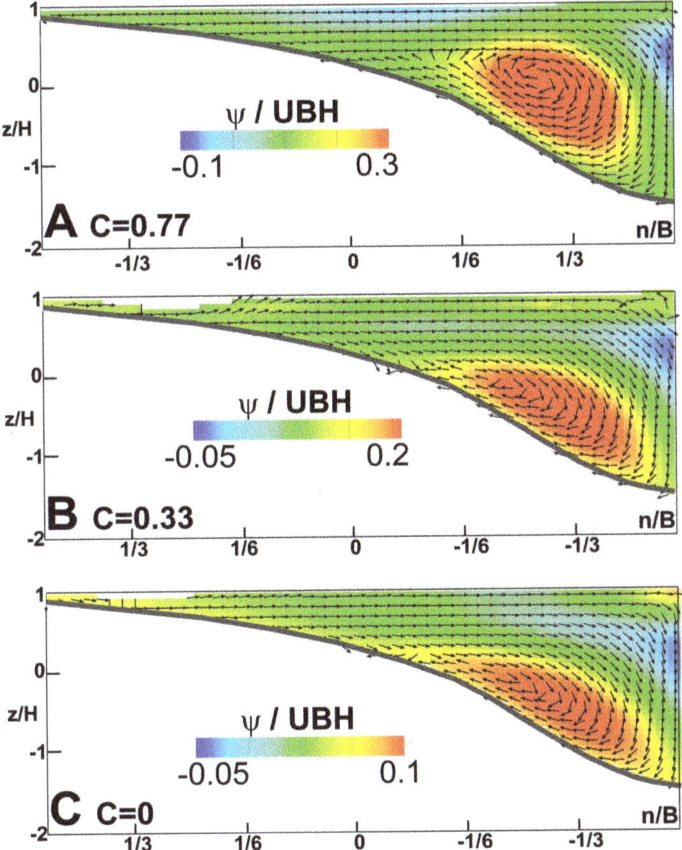

Figure 12. Transverse stream function distribution ψ/UBH reveals the secondary circulation of transverse flow cells rotating at the meander apex 1.5 s/B for channel curvature $C = 0.77$ (**A**), $C = 0.33$ (**B**), and $C = 0$ (**C**), with positive values representing clockwise rotation direction when facing upstream, and negative values representing counter-clockwise rotation when facing upstream.

5. Discussion and Conclusions

This study has quantified how straightening a previously sinuous river channel impacts the spatial variation in water depth, velocity, and their products which create hydraulic complexity. As a river meander is straightened, the loss to hydraulic complexity is initially gradual and then is abrupt when the river curvature decreased below a threshold of $C = 0.33$, i.e., the radius of the channel is three times of the channel width. Leopold and Wolman [2] also identified a planform curvature of approximately $C = 0.33$ as a threshold below which the river meanders no longer support the process of flow separation and the associated hydraulics of wake zones and eddies.

River management can refer to these findings as a guide to hydraulic patterns about meander bends at different river curvature. Such a guide could direct data collection to document and confirm spatial patterns of depth, velocity, and the hydraulic complexity metrics. In cases where river meanders were straightened and bedform topography were degraded, river managers might use this study to consider at least two approaches to restoration. If floodplain encroachment prohibits restoration of full curvature, managers might try restoration of the bedform topography associated with full curvature. Such bedform manipulation can be supported with boulders, submerged logs, vegetation, and other techniques (discussed below). Without constraints of floodplain encroachment, managers might be able to restore channel curvature along with its bedform topography. Given the sudden loss to hydraulic complexity when curvature drops below the threshold of $C = 0.33$, rivers should be managed to maintain a range of meander curvatures that exist above this threshold in order to support the morphodynamic and associated biotic functions of rivers. While this is the first study to systematically document how the loss of curvature leads to progressive and then rapid loss of hydraulic complexity, other studies have documented the links between curvature and river functions.

The morphodynamic connections to river curvature include the generation of transverse, or secondary circulation patterns [31,32], which establish feedback loops between the dynamics of sediment transport and morphology of bed topography [33,34]. Our experiments documented how the loss of secondary circulation cells was associated with changes patterns of maximum velocities in three directions and a loss of vortices others have shown useful for biota. Liao et al. [35] used high-speed video to show how fish benefit from the rich 3D flows common to high curvature meanders, using vortices to decrease energy expenditure as they migrate up-river.

As these findings are referred to for guidance in managing river meanders, it is important to reiterate this study used a fixed-bed experimental design, while most natural rivers will have mobile-beds where the substrate can move in larger flows. The fixed-bed condition allowed for establishing bed topography associated with high curvature meanders, even in lower curvature conditions. To achieve this in the field, river restoration uses fixed, in-channel forms such as boulders to steer the river water and achieve multiple goals of reducing channel bank erosion and enhancing hydraulic complexity and ecosystem services such as habitat and hyporheic exchange [36,37]. In most alluvial rivers sediment ultimately deposits and scours around these forms, in complex ways that extend beyond the reach of this study. To make progress in this complex environment of river research, we controlled the covariate of the bedform to understand the role of curvature. To further address the issues of mobile bed dynamics, we recommend follow on research to provide detailed estimates of equilibrium bed scour and deposition caused by different channel curvatures. Another limitation is that the bed topography is defined as equivalent across the experiments. However, the decreasing of the channel curvature also related to bedform changes with a shrinking outer bank and extending inner bank, which may also contribute to the flow pattern changes.

In conclusion, this study detailed where river flow about a meander bend loses hydraulic complexity when the river is straightened. This has implications for protecting and restoring river curvature to manage river ecosystem services. Below is a summary of changes to river flow complexity when a river meander is modified from high curvature to straightened:

1. A weakening of the superelevated water surface along the outer bank at the meander apex (see Figures 2 and 3) , and a diminishing transverse unit discharge (Figures 5 and 6).
2. The creation of an uninterrupted band of high magnitude streamwise unit discharge along the outer bank, which had previously been limited to an area about the meander apex (Figures 4 and 8), and associated loss of high flows along the inside bank just upstream of the point bar (Figure 7).
3. The loss of large magnitude bed shear stresses in the pool at the meander apex and establishment of a larger zone of near-zero bed shear about the point bar (Figure 9), and the related loss of large swaths of upwelling flow of groundwater into the channel at the apex along the outer bank and downstream of the apex along the inner half of the channel (Figure 10).
4. The loss of a tightly coupled upward and downward vertical velocities along the outer bank at the meander apex (Figure 11) and the associated weakening and shifting of secondary circulation transverse flow cells (Figure 12).

Author Contributions: T.Z. and T.E. have equally contributed to the conceptualization, methodology, analysis, and writing. All authors have read and agreed to the published version of the manuscript.

Funding: This research was funded by NSF grant number EAR 09-11547.

Acknowledgments: The authors thank the data sharing provided by K. Blanckaert of Ecole Polytechnique Federale Lausanne (EPFL) as well as experiment support by B. Han of Ball State University.

Conflicts of Interest: The authors declare no conflict of interest.

References

1. Langbein, W.; Leopold, L. *River Meanders-Theory of Minimum Variance (Geological Survey Professional Paper 422-H)*; U.S. Government Printing Office: Washington, DC, USA, 1966.
2. Leopold, L.B.; Wolman, M.G. River meanders. *Bull. Geol. Soc. Am.* **1960**, *71*, 769–793. 0016-7606(1960)71. [CrossRef]
3. Wohl, E. *Rivers in the Landscape*; John Wiley & Sons: Hoboken, NJ, USA, 2020.
4. Dietrich, W.E.; Smith, J.D. Influence of the point bar on flow through curved channels. *Water Resour. Res.* **1983**, *19*, 1173–1192. [CrossRef]
5. Harvey, J.W.; Bencala, K. The effects of streambed topography on surface-subsurface water exchange in mountains catchments. *Water Resour. Res.* **1993**, *29*, 89–98. [CrossRef]
6. Bridge, J.S. *Rivers and Floodplains: Forms, Processes, and Sedimentary Record*; John Wiley & Sons: Hoboken, NJ, USA, 2009.
7. Schumm, S.A. Patterns of alluvial rivers. *Annu. Rev. Earth Planet. Sci.* **1985**, *13*, 5–27. [CrossRef]
8. Vermeulen, B.; Hoitink, A.J.F.; Labeur, R.J. Flow structure caused by a local cross-sectional area increase and curvature in a sharp river bend. *J. Geophys. Res. Earth Surf.* **2015**, *120*, 1771–1783. [CrossRef]
9. Konsoer, K.M.; Rhoads, B.L.; Best, J.L.; Langendoen, E.J.; Abad, J.D.; Parsons, D.R.; Garcia, M.H. Three-dimensional flow structure and bed morphology in large elongate meander loops with different outer bank roughness characteristics. *Water Resour. Res.* **2016**, *52*, 9621–9641. [CrossRef]
10. Li, B.D.; Zhang, X.H.; Tang, H.S.; Tsubaki, R. Influence of deflection angles on flow behaviours in openchannel bends. *J. Mt. Sci.* **2018**, *15*, 2292–2306. [CrossRef]
11. Gualtieri, C.; Abdi, R.; Ianniruberto, M.; Filizola, N.; Endreny, T.A. A 3D analysis of spatial habitat metrics about the confluence of Negro and Solimões rivers, Brazil. *Ecohydrology* **2020**, *13*, e2166. [CrossRef]
12. Gualtieri, C.; Ianniruberto, M.; Filizola, N.; Santos, R.; Endreny, T. Hydraulic complexity at a large river confluence in the Amazon basin. *Ecohydrology* **2017**, *10*, e1863. [CrossRef]
13. Kozarek, J.; Hession, W.; Dolloff, C.; Diplas, P. Hydraulic complexity metrics for evaluating in-stream brook trout habitat. *J. Hydraul. Eng.* **2010**, *136*, 1067–1076. [CrossRef]
14. McCoy, E.D.; Bell, S.S.; Mushinsky, H.R. Habitat structure: Synthesis and perspectives. In *Habitat Structure*; Springer: Berlin, Germany, 1991; pp. 427–430.
15. Re-Engineering Britain's Rivers. *The Economist*. 6 March 2020. Available online: https://www.latestnigeriannews.com/news/8279579/reengineering-britains-rivers.html (accessed on 12 April 2020).
16. Palmer, M.A.; Bernhardt, E.; Allan, J.; Lake, P.S.; Alexander, G.; Brooks, S.; Carr, J.; Clayton, S.; Dahm, C.; Follstad Shah, J.; et al. Standards for ecologically successful river restoration. *J. Appl. Ecol.* **2005**, *42*, 208–217. [CrossRef]
17. Abad, J.D.; Rhoads, B.L.; Güneralp, İ.; García, M.H. Flow structure at different stages in a meander-bend with bendway weirs. *J. Hydraul. Eng.* **2008**, *134*, 1052–1063. [CrossRef]
18. Blanckaert, K.; Schnauder, I.; Sukhodolov, A.; van Balen, W.; Uijttewaal, W. Meandering: Field Experiments, Laboratory Experiments and Numerical Modeling. Technical Report. 2009. Available online: https://infoscience.epfl.ch/record/146621/files/2009-695-Blanckaert_et_al-Meandering_field_experiments_laboratory_experiments_and_numerical.pdf (accessed on 12 April 2020).
19. Constantinescu, G.; Koken, M.; Zeng, J. The structure of turbulent flow in an open channel bend of strong curvature with deformed bed: Insight provided by detached eddy simulation. *Water Resour. Res.* **2011**, *47*, doi:10.1029/2010WR010114. [CrossRef]
20. Sawyer, A.H.; Bayani Cardenas, M.; Buttles, J. Hyporheic exchange due to channel-spanning logs. *Water Resour. Res.* **2011**, *47*, doi:10.1029/2011WR010484. [CrossRef]

21. Zhou, T.; Endreny, T. Meander hydrodynamics initiated by river restoration deflectors. *Hydrol. Process.* **2012**, *26*, 3378–3392. [CrossRef]

22. Hirt, C.W.; Nichols, B.D. Volume of fluid (VOF) method for the dynamics of free boundaries. *J. Comput. Phys.* **1981**, *39*, 201–225. [CrossRef]

23. Van Balen, W.; Uijttewaal, W.; Blanckaert, K. Large-eddy simulation of a curved open-channel flow over topography. *Phys. Fluids* **2010**, *22*, 075108. [CrossRef]

24. Blanckaert, K. Topographic steering, flow recirculation, velocity redistribution, and bed topography in sharp meander bends. *Water Resour. Res.* **2010**, *46*.

25. Zeng, J.; Constantinescu, G.; Blanckaert, K.; Weber, L. Flow and bathymetry in sharp open-channel bends: Experiments and predictions. *Water Resour. Res.* **2008**, *44*, doi:10.1029/2007WR006303. [CrossRef]

26. Elliott, A.H.; Brooks, N.H. Transfer of nonsorbing solutes to a streambed with bed forms: Laboratory experiments. *Water Resour. Res.* **1997**, *33*, 137–151. [CrossRef]

27. Zhou, T.; Endreny, T.A. Reshaping of the hyporheic zone beneath river restoration structures: Flume and hydrodynamic experiments. *Water Resour. Res.* **2013**, *49*, 5009–5020. [CrossRef]

28. Lane, S.; Bradbrook, K.; Richards, K.; Biron, P.; Roy, A. The application of computational fluid dynamics to natural river channels: Three-dimensional versus two-dimensional approaches. *Geomorphology* **1999**, *29*, 1–20. [CrossRef]

29. Vardy, A. *Fluid Principles*; McGraw-Hill International Series in Civil Engineering; McGraw-Hill: London, UK, 1990.

30. Rozovskii, I.L. *Flow of Water in Bends of Open Channels*; Academy of Sciences of the Ukrainian SSR: Kiev, Ukraine, 1957.

31. Blanckaert, K.; De Vriend, H.J. Secondary flow in sharp open-channel bends. *J. Fluid Mech.* **2004**, *498*, 353–380. [CrossRef]

32. Johannesson, H.; Parker, G. Linear theory of river meanders. *River Meand.* **1989**, *12*, 181–213.

33. Camporeale, C.; Perona, P.; Porporato, A.; Ridolfi, L. Hierarchy of models for meandering rivers and related morphodynamic processes. *Rev. Geophys.* **2007**, *45*, doi:10.1029/2005RG000185. [CrossRef]

34. He, L. Distribution of primary and secondary currents in sine-generated bends. *Water SA* **2018**, *44*, 118–129. [CrossRef]

35. Liao, J.C.; Beal, D.N.; Lauder, G.V.; Triantafyllou, M.S. Fish exploiting vortices decrease muscle activity. *Science* **2003**, *302*, 1566–1569. [CrossRef]

36. Crispell, J.K.; Endreny, T.A. Hyporheic exchange flow around constructed in-channel structures and implications for restoration design. *Hydrol. Process.* **2009**, *1168*, 1158–1168. [CrossRef]

37. Hester, E.T.; Gooseff, M.N. Moving Beyond the Banks: Hyporheic Restoration Is Fundamental to Restoring Ecological Services and Functions of Streams. *Environ. Sci. Technol.* **2010**, *44*, 1521–1525. [CrossRef]

Article

Dynamic Evapotranspiration Alters Hyporheic Flow and Residence Times in the Intrameander Zone

James Kruegler [1,*], **Jesus Gomez-Velez** [2], **Laura K. Lautz** [3] **and Theodore A. Endreny** [4]

1   The Davey Institute, 1500 North Mantua Street, Kent, OH 4424, USA
2   Department of Civil and Environmental Engineering, Vanderbilt University, 400 24th Avenue South,
     267 Jacobs Hall, Nashville, TN 37212, USA; jesus.gomezvelez@vanderbilt.edu
3   Department of Earth Sciences, Syracuse University, 204 Heroy Geology Laboratory, Syracuse, NY 13244,
     USA; lklautz@syr.edu
4   Department of Environmental Resources Engineering, State University of New York,
     College of Environmental Science and Forestry, 1 Forestry Drive, 402 Baker Labs, Syracuse, NY 13210, USA;
     te@esf.edu
*   Correspondence: JKruegler@gmail.com

Received: 30 December 2019; Accepted: 1 February 2020; Published: 5 February 2020

Abstract: Hyporheic zones (HZs) influence biogeochemistry at the local reach scale with potential implication for water quality at the large catchment scale. The characteristics of the HZs (e.g., area, flux rates, and residence times) change in response to channel and aquifer physical properties, as well as to transient perturbations in the stream–aquifer system such as floods and groundwater withdraws due to evapotranspiration (ET) and pumping. In this study, we use a numerical model to evaluate the effects of transient near-stream evapotranspiration (ET) on the area, exchange flux, and residence time (RT) of sinuosity-induced HZs modulated by regional groundwater flow (RGF). We found that the ET fluxes (up to 80 mm/day) consistently increased HZ area and exchange flux, and only increased RTs when the intensity of regional groundwater flow was low. Relative to simulations without ET, scenarios with active ET had more than double HZ area and exchange flux and about 20% longer residence times (as measured by the median of the residence time distribution). Our model simulations show that the drawdown induced by riparian ET increases the net flux of water from the stream to the nearby aquifer, consistent with field observations. The results also suggest that, along with ET intensity, the magnitude of the HZ response is influenced by the modulating effect of both gaining and losing RGF and the sensitivity of the aquifer to daily cycles of ET withdrawal. This work highlights the importance of representing near-stream ET when modeling sinuosity-induced hyporheic zones, as well as the importance of including riparian vegetation in efforts to restore the ecosystem functions of streams.

Keywords: hyporheic zone; hyporheic exchange; evapotranspiration; groundwater modeling; riparian vegetation

---

1. Introduction

The hyporheic zone (HZ) plays a crucial role in basic ecosystem functions in riparian corridors, being the region of an aquifer where there is some degree of mixing between stream water and groundwater. Focusing at the local reach scale, the exchange of water, solutes, and biota to and from the stream gives the HZ unique hydrodynamic and chemical properties [1], as well as a rich diversity of microbial communities [2]. Hyporheic exchange allows solutes carried by a stream to temporarily reside in the geochemically and microbially active streambed and banks [3], which can create the microfauna and solute residence times (RTs) necessary for critical biogeochemical transformation [4]—for example, retention of nutrients [5] and metals [6]. Thus, hyporheic exchange influences major environmental engineering problems such as degradation of water quality [7], stream restoration [8], and the integrity of riparian ecosystems [9].

Spatiotemporal changes of the HZ's characteristics such as area, fluxes, and residence times can have a significant impact on its potential for biogeochemical transformation. For example, changes in the areal extent of the hyporheic zone can dictate the location of hotspots for biogeochemical transformations [10,11]. Similarly, changes in hyporheic flux strongly constrain the mass and spatial distribution of reactants available for transformations within the hyporheic zone. Lastly, hyporheic residence times can serve as a proxy for the likelihood that a solute will be consumed during a biogeochemical reaction (e.g., denitrification), and therefore to quantify the HZ's biogeochemical potential [6,12]. This can be done by comparing the HZ's RTs with a characteristic timescale for the reaction of interest, typically defined as the reciprocal of a reaction rate constant [13,14].

In general, the hyporheic zone hydrodynamics and its associated characteristics are defined by the porous media properties (e.g., permeability [15], porosity, specific yield, and dispersivity [14]); modulators such as regional groundwater flow [16], pumping [17], and evapotranspiration fluxes [18]; and drivers such as pressure gradients induced by interactions with geomorphic features (e.g., bedforms and meanders) [11,19–22]. In this work, we focus on the case of lateral hyporheic exchange driven by channel sinuosity (exchange between the channel and its banks) [11,19,22,23] as it is modulated by regional groundwater flow and riparian evapotranspiration.

Near-stream vegetation (e.g., phreatophytes) and diel cycles of radiation influence the hydrodynamics of the HZ though the combined effect of transpiration (root uptake) and evaporation of surface water, soil water, and groundwater within the alluvial valley. The evapotranspiration process results in a diurnal cycle of alluvial valley water withdraw, where the local water table declines throughout the day and recovers at night [24]. Lowering of the water table induces additional hydraulic gradients from the channel to the alluvial valley, potentially enhancing the hyporheic exchange process [23,25]. Patterns of fluctuations similar to the riparian ET diel cycles are commonly observed in near-stream well hydrographs [26] and stream discharge measurements [27]. These fluctuations have been attributed directly to evapotranspiration (ET) in areas where other potential perturbations on water table fluctuations (e.g., well pumping and barometric pumping) were considered negligible [24,28].

Despite the effects riparian ET has on fundamental aspects of the HZ's hydrodynamics, relatively few numerical models of the HZ account for any amount of ET, and fewer have attempted to quantify the relationships between ET and metrics such as HZ area, exchange flux, and RTs. With this in mind, the goal of this work is to identify the magnitudes of ET withdrawal necessary to alter major characteristics of lateral reach scale hyporheic exchange, across a range of regional groundwater conditions. This goal is central to our ongoing effort to identify and maximize the potential for environmentally beneficial hyporheic activity along river corridors.

To meet our goal, we modeled and evaluated the effects of transient near-stream ET in a numerical model for lateral reach scale hyporheic exchange. To put the results in proper context, the effects of ET were compared with results from previous modeling studies (e.g., [11,14,19]) that have simulated a range of stream morphologies and ambient groundwater conditions to identify maximum values for HZ area, hyporheic exchange flux, and RTs.

Using a finite-element model, Cardenas [11,19,20] explored the relationship between sinuosity-driven hyporheic exchange, channel sinuosity, and regional groundwater flow. Their simulations showed that channel morphology, as represented by sinuosity, exerts a dominant control in the shape and fluxes within the HZ. The exchange through homogeneous meanders is characterized by broad power law residence time distributions (RTDs) extending across several orders of magnitude, from minutes to decades. This is an important finding that suggests prolonged effects of hyporheic exchange on the biogeochemical transformation potential at scales that range from the reach to the watershed [20]. Specifically, gaining or losing regional groundwater flow (RGF) conditions compress the HZs toward meander apexes and shorten residence times [11]. As sinuosity increases, the sensitivity to RGF decreases and steeper transverse regional water table gradients are needed to compress the HZ [11].

Boano et al. [13] used a similar modeling framework to make a direct connection between hyporheic flow, RTs, and biogeochemical transformation in the intra-meander zone. Hyporheic flow was simulated in meanders with a wide range of channel sinuosities, producing flow paths and residence time distributions that vary widely between scenarios. Then, a reactive solute transport model was used to simulate a sequence of redox reactions as a steady supply of dissolved organic carbon was supplied by the channel to the HZ. These biogeochemical simulations show the tight connection between residence time distributions and the potential for denitrification. In particular, Boano et al. [13] highlighted the usefulness of comparing threshold biogeochemical timescales (associated to the reaction of interest) with the HZ's residence times in order to quantify the effects of stream morphology on the biogeochemical potential of the HZ. Expanding upon this work, Gomez et al. [14] used the biogeochemical model from Boano et al. [13] with the numerical flow model from Cardenas [11,19] to establish numerical relationships between sinuosity, water table gradient, hydraulic conductivity, aquifer dispersivity, and RTDs. Keeping aquifer dispersivity constant, changes in the other aquifer and morphological parameters either stretched or compressed the RTD for each simulation. By comparing those RTDs with biogeochemical timescales [13], each set of control parameters could be used to classify a meander as a net sink or source of nitrate.

In an investigation of the relationship between riparian ET and streamflow diel cycles, Wondzell et al. [23] examined ET-induced fluctuations in streamflow at the mouth of a watershed. When compared to data from previous stream-tracer experiments performed throughout the watershed, they concluded that the combination of ET signals from local and watershed scales produced a nonlinear relationship between ET and streamflow. The study results also suggested that some observed water table fluctuations were in response to ET perturbation effects being transported along hyporheic flow paths. In other words, the effects of ET on hyporheic exchange can persist in the HZ beyond the time scale associated with the daily cycle. This is consistent with the memory effects found by Gomez-Velez et al. [29] in the context of flood event perturbations.

Efforts to characterize the effects of transient model boundaries on HZ exchanges and RTs, and in turn on the biogeochemical potential of HZs, have been limited and remain inconclusive. Gomez-Velez et al. [29] explored the effect of dynamic (or transient) perturbations due to flood events on lateral hyporheic exchange. These perturbations result in a wide variation in exchange fluxes and residence times. The variable perturbation in the model took the form of transient stream discharge, representing flood pulses of varying duration and intensity. Gomez and Wilson [30] drew similar conclusions after modeling transient flow in a generic flow system, where large enough fluctuations in the model perturbations create entirely new modes in the resulting distributions of residence time distributions. The results of Larsen et al. [31], however, suggest that dynamic perturbations from flood pulses and ET affect HZs on different spatial and temporal scales. The study ran multi-year simulations of hyporheic exchange, all with seasonal-timescale flood pulses and some with ET changing on a daily cycle. The ET boundary had different effects on HZ metrics at different temporal scales: on a monthly timescale, hyporheic exchange fluxes increased by several orders of magnitude; six-year mean residence times decreased by over half compared to no ET. The authors concluded that seasonal flood pulses produced separate characteristic RTDs from the ET effects taking place at shorter timescales.

The limited studies of transient conditions in hyporheic models leave some aspects of the research in a state of debate. Characteristic hyporheic RTs have been used to estimate the potential for biogeochemical transformation in HZs [13,32], but many studies have reported multimodal RTDs are inadequately described by a single characteristic RT [20,21]. Wondzell et al. [23] and Larsen et al. [31] suggest that transient model conditions produce complex multimodal distributions because they affect HZ transport and storage dynamics on multiple timescales. This is particularly important given than biogeochemical timescales for oxygen and nitrate consumption in meanders can vary over nine orders of magnitude from 10^{-2} to 10^6 days [14].

2. Materials and Methods

2.1. Conceptual Model

We used MODFLOW-2005 to build a finite-difference model of the hyporheic exchange process modulated by ET and regional groundwater flow (RGF) (Figure 1). Our conceptual model is based on the model previously proposed by Cardenas [11,19] and revisited by Gomez et al. [14] and Gomez-Velez et al. [29]. The generic domain represents one side of an alluvial valley overlying an impermeable surface. The valley is constrained on the x-y plane by the outer edge of the valley on one side and the valley's stream on the opposite side. The left and right (upstream and downstream, respectively) domain edges are periodic boundaries, i.e., they are boundaries beyond which the stream meanders repeat periodically and indefinitely. The model domain is vertically-integrated in a 2 m-thick layer, the stream fully penetrates that layer, and the alluvial aquifer is homogenous and isotropic. The stream channel is sinusoidal, with fixed head values varying linearly along the arc-length of the channel.

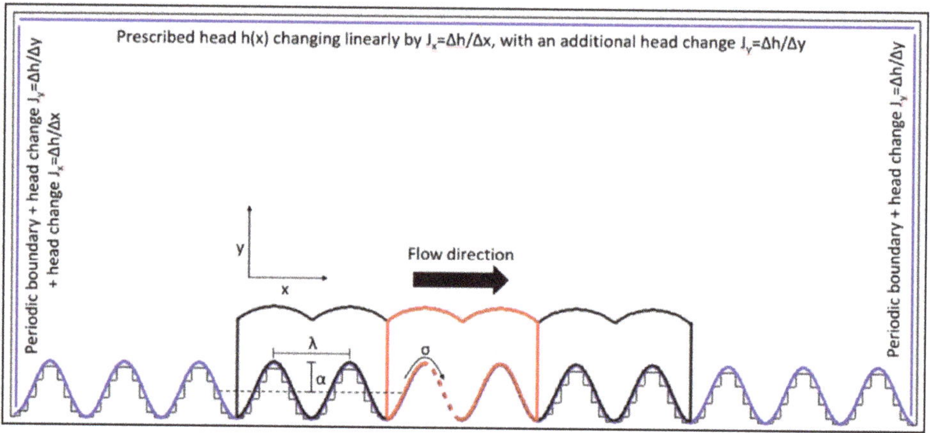

Figure 1. Overview of the finite-element model domain, with all boundary conditions labeled. The near-stream area outlined in black, encompassing the smaller area outlined in red, is the full extent of the vegetated riparian area, simulated with MODFLOW's EVT boundary condition. The area outlined in red is the area of interest pictured in proceeding figures. The dashed red line corresponds to the segment of the stream–aquifer interface along which MODPATH particles were released. Blue lines correspond to constant head (CHD) boundaries.

All four edges of the model domain were represented by the MODFLOW Time-Variant Specified Head package (constant head, or CHD). These constant heads varied linearly along a single border to produce the desired mean head gradients in the x- and y-directions and thereby control mean water table configurations before the addition of ET [11]. The near-stream ET was generated by MODFLOW's Evapotranspiration package (EVT) and the riparian vegetation was represented as a band of EVT cells running adjacent and parallel to the stream. The EVT cells ran on a daily cycle of hourly pumping rates (Figure 2), to reflect the observed sub-daily effects of solar radiation on groundwater ET [27]. Transport of stream water through the aquifer was modeled both as a conservative solute using the Modular Three-Dimensional Multispecies Transport Model (MT3DMS) [33], and as a series of discrete particles using the particle-tracking program MODPATH [34]. Observations from these transport software packages were used to calculate different metrics (described in Section 2.3) and did not interact or exchange data.

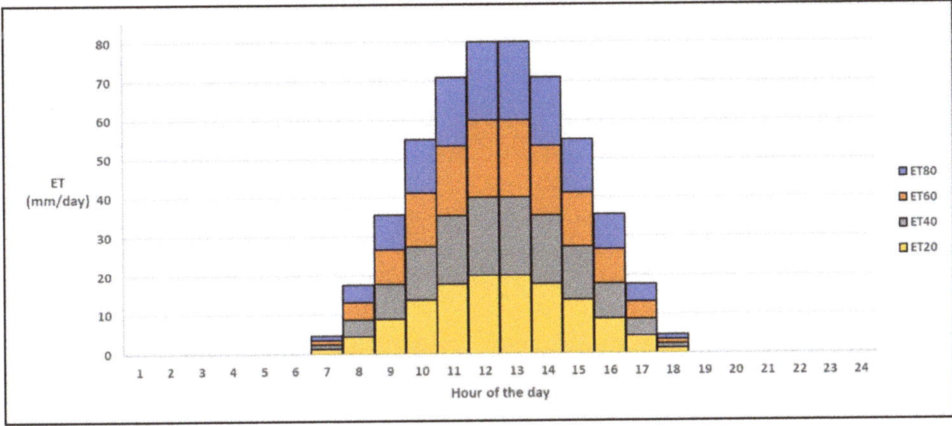

Figure 2. Daily schedules of evapotranspiration pumping rates in the vegetated zones of the model. Schedules are color-coded according to the maximum pumping rate (mm/day) achieved in each schedule, and labeled in shorthand (e.g., ET80 = maximum rate of 80 mm/day).

To represent an infinitely long stream and vegetated zone, the domain was expanded beyond the central 2 wavelengths of the stream that make up the representative area of interest (Figure 1, in red). Measurements of all metrics aside from HZ area come exclusively from this area. The top, right, and left edges of the domain (in plan view, Figure 1) were expanded until the total steady-state fluxes entering or leaving those edges during active ET (80 mm/day, the maximum withdrawal rate simulated) were within 5 percent of the same fluxes from simulations without active ET (see the description of scenarios below). Then, wavelengths with active ET zones were added upstream and downstream of the area of interest until, under transient conditions with daily ET cycles, drawdown and head within the area of interest experienced a maximum change of <5% due to the additional ET wavelengths.

The top boundary of the vegetated zone ET was established in a way that places it at least 30 m away from the stream at all points along the stream. This was done to maintain a more consistent thickness to the vegetated zone ET, in sync with the sine wave of the stream, than would have been produced with a straight line running along the x-axis of the domain. The 30 m vegetation width is consistent with recommendations from the USDA [35] for riparian buffers. A more detailed description of how this boundary was created in ModelMuse [36] is provided in the Supplementary Materials.

Some aspects of the model were chosen with the goal of avoiding numerical instability and water budget imbalances in the finite-difference solution. The model's x-y grid was initially given a regular cell size of 4 × 4 m, but the cells in the area of interest were refined to 0.4 × 0.4 m, with several rows and columns of transitional cells in-between. This gives the model over 47,000 active cells total and over 22,000 active cells in the area of interest alone. The MODFLOW Geometric Multigrid (GMG) solver package was chosen because it kept water budget discrepancies below 0.1% for all steps in all simulations. All solver packages solve the same equations for groundwater flow, but with different algorithms that allow some to arrive at satisfactory solutions when others will not [37].

2.2. Modeling Scenarios

Five different scenarios for ET withdrawal rates were applied, based on the daily maximum ET rate (ET_{max}): 0, 20, 40, 60, and 80 mm/day. These scenarios will be abbreviated from here onward as ET0, ET20, ET40, ET60, and ET80, respectively. These values cover the full range of ET activity found in the literature [26,27,38,39]; but it should be noted the ET withdrawal rates produced in the ET80 scenarios will be impossible for phreatophytes to achieve in most climate zones. The mean head gradient of the aquifer in the y-direction (J_y) was also altered between simulations to produce different scenarios of ambient groundwater flow. The ratio of the regional groundwater flow gradient in the y-direction (J_y) to the one in the x-direction (J_x) is given by $J_{y/x} = J_y/J_x$. This ratio represents the magnitude of the regional groundwater flux constraining the hyporheic exchange, and it is positive for gaining conditions, negative for losing conditions, and zero for neutral conditions. We explored five different scenarios for $J_{y/x}$, with values of 2, 1, 0, −1, and −2. These scenarios will be abbreviated from here onward as J + 2, J + 1, J0, J − 1, and J − 2, respectively. Most model parameters were kept constant across all simulations, to reduce the total number of simulations and simplify the analysis of the parameters that were changed. The main model constants can be found in Table 1.

Table 1. Constant values for major geomorphic and aquifer parameters.

Parameter	Symbol and Units	Constant Value
Sinuosity	S, (-)	1.87
Wavelength	λ, m	40
Down-valley water table gradient	J_x, (-)	0.00125
Hydraulic conductivity	K, m/h	3.5
Porosity	φ, (-)	0.25
Specific yield	S_y, (-)	0.20
Longitudinal dispersivity	α_L, m	10

The EVT cells were set with extinction depths at the bottom of the model layer (2 m) and withdrawal coming from the top of the model layer. As a result, the real withdrawal rate of an active EVT cell was a fraction of the maximum possible withdrawal rate of a given time step (the hly values of pumping rates in Figure 2), with the fraction being proportional to the head in the cell:

$$ET_R = ET_P \times (h/2) \tag{1}$$

where ET_R is the real withdrawal rate (mm/day) of a cell, ET_P is the maximum potential withdrawal rate (mm/day) for that time step, and h is the head (m) in the cell. An extinction depth of 2 m is a plausible value for riparian phreatophytes, based on ranges of values produced in previous modeling studies [40–42].

Each simulation was run with one steady-state flow step, one steady-state solute transport step, and then 744 steps (31 days) of hly ET and transient solute transport (Table 2). Representative measurements of stream water concentrations and flow terms needed for area and flux metrics were taken from the 31st day of the transient simulation. The particle tracking used for residence time calculations ran beyond the first 31 days for reasons described below. Post-processing of model data was completed with a combination of tools from ModelMuse and scripts written in R [43]. For more detailed explanations of the model setup and execution, see the Supplementary Materials.

Table 2. The changing status and activity of the MODFLOW, MT3DMS, and MODPATH software through a simulation's time steps.

Start Time Step	End Time Step	MODFLOW	MT3DMS	MODPATH
−1	0	Steady-state	Inactive	Inactive
0	1	Transient	Steady-state	Inactive
1	720	Transient	Transient	Inactive
720	721	Transient	Transient	Particles released
721	744	Transient	Transient	Particles travel
744	variable	Transient	Transient	Particles travel

2.3. Characterization of the Hyporheic Exchange

The areal extent of the HZ was evaluated using a geochemical definition proposed by Triska et al. [44] and used by Gomez-Velez et al. [29], where the HZ is the area within the aquifer composed of more than 50% stream water. The geochemical definition of the hyporheic zone has been identified in previous research [13,14,32] as an effective method of predicting the potential for biogeochemical reactions. At the beginning of each simulation the stream water, with an initial concentration of 1 g/m^3, was released from the upstream half of the central wavelength of the stream into the aquifer with an initial concentration of 0. At the end of each simulation, each cell with a final stream water concentration greater than 0.5 g/m^3 was added to the HZ area (A_{HZ}). The percent increases in A_{HZ} relative to the corresponding ET0 scenarios (i.e., with the same $J_{y/x}$) were also calculated.

The active ET in this model acted as a sink distributed across a wide band of cells, some of them up to 30 m away from the stream. Two flux metrics were developed to provide a look at fluxes throughout that distributed sink as well as fluxes exchanged along the stream–aquifer interface. The first metric was a net flux term in the y-direction (Q_y), evaluated cell by cell with positive terms flowing toward the stream. Cross-sectional profiles of Q_y values in the central wavelength were plotted for direct comparison between simulations.

The second flux metric was a normalized dimensionless flux (F) describing exchange along the stream–aquifer interface. On the 31st day of each simulation, a daily average flux was calculated for the stream–aquifer interface of the central two stream wavelengths using ZoneBudget. This daily average flux was made dimensionless (Q^*):

$$Q^* = Q_{ave}/(Kh_c{}^2) \tag{2}$$

where Q_{ave} is the calculated daily average flux (m^3/h), K is hydraulic conductivity (m/h), and h_c is a characteristic head value (1 m). Then the Q^* terms were normalized to the ET0 simulation with the same $J_{y/x}$:

$$F = Q^*/Q_0 \tag{3}$$

where F is the normalized dimensionless flux and Q_0 is the dimensionless daily average flux from the respective ET0 simulation.

Because the model was transient and there were changes in model storage, aquifer residence times (RTs) were estimated with a Lagrangian approach, which involved tracking individual particles through time as they traveled through the area of interest. Along the upstream half of the central wavelength, a particle was placed on every cell face along the stream–aquifer interface at a depth of 1m, for a total of 125 particles. Before release, MODFLOW was run with a wind-up time that replicated the conditions imposed on the model for the other metrics: one steady-state flow step with no ET and then 720 transient time steps (30 days) with daily cycles of ET withdrawal. After this wind-up time, the particles were released and MODFLOW was run with the same daily ET cycle until all particles had exited the area of interest. Sets of particles were released during the 1st, 6th, and 12th hours of the 31st day, to explore the sensitivity of the metrics described below to particle release time.

The RT (h) of each particle was weighted according to the volumetric fluxes of stream water entering the HZ at the time the particles were released, along the section of the stream–aquifer interface where particles were released. Flux-weighted RT values (RT_{FW}, h) are defined as:

$$RT_{FW} = RT^*(Q_P/Q_T) \tag{4}$$

where Q_P is the flux represented by a particle (m^3/h) entering the aquifer at the cell face corresponding to the starting location of that particle, and Q_T is the total flux of stream water entering the aquifer (m^3/h) along all cell faces on the upstream half of the central meander bend. Flux-weighted RTs were divided by a characteristic timescale of 24 h to produce dimensionless RT (RT^*) values reflecting the 24 h-long cycle of ET withdrawal. For each simulation, these dimensionless values were arranged in histograms, cumulative distribution functions (CDFs), and plotted as a function of dimensionless arc-length distance (σ^*):

$$\sigma^* = \sigma/\sigma_T \tag{5}$$

where σ is the distance (m) along the arc-length of the upstream half of the central meander bend, increasing when moving downstream, and σ_T is the total arc-length of the upstream half of the bend. Median RT^* values from all scenarios were plotted once as a function of $J_{y/x}$ and ET_{max}, and then median RT^* values from active ET scenarios were normalized to their respective ET0 scenarios and plotted again.

To provide mechanistic explanations of how hydraulic head influences the above metrics, contour maps of drawdown and head were interpreted in terms of how they affected the velocity and direction of groundwater flow paths relative to ET0 scenarios. Snapshots of drawdown and head contours were taken on the 31st day of the five ET80 simulations. On the head contour maps, the general orientation of the contours was estimated by drawing a straight line between two points: the top apex of the left meander bend in the area of interest and the cell with the lowest head value in the area of interest (Figure 1). The orientation of this line was recorded as an azimuth, or degrees clockwise from north. The average head gradients of these lines were also calculated. For each of the five simulations, these orientations and gradients were compared between the 6th and 12th h snapshots. These metrics were used to compare conditions just before the daily cycle of active ET with conditions while $ET = ET_{max}$. The patterns in the drawdown maps were not quantified but used to support conclusions drawn from other metrics.

Time sensitivity was quantified using earlier methods. The ET in this study was expected to affect the stream–HZ–alluvial aquifer system similar to the use of flooding events in Gomez-Velez et al. [29], in that dynamic ET rates would produce dynamic responses in all of the above metrics. The intensity of change in these responses depend on the sensitivity of the aquifer to the daily cycles of ET withdrawal, and this sensitivity was quantified with an adapted version of a hydraulic time constant (t_h) used in Gomez-Velez et al. [29]:

$$t_h = (S_y \lambda^2)/(Kb) \tag{6}$$

where S_y is specific yield, λ is meander wavelength, and b is aquifer layer thickness. The hydraulic time constant was compared to the time scale of one cycle of ET withdrawal (24 h) to establish the relative importance of ET perturbations on the near-stream aquifer:

$$\Gamma = t_h/24 \tag{7}$$

If $\Gamma < 1$, the aquifer was considered insensitive and the water table configuration would likely not change in response to the daily cycle of ET withdrawal; if $\Gamma > 1$, the opposite was true.

3. Results

3.1. Hyporheic Zone Area

Compared to their base ET0 simulations, all regional groundwater flow (RGF) scenarios (intensity of RGF is proportional to $J_{y/x}$) showed relative increases in the hyporheic zone area (A_{HZ}) during simulations with active ET (Figure 3). The greatest relative increases were found in gaining scenarios ($J_{y/x} > 0$), where A_{HZ} increased over 100% from ET0 to ET80. The strongest losing scenario ($J_{y/x} < 0$) had markedly smaller growth, with just over a 6% increase from ET0 to ET80.

Active ET simulations also showed altered HZ geometry. For gaining and neutral scenarios, the HZ expanded primarily along the y-axis of the domain. In losing scenarios, any growth or movement of the HZ was along the x-axis (see the fourth and fifth rows in Figure 3). After the steady-state transport step in each simulation, the shapes and sizes of the HZ did not change on daily timescales, and so only a single final snapshot was taken from each simulation.

3.2. Net Groundwater Flux

For the net flux in the y-direction (Q_y), gaining scenarios were the only scenarios to show any movement in the divide between positive terms representing net flux away from the stream and negative terms representing net flux toward the stream (Figure 4a,b top). Increasing ET rates, whether through the hly progression of one simulation or between simulations with different ET_{max}, corresponded to the divide moving away from the stream and a greater fraction of the active ET zone showing a net flux away from the stream. Although this divide did not move in neutral and losing scenarios, these scenarios' profiles of discrete Q_y values (Figure 4c–e, bottom) still showed greater flux away from the stream in scenarios and time steps with higher ET rates. This increased flux away from the stream could be seen in lower average values, lower minimum values, and lower values at the stream–aquifer interface (A' location on each profile). Of those three indicators, the values at the stream–aquifer interface may be the most crucial because they were direct measurements of exchange between stream water and hyporheic water. On average, when comparing ET0 scenarios to corresponding 12th h ET80 scenarios, Q_y values at the stream–aquifer interface decreased by 117%, with a maximum difference of 11% between $J_{y/x}$ scenarios. All active ET scenarios displayed increased flux away from the stream regardless of ET_{max}, with the only difference being that smaller ET_{max} corresponded to smaller Q_y values. To simplify the snapshots the ET0 scenarios were only compared to ET80 scenarios. Q_y values from the last 12 h of the daily ET curve were identical and symmetrical to those from the first 12 h, so only snapshots from the first 12 h are shown.

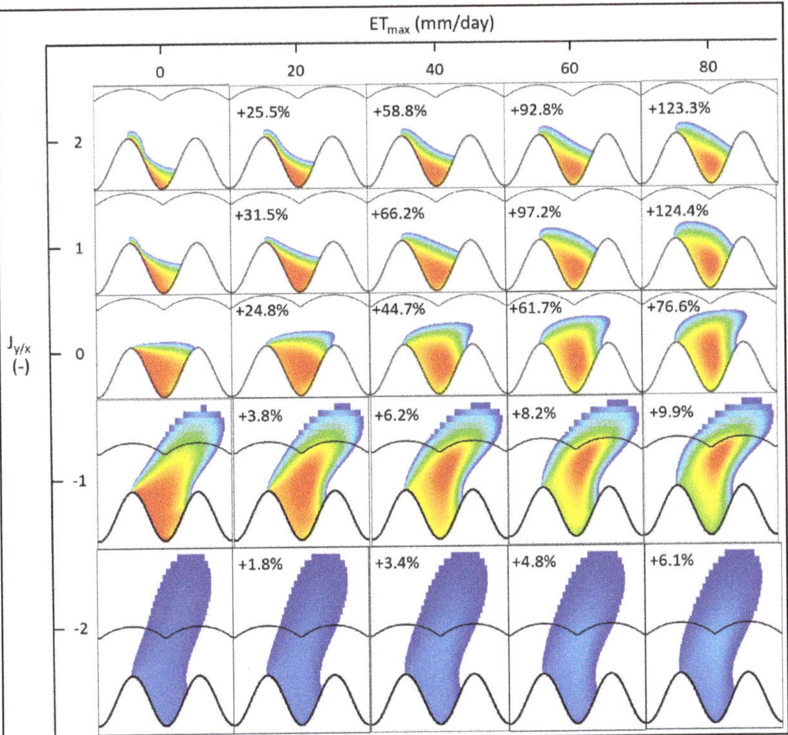

Figure 3. Maps of hyporheic zone area (A_{HZ}) as a function of maximum daily ET (ET_{max}) (columns) and regional groundwater flux ($J_{y/x}$; positive for gaining, negative for losing) (rows). Each map is a snapshot taken at the 12th h of the last day of the simulation. The percentages listed are the percent increases in A_{HZ} relative to the corresponding ET0 scenarios (left column). The colors on each map correspond to concentrations of stream water that have entered the aquifer; the range of concentrations are different for each simulation and so the colors cannot be compared between maps. The colors follow a rainbow gradient where blue indicates lower concentrations and red indicates higher concentrations.

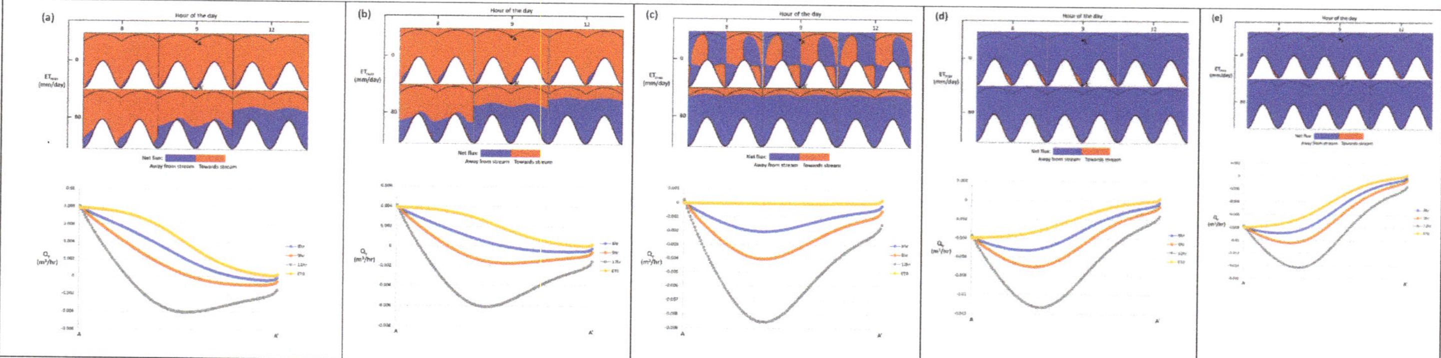

Figure 4. Snapshots of net Q_y values at different h of the day (**top row**), as a function of maximum daily ET and regional groundwater flux ($J_{y/x}$; positive for gaining, negative for losing). Profiles of the above Q_y (positive towards the stream) values along the middle of the vegetated area of interest (**bottom row**). Note that the no-ET scenario is unchanging with time and plotted only once on each profile. Columns correspond to simulations where the magnitude of $J_{y/x}$ was, respectively, (**a**) 2, (**b**) 1, (**c**) 0, (**d**) −1, and (**e**) −2.

Normalized dimensionless flux (F) increased across all $J_{y/x}$ scenarios when directly comparing ET0 simulations with their respective ET80 simulations (Figure 5). In other words, all ET80 simulations produced $F > 1$. For neutral and losing scenarios, all increases in ET_{max} corresponded with increases in F, and ET80 F values fall between 1.5 and 1.7. Gaining scenarios produced $F < 1$ at lower ET_{max} and eventually $F > 1$ starting at ET60 for J + 1 and only at ET80 for J + 2. The lowest F values, both overall (0.87) and on average (0.94), were produced in J + 2 scenarios.

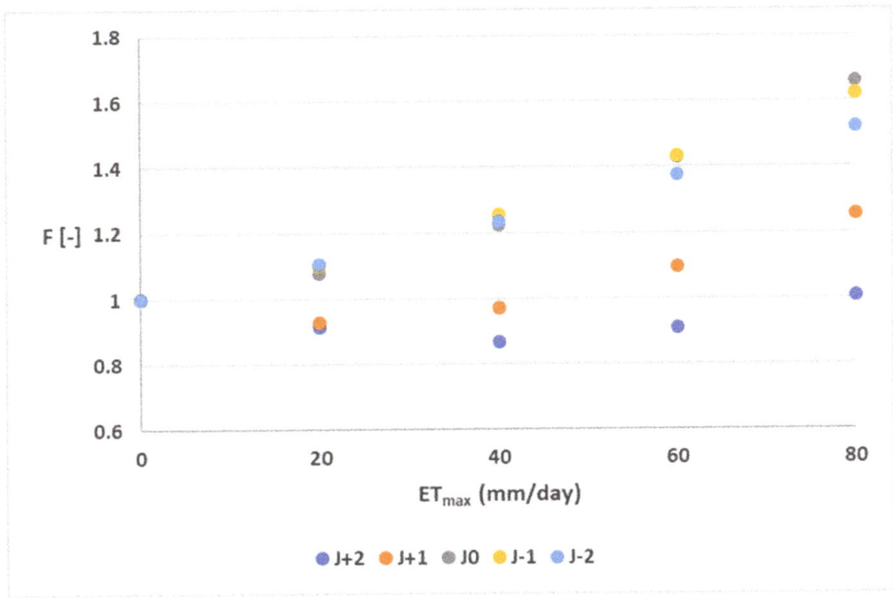

Figure 5. Normalized dimensionless flux (F) for each regional groundwater flux ($J_{y/x}$; positive for gaining, negative for losing) scenario, plotted as a function of ET_{max}.

3.3. Residence Time

For all simulations, the difference in median RT^* between particle release times was less than 5%. The RT^* values used for analysis came from the particles released during the 1st h of the 31st day of each simulation.

The residence time distributions (RTDs) displayed in the histograms and cumulative residence time distributions (CRTDs) in the CDFs developed unique shapes in response to both $J_{y/x}$ and ET_{max}. Gaining simulations displayed a strong early mode and long tail (Figure 6a,b), and this basic RTD shape did not change between ET0 and active ET scenarios. Neutral simulations produced a similar early mode distribution with no ET (Figure 6c, top), but with a less severe peak and a more gradual decline in the tail. With active ET, the early mode remained but a larger number of particles clustered around a range of high RT^* values, eventually producing a smaller secondary mode in that range at the highest ET_{max} (Figure 6c, bottom). Compared to ET0, median RT^* and standard deviation more than doubled in ET80. Progressing from ET0 to ET80, losing scenarios (Figure 6d,e) began with large early modes that shifted to higher and higher RT^* values, eventually settling close to the median RT^*. J − 2 scenarios (Figure 6e) began with several smaller secondary modes that shrank and disappeared as ET_{max} increased. Active ET decreased standard deviations and compressed RTDs for all losing scenarios.

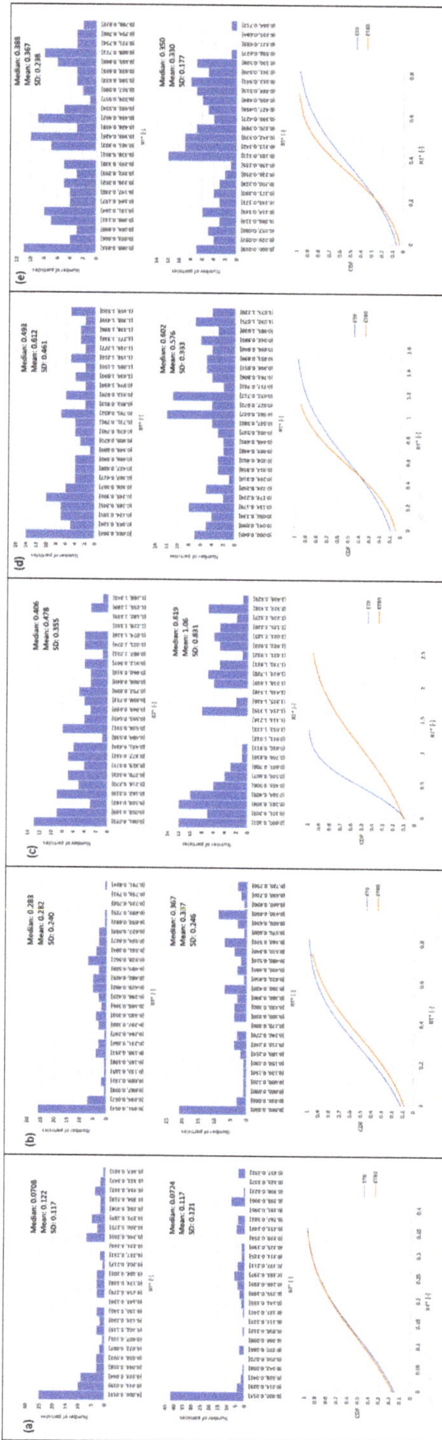

Figure 6. Residence time distributions (RTDs) from no-ET (ET0) scenarios (**top row**). RTDs from scenarios where maximum daily ET (ET_{max}) is 80 mm/day (**middle row**). Comparison of cumulative RTDs from ET0 and ET80 scenarios (**bottom row**). Columns correspond to simulations where the magnitude of regional groundwater flux ($J_{y/x}$; positive for gaining, negative for losing) was, respectively, (**a**) 2, (**b**) 1, (**c**) 0, (**d**) −1, and (**e**) −2.

Changes in median RT^* as a response to ET_{max} corresponded to the absolute magnitude of $J_{y/x}$. As ET_{max} increased, intensely gaining and losing scenarios ($|J_{y/x}| = 2$) produced mostly decreases in median RT^*, and less intense and neutral scenarios ($|J_{y/x}| = \{0, 1\}$) produced mostly increases in median RT^* (Figure 7). The largest changes in median RT^* were over 20%: ET40 to ET60 and ET60 to ET80 in neutral scenarios produced >20% higher median RT^*, and ET20 to ET40 in strongly gaining (J + 2) scenarios resulted in >20% lower median RT^*.

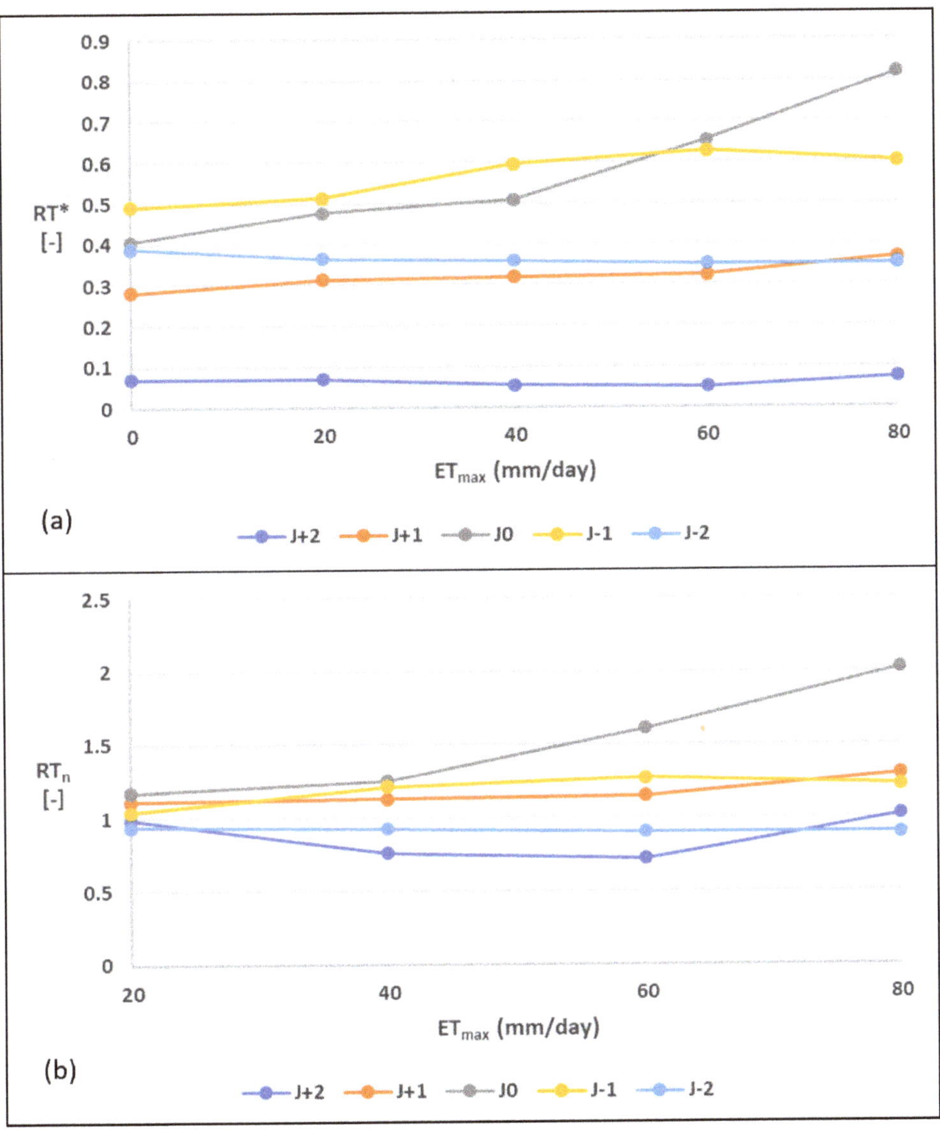

Figure 7. (**a**) Median dimensionless RT (RT^*) for all regional groundwater flux ($J_{y/x}$; positive for gaining, negative for losing) scenarios as a function of ET_{max}. (**b**) Median RT^* values from active ET scenarios, normalized to their respective ET0 simulations.

Regional water table slope also affected the spatial distribution of RTs along the stream–aquifer interface. Gaining scenarios (Figure 8a,b) produced near-normal distributions of RT^*, with the highest values close to $\sigma^* = 0.5$ and the lowest values close to $\sigma^* = \{0, 1\}$. The greatest changes in RT^* from ET0 to ET80 occurred near $\sigma^* = 0$, with decreases of over 90% in J + 2 and increases of over 2000% in J − 1. Neutral scenarios (Figure 8c) had comparatively flat spatial distributions, with slightly higher RT^* closer to $\sigma^* = 0$. Nearly all RT^* values increased from ET0 to ET80, with maximum increases of over 300% around $\sigma^* = 0.65$. With no ET, losing scenarios (Figure 8d,e) displayed two spatially distinct modes of RT^*, where average RT^* was about four to five times higher where $\sigma^* < 0.5$ compared to $\sigma^* > 0.5$. At ET80, the high RT^* modes dropped and the low RT^* modes rose to produce a flat spatial distribution like the ones seen in neutral scenarios. The greatest changes in RT^* were from increases in RT^* where $\sigma^* < 0.5$; these increases were >300% in J − 1 and >200% in J − 2.

The point-to-point noise on the spatial distribution graphs is due to the fine-scale jagged edge of the stream–aquifer interface. This resulted in alternating groups of particles being initially placed on either side of the ideal, perfectly sinuous stream–aquifer interface being approximated by the model domain. This, in turn, produced RTs based on flow paths that are slightly too long or too short. To reduce some of this noise and make the spatial distributions more legible, the values plotted are moving five-point averages of the original RT^* values.

3.4. Drawdown and Head

For all scenarios, the maximum observed drawdown was close to 0.1 m and centered on the top edge of the vegetated zone farthest from the stream (Figure 9), indicating the stream was the major source of recharge to the riparian aquifer. If the ET zone was drawing in a similar quantity of water from the alluvial valley, maximum drawdown values could be expected along the center of the ET zone, equidistant from the stream and alluvial valley.

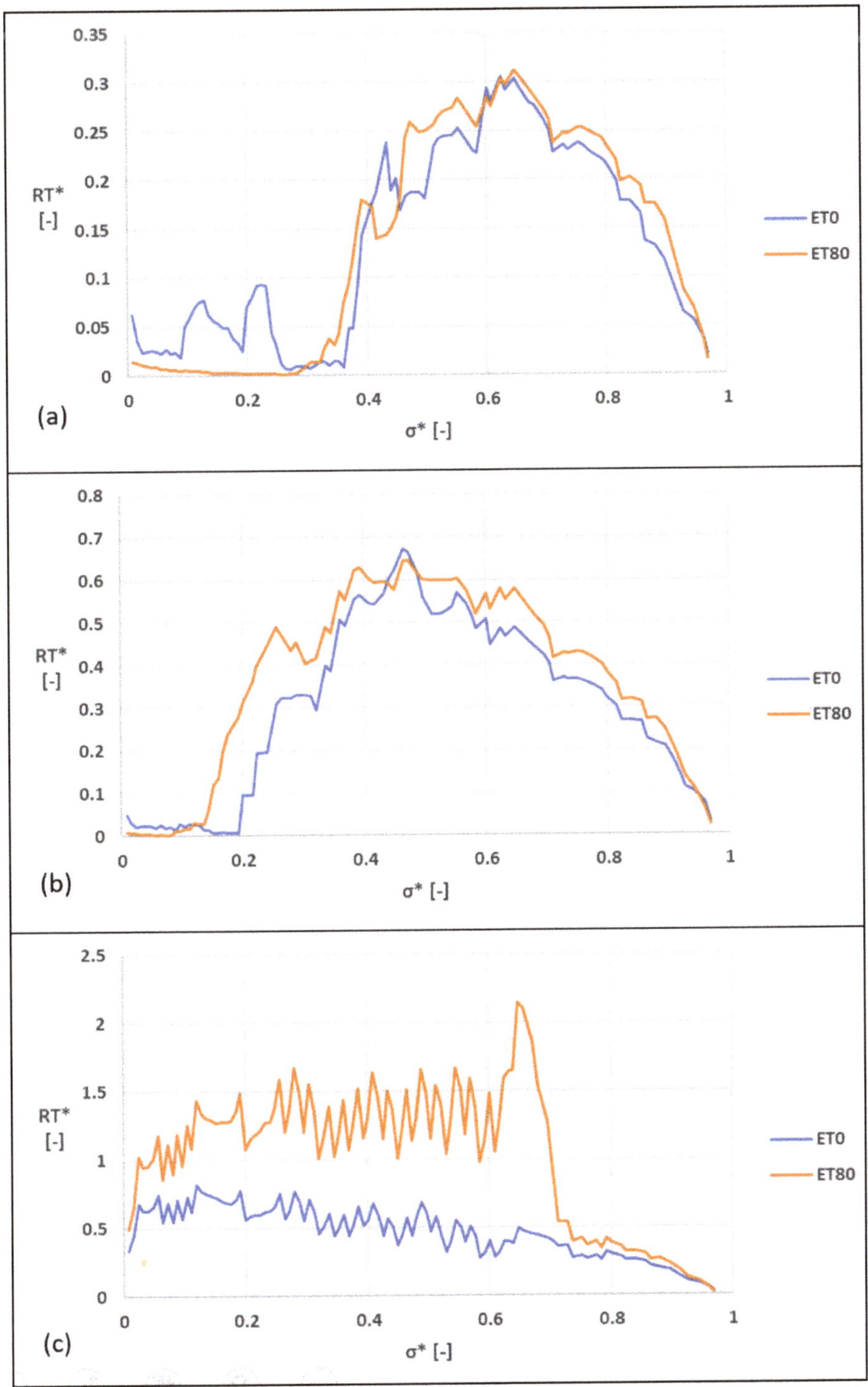

Figure 8. *Cont.*

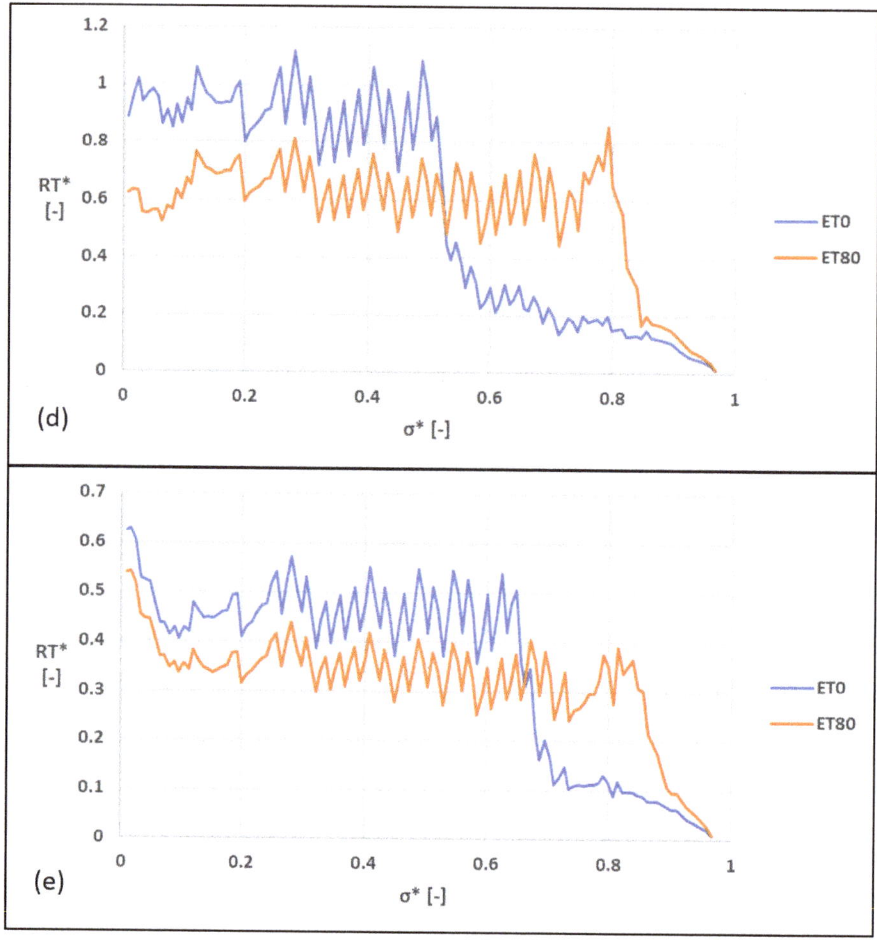

Figure 8. Moving 5-particle average *RT** as a function of regional groundwater flux ($J_{y/x}$; positive for gaining, negative for losing) and initial particle placement along the upstream half of the central meander bend. Comparing ET0 and ET80 scenarios for (**a**) J + 2, (**b**) J + 1, (**c**) J0, (**d**) J − 1, and (**e**) J − 2.

For the head distributions (Figure 10), increased ET rates resulted in contour lines in the ET zones orienting themselves away from the stream and forming steeper gradients in some locations (see Table 3). Contour lines in gaining and neutral scenarios reoriented so the lowest head value in the area of interest moved away from the stream. In losing scenarios, the lowest head value in the area of interest did not reorient at all relative to the point placed at the meander apex. All scenarios produced a steeper head gradient when ET was active, with neutral and losing scenarios producing slightly higher changes in gradient.

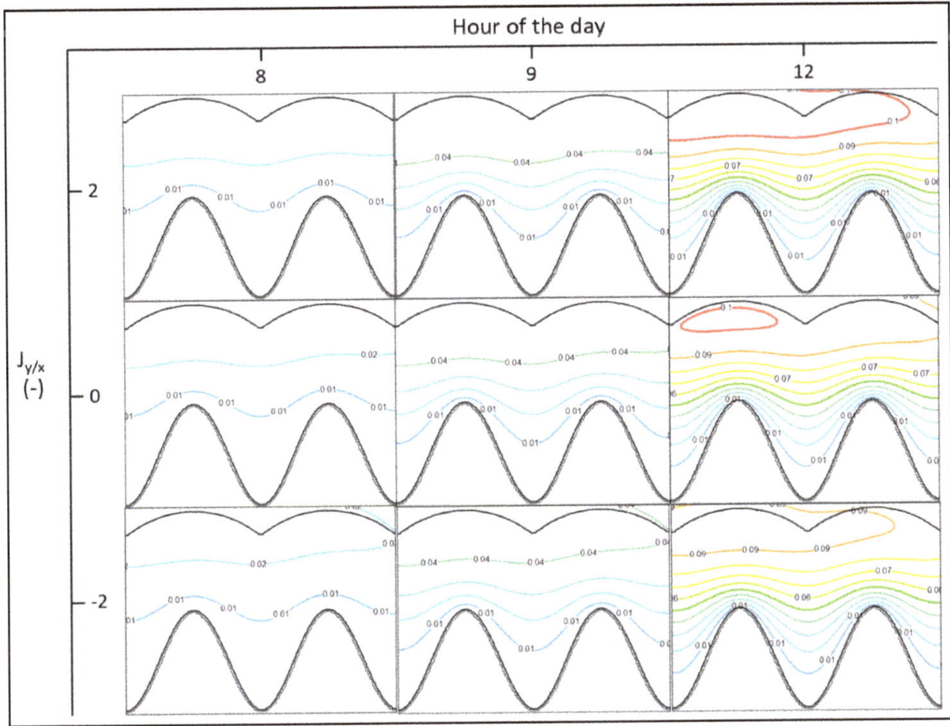

Figure 9. Snapshots of drawdown (m) at different h of the day, as a function of regional groundwater flux ($J_{y/x}$; positive for gaining, negative for losing) (rows). All snapshots came from simulations with ET_{max} values of 80 mm/day. Drawdown values are relative to the head values from the first h of the day of each respective simulation.

Table 3. Contour line orientations and average head gradients for all regional groundwater flux ($J_{y/x}$) scenarios at ET80. All delta values are 12th h conditions −6th h conditions.

$J_{y/x}$	Head Contour Azimuth, °			Head Gradient, %		
	6th h	12th h	Δ	6th h	12th h	Δ
2	117	82	35	0.115	0.179	0.064
1	117	75	42	0.114	0.213	0.099
0	96	70	26	0.125	0.253	0.128
−1	65	65	0	0.185	0.307	0.122
−2	65	65	0	0.256	0.376	0.120

The reoriented head contours formed unique patterns depending on the $J_{y/x}$ scenario. In gaining and neutral scenarios, the contours formed a trough of lower heads parallel to the main axis of the stream; losing scenario contours did not form a trough but compressed towards the stream without a major change in their orientation. When $ET = ET_{max}$, the greatest steepening of head gradients was consistently at the apexes of meander bends where the distance between the stream and maximum drawdown values was smallest. In Figures 9 and 10, only ET80 snapshots are shown, since all active ET scenarios displayed similar patterns. Likewise, $J + 1$ and $J − 1$ scenarios were omitted because they showed trends like those of $J + 2$ and $J − 2$, respectively.

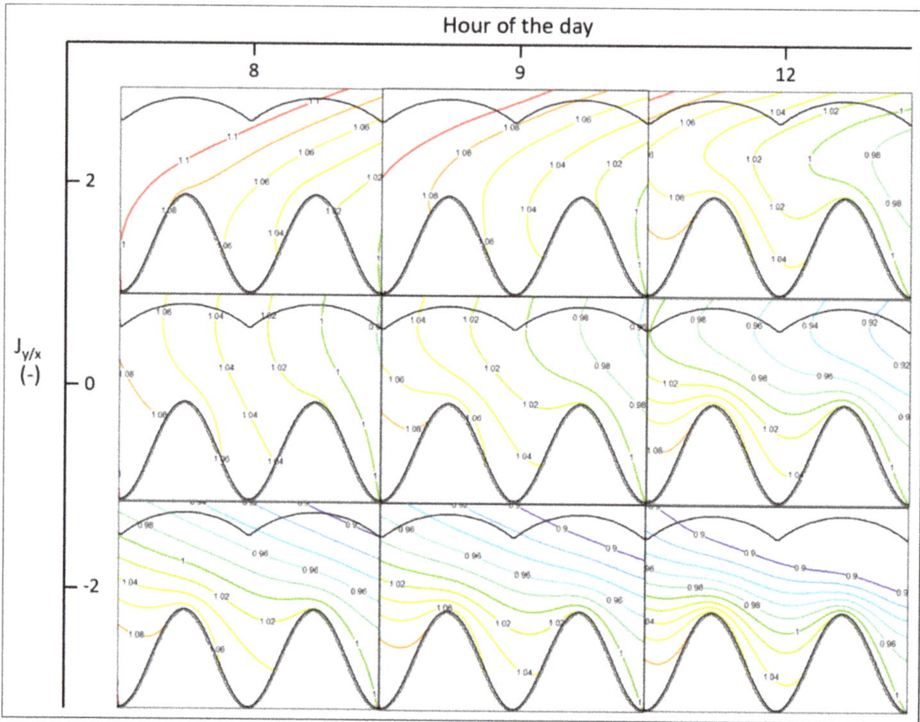

Figure 10. Snapshots of head (m) at different h of the day, as a function of regional groundwater flux ($J_{y/x}$; positive for gaining, negative for losing) (rows). All snapshots came from simulations with ET_{max} values of 80 mm/day.

3.5. Aquifer Sensitivity

For all model simulations, the aquifer sensitivity parameter Γ was constant at 1.9. This means the duration of effects produced by a daily cycle of ET was almost twice as long as the daily cycle itself. With $\Gamma > 1$, perturbations in the water table from ET were not expected to significantly affect the aquifer at sub-daily timescales [28].

4. Discussion

4.1. Hyporheic Zone Area

The patterns of changes in HZ area sizes and shapes can be explained as the result of active ET altering head distributions and flow paths. In gaining and neutral scenarios, ET temporarily reoriented groundwater flow paths away from the stream and towards the trough of low heads that developed. This means groundwater was flowing away from the stream for a period of h every day, compared to ET0 scenarios where flow paths toward the stream were constant. This change was what gave gaining and neutral HZs larger areas growing primarily in the y-direction when ET was active. Losing scenarios experienced much smaller changes in the orientation of head contours, and so their HZs did not undergo much growth or change in shape.

The width of the vegetated belt (30 m) also had a strong influence on HZ geometry because it controlled how head distributions were reoriented. The vegetated belt, represented in this study by the cells with active ET, was wide enough to reach into the alluvial valley well beyond the edge of the stream. This was where troughs of low head values formed in gaining and neutral scenarios, and therefore where stream water particles were drawn during h of the day with high ET rates. Preliminary tests of the model domain, not detailed in this study, found that a much narrower vegetated belt led to smaller A_{HZ} values relative to no-ET scenarios. The preliminary model had identical values for all the model constants listed in Table 1, but a vegetated belt width of only 2 m. Water table depressions from these ET cells were focused in a narrow band close to the edge of the stream, and stream water particles entering the aquifer tracked more closely to the stream and exited the aquifer more quickly than without ET.

With the addition of ET, HZ areas grew on a scale comparable to the changes produced in other studies by varying basic geomorphic and aquifer parameters and/or adding a dynamic perturbation like ET. In identically-shaped ($S = 1.87$, $\lambda = 40$ m) intrameander zones, Cardenas [12] reduced HZ areas by about 80% when increasing and decreasing $J_{y/x}$ (J0 base case, increased to J + 2 and decreased to J − 2). At higher ET_{max}, the neutral and gaining scenarios in this study increased the HZ area close to or surpassing that amount. Gomez-Velez et al. [29] mapped the growth of HZs in response to flood pulses across a range of aquifer sensitivities (Γ values). Comparing those maps to the growth seen in this study, the results of gaining and neutral scenarios were similar to what Gomez-Velez et al. [29] produced in less sensitive ($1 < \Gamma < 10$) simulations.

With $\Gamma > 1$ for this study, A_{HZ} was insensitive to hly changes in ET because the hydraulic signal produced by ET took much longer than one h to propagate through the area of interest (i.e., $t_h > 1$ h). Keeping $J_{y/x}$ constant, it is also clear the HZ did respond to changes in ET_{max} between simulations. In the field, ET_{max} is primarily dependent on the intensity of solar radiation, especially in energy-limited environments like riparian zones [27,38]. This is also true for the h-to-h changes in ET withdrawal rates. Daily maximum solar radiation follows long-term cyclic (seasonal) fluctuations [38,39], so it follows that active riparian ET could affect A_{HZ} on primarily seasonal timescales.

4.2. Net Groundwater Flux

Like A_{HZ}, net flux in the y-direction (Q_y) and normalized dimensionless flux (F) were dependent on the head patterns that developed with active ET. Active ET rearranged head contours so that more water was drawn from the stream into the aquifer, regardless of whether the simulation was originally gaining or losing. All profiles of discrete Q_y values (Figure 4, bottom) captured this: values decreased across the profile, and gaining scenarios showed some cells reversing from positive to negative. When ET = ET_{max}, the divide between positive and negative Q_y cells coincided with the troughs of low head values that developed in gaining scenarios.

In gaining scenarios with lower ET_{max}, F values dropped below 1 because daily average flux from the aquifer to the stream decreased more than daily average flux from the stream to the aquifer increased (Figure 5). At higher ET rates, both trends continued, but eventually, the decrease in flux to the stream was outpaced by increases in flux to the aquifer, resulting in more total exchange flux than at ET0 ($F > 1$). Looking at gaining scenarios only, the ET rates at which F dropped below 1 correspond to the ET_{max} values for simulations where head contours either did not reorient themselves beyond the x-axis or did so only briefly. In other words, even during ET_{max} in the ET20 and ET40 gaining simulations, head contour trendlines never produced an azimuth below 80 degrees (Table 3). Neutral and losing scenario F values only increased with ET because their head gradients did not reorient as strongly or at all in response to ET. These results support the idea that increases in F were due to increases in exchange flux from the stream to the aquifer. The process captured here by which ET increased net flux to the aquifer has been observed in field studies of lateral hyporheic exchange [23,25], suggesting the ET in this model functioned in a way comparable to conceptual models developed from field data.

Increasing ET did draw more water into the ET zone from the alluvial valley, but it was less than the water drawn in from the stream. Comparing ET0 results with ET80 results for all $J_{y/x}$ scenarios, active ET produced an average 0.024% increase in volumetric flux from the alluvial valley, measured at the A location cell in the Q_y profiles. Using the same scenarios, active ET produced an average 0.17% increase in flux from the stream, measured at the A' location cell.

The flux metrics developed in previous modeling studies focused on the exchange at the stream–aquifer interface, and therefore can be more easily compared to F than Q_y. In general, previous flux metrics were a variation on a Darcy flux term, where the total or average volumetric flux entering or leaving the HZ was divided by the surface area of the stream–aquifer interface. Some of these Darcy fluxes were then made dimensionless by dividing by hydraulic conductivity and two characteristic length terms (see Equation (2)). Each study depicted a slightly different hyporheic exchange flux; but the patterns the studies observed in response to changes in stream morphology, aquifer characteristics, and transient perturbations can still be compared to the effects of ET in this study.

In Cardenas [12], a dimensionless hyporheic flux term was established by dividing the total Darcy flux entering the HZ by hydraulic conductivity. This made the term insensitive to changes in flux leaving the HZ, which this study's F term accounted for. Since the dimensionless values were then normalized to neutral ($J_{y/x} = 0$) scenarios, however, it still describes how a part of the hyporheic exchange flux would respond to changes in ambient groundwater flow. Cardenas [12] showed exponential decreases in their flux term with increases in $|J_{y/x}|$. At $|J_{y/x}| = 2$ and $S = 1.87$, flux was about 70% reduced; in this study, changes in F ranged from -15% to +65%.

Gomez et al. [14] developed two interrelated metrics: a dimensionless Darcy flux normal to the stream, and a dimensionless volumetric exchange flux per unit stream length. Both metrics were used to describe flux leaving the HZ and entering the stream. Since ambient groundwater flow was kept neutral ($J_{y/x} = 0$) in this model, these flux terms captured all exchange at the stream–aquifer interface on the downstream half of the meander bend of interest, and this outgoing flux was symmetrical to the incoming flux on the upstream half of the meander bend. As stream sinuosity (measured here as the ratio of meander amplitude to wavelength) was increased from 1/16 to 6/16, the maximum dimensionless Darcy flux increased over 150% and volumetric exchange flux increased 200% [14]. Neutral scenarios from this study with a similar sinuosity (15/40, the same ratio as 6/16) achieved a maximum increase in F of about 65%. It is likely this study's simulations produced localized changes in exchange flux higher than 65%, since F values were based on flux magnitudes averaged across 24 h and two full stream wavelengths, unlike the metrics used in Gomez et al. [14].

Gomez-Velez et al. [29] established separate metrics for dimensionless fluxes entering the HZ and those leaving the HZ, measured along sections of the stream where the net movement of water was out of and into the stream, respectively. The difference between the fluxes became the net flux leaving the HZ, which was then made dimensionless. In response to a flooding event, dimensionless net flux first decreased and turned negative as stream stage rose and pushed water into the aquifer. Once the stream stage dropped, head gradients reversed and the net flux term increased [29]. These net flux values were not normalized, but all modeled flooding events started at a neutral base case where net flux = 0. Net flux and the total flux used in this study could give very different results for the same scenario, if that scenario had ≈ 0 net flux. The maximum F values in this study were from a scenario comparable to the minimum net flux values produced in Gomez-Velez et al. [29] because they both represented moments where the majority of exchange flux was moving from the stream to the HZ, meaning the net flux /≈ 0. When directly comparing the total dimensionless flux in this study (before being normalized to ET0 scenarios; values not shown) to the minimum dimensionless flux values in Gomez-Velez et al. [29] from simulations with comparable Γ, ET produced responses an order of magnitude lower than those produced by flood events.

4.3. Residence Time

The patterns of flux-weighted RTDs can be explained by comparing them to the changes seen in spatial distributions of RT^* (Figure 8). Because these RT^* values were flux-weighted, they reflect changes in a combination of particle travel times and spatially-varying flux at the stream–aquifer interface. Higher RT^* values were a result of longer particle travel times in the area of interest, higher flux entering the aquifer where and when the particle was released, or a combination of both; the opposite is true for lower RT^* values. The strong early mode observed in gaining scenarios was produced by stream water originating near meander apexes. The shape of these RTDs did not fundamentally change from ET0 to ET80 because active ET did not alter the general shape of RT^* spatial distributions. For neutral scenarios, the early mode in the ET0 RTD came primarily from stream water close to $\sigma^* = 1$. With the addition of active ET, RT^* values for that section of the stream–aquifer interface remained largely unchanged, but the RT^* values further upstream increased and eventually formed a second mode much higher than the first. The average change in RT^* from ET0 to ET80 where $\sigma^* \geq 0.8$ was 19.9; the same average change where $\sigma^* < 0.8$ was 135.9. In losing scenarios, early modes in the RTDs for ET0 came from stream water close to $\sigma^* = 1$ and any later secondary modes present represented water from further upstream. Adding and then intensifying active ET flattened out spatial distributions of RT^*, shifting all modes toward the median and reducing standard deviations.

The trends of median RT^* values also correspond to changes in spatial distributions of RT^*. Neutral scenarios produced the largest increases in median RT^* because increases in ET_{max} led to larger RT^* for all stream-origin water. For weakly gaining and losing scenarios, increasing ET_{max} caused some sections of σ^* to produce lower RT^*, but those decreases were outpaced by increased RT^* elsewhere along σ^*, resulting in small but consistent increases in median RT^*. In strongly gaining and losing scenarios, the opposite was almost always true: decreases in RT^* for some stream-origin water outpaced increases in RT^* for other stream-origin water. The sole exception can be seen in J + 2 results when moving from ET60 to ET80. Here, RT^* values from $\sigma^* < 0.3$ had already dropped to ~0 by ET60, but RT^* values from $\sigma^* > 0.3$ still substantially increased.

Previous studies of RTDs in lateral HZs have used mean [29] and mode [13,14] RTs as characteristic timescales for further comparison and analysis. Their choices reflect the model domains that produced particular RTD shapes as well as the subsequent analyses performed. This study used (flux-weighted, dimensionless) median RT values so comparisons to timescales in other studies could be made with the understanding that half of the particles modeled in this study were reaching or exceeding some RT. This is a straightforward way to conceptualize whether simulations were allowing HZ water to reach the timescales necessary for biogeochemical transformation reactions.

Boano et al. [13] produced strongly bimodal probability distributions of flow paths generated in intrameander zones of increasing sinuosity, the lowest sinuosity being higher than the $S = 1.87$ used in this study. The two modes of each distribution reflected different travel times at the neck and apex of meander bends, with the difference between them increasing with sinuosity. These modes were then used as minimum and maximum RTs and compared to timescales for the biogeochemical transformation of organic carbon in order to map the resulting chemical zonations of the HZ. Increasing their meander sinuosity from 2.5 to 5.0, Boano et al. [13] decreased their minimum RT mode by about 200% and increased their maximum RT mode by about 140%. As mentioned in Section 3.3, this study produced maximum increases in median RT^* of about 20% when comparing ET0 scenarios to their respective ET80 scenarios.

Building on this work, Gomez et al. [14] compared their own RTDs to the biogeochemical timescales proposed by Boano et al. [13], but within a model based on the domain used in Cardenas [11] and this study. Gomez et al. [14] chose the first and primary mode of their RTDs as a characteristic timescale. A dimensionless version of the characteristic mode (τ^*) increased as sinuosity increased and as dispersivity decreased. In response to increasing ET_{max}, median RT^* increased at a rate comparable to τ^* responding to sinuosity, but at a much lower rate than τ^* responding to dispersivity. In Gomez et al. [14], modes other than the characteristic mode also shrank and disappeared from probability distributions with higher sinuosity and lower dispersivity. Increasing ET_{max} was demonstrated to have a similar effect for J − 2 scenarios. Based on characteristic timescales developed in Gomez et al. [14] by applying the definitions of Boano et al. [13] to the data of Zarnetske et al. [14], the simulations in this study produced median RT^* values below the threshold timescales for the transformation of organic carbon ($t_{DOC} = 2.10$ days) and nitrate ($t_{NO3} = 0.92$ days), but above the threshold timescale for oxygen ($t_{O2} = 0.20$

days), discounting median RT^* values from $J - 2$ scenarios. RT^* values are dimensionless, but since they were made dimensionless by dividing RTs of h by a characteristic 24 h, they are equivalent to RTs in units of days.

Gomez-Velez et al. [29] tracked the dimensionless mean RT of all flow paths discharging to the stream from the HZ at a given time during and after flood events. These dimensionless values were normalized to mean RTs from base flow conditions, and these normalized dimensionless values were plotted across time to show relative changes in mean RT as flood pulses interacted with the HZ. For this study, a similar plot of RT changes through time was unnecessary because RT^* values demonstrated almost no change in response to varying release time for water particles. Even in simulations with comparable Γ values, the RTs from Gomez-Velez et al. [29] were variable in response to flooding whereas RTs from this study did not respond to ET. This difference in response may be because a single daily cycle of ET drew much less water into the aquifer than a flood simulated with similar Γ values. The greater exchange flux from the flood could penetrate farther into the alluvial valley and subsequently take much longer to return to the stream. In some simulations, Gomez-Velez et al. [29] had a part of the HZ detach itself from the intrameander zone and travel down-valley, similar to how particles were pulled into the trough of depression in neutral scenarios of this study. Since this study's particle flow paths did not change if the daily cycle of ET withdrawal rates remained constant, neutral field conditions with a similar aquifer sensitivity could possibly provide a steady supply of stream-origin water to the alluvial valley beyond the intrameander zone. Even outside the geochemical limit of the HZ, this could enhance reaction potential in areas of the aquifer further from the stream by providing aquifer environments reactants unique to the stream.

4.4. Model Assumptions and Limitations

The MODFLOW and related models used in this study made assumptions about aspects of the hyporheic system that have been demonstrated to affect hyporheic form and function. The assumptions were made either to keep the study focused on the addition of the dynamic ET sink, so the results could clearly answer the primary research question; or they were made due to the limited time available to run, process, and analyze additional scenarios.

The stream modeled here had an idealized channel morphology, but most of its major characteristics were representative of what can be found in both natural and engineered channels. The stream was assumed to be perfectly sinusoidal, with a constant sinuosity of 1.87 extending indefinitely both upstream and downstream of the domain. Natural channel meanders are often asymmetrical and irregular because of fluctuations between bends in meander wavelength and radius of curvature [45]. Even so, sinusoidal channels can approximate symmetrical and regular meanders produced naturally [45] as well as those engineered to promote meandering and intrameander HZs [46]. Variation in sinuosity has been demonstrated to control hyporheic exchange fluxes [11,14], RTs [13,14], and potential for biogeochemical transformation [13]. Sinuosity was kept constant in order to limit the number of simulations, and a constant value was chosen that fell within the ranges of S explored by studies modeling similar domains [11,14,19,29], for ease of comparison. A sinuosity on the higher end of these ranges was chosen to represent a possible situation where a stream is engineered to enhance meandering, and thereby enhance hyporheic exchange flux [47]. Models of slow, meandering, high-order streams from Gomez-Velez and Harvey [48] indicate they produce higher RTs in lateral HZs, and therefore have enhanced potential for biogeochemical processing through lateral exchange compared to higher-relief and lower-order streams.

Taken together, the static aquifer parameters (Table 1) describe a homogeneous aquifer of sand to silty sand. These parameter values were chosen to remain consistent with the ranges of values explored in previous studies [11,13,14,19,29], for ease of comparison. If the model had been designed to represent a gravel aquifer by increasing hydraulic conductivity, specific yield, and porosity, it is possible the time constant Γ would have dropped below 1 and the aquifer would have been more sensitive to hly changes in ET. Studies exploring the effects of median grain size on lateral hyporheic exchange indicate that a gravel aquifer also would have likely produced higher exchange fluxes at the stream–aquifer interface [12,48].

Some parameters of the model domain were made homogeneous to reduce the number of parameters altered between simulations, but are known to be widely heterogeneous in the field. Assuming the aquifer was completely homogeneous ignored the effects of common paleochannels in intrameander zones, where they have been observed to significantly alter RTDs by providing preferential flow paths for water and solutes [49]. Riverbed and bank sediments in the stream–aquifer interface can hold and transmit water very differently than surrounding aquifer sediment [50]. Since the stream was the only source of water in this model, changing the basic characteristics of this liminal boundary could have strongly influenced the flux of stream water to the aquifer both overall and between hly time steps. Altering the model to represent these elements as heterogeneous would have made results more realistic, but also would have added a complex parameter (degree of heterogeneity) to be accounted for when attempting to interpret the effects of ET on this model.

The boundary conditions of this model were central to its representation of the stream–aquifer–ET system, and some of the assumptions of those boundary conditions limited the realism and variety of situations explored. The model stream, represented with MODFLOW's CHD package, maintained static head despite the adjacent ET sink. This implies the stream was large enough to act as an infinite supply of water to the aquifer. If stream heads had been allowed to fluctuate in response to drawdown in the vegetated zone, the feedback of hydraulic signals could have produced more complex patterns of results like those seen in Gomez-Velez et al. [29]. The static CHD values enforced quasi-steady-state regional groundwater fluxes (RGFs) throughout the rest of the model domain. Fluxes between cells (and by extension, heads) could vary between timesteps, but still generally oscillated around average values, collectively producing a simulation-average water table. These quasi-steady-state groundwater tables helped clarify the hydraulic signals produced by transient ET, but also preclude modeling long-term dynamic equilibrium trends such as regional water table decline and the subsequent response of phreatophyte roots. Long-term decline of average groundwater levels would need to be modeled across a series of simulations where all CHD boundary conditions were altered to enforce new RGF regimes that produced lower water tables.

Many of the assumptions made about the EVT boundary simplified the physiology of the riparian vegetation being modeled. The extinction depth of 2 m may not be appropriate to represent many riparian phreatophyte species, at least in part because of the group's wide variability in extinction depths [42]. Considering that Carroll et al. [42] reported 90% confidence interval values (0.3 to 9 m), no single value would have been fully representative, and 2 m appears to be plausible for riparian phreatophyte species with shorter rooting depths. The model also assumed the phreatophytes were evenly spaced in the vegetated zone and withdrew water at an identical rate set for all EVT cells, before the variation due to head differences. Even spacing is possible when a restored intrameander zone is first being planted, but many field conditions will deviate. Spatial variations in ET could make patterns of drawdown and head less uniform, leading to less predictable flow paths and RTDs.

The EVT package also produced ET withdrawal rates that varied linearly with head. As Shah et al. [41] indicate, an exponential decay relationship more closely matches field observations of ET withdrawal for forested cover. Unlike the EVT package, the Evapotranspiration Segments (ETS) package can have its formula adjusted to account for the exponential and otherwise nonlinear relationships between water table depth and parameters such as S_y, the fraction of vadose zone contribution to ET, and rooting density [41]. The EVT package cannot adjust its linear decay of ET and therefore cannot represent these parameters with the same degree of realism. Despite this, the EVT package is adequate for its purpose in this work. The low-order approach of the model included simplifications to the simulation of ET and basic aquifer characteristics, to the point that parameters the ETS package could more accurately represent were already simplified or did not apply. For example, the model aquifer's S_y was constant at all depths; MODFLOW models only saturated flow and therefore could not capture the depth-dependent contribution of the vadose zone to ET; and the root network density of hyporheic trees was assumed to be completely homogeneous with depth. The EVT package still captured the essential aspect of the relationship between ET and water table depth: a model cell achieved ET_{max} at depth = 0, ET = 0 at the extinction depth, and some fraction of ET_{max} in-between.

As was noted in Section 2.2, the most extreme ET withdrawal rates produced in the ET80 scenarios lie outside what is physically possible in most climate zones [26,27,38,39]. This limits the usefulness of some results in understanding the field conditions of sinuosity-induced hyporheic flow. However, results from ET80 scenarios still contribute to a more thorough exploration of the relationship between the magnitude of ET perturbances and the magnitude of hyporheic zone response. This exploration is central to the goal of this work stated in Section 1.

The temporally symmetrical results for head, drawdown, and flux may be due in part to the assumption of a symmetrical ET curve (Figure 2). This curve was based on the hydrograph used to control the flood stage in Gomez-Velez et al. [29], but it was also a fair approximation of ET estimated in the field [51] and calculated in models with ideal "clear sky" conditions [38]. Despite this, all field estimates of ET reveal day-to-day variation in ET_{max} due to fluctuations in solar radiation [26,38,51], or deviations from the clear sky ideal. Including this variation in the diel ET schedule could be crucial for accurately modeling ET in areas with high cloud cover.

5. Conclusions

Hyporheic zones provide unique and critical environments for the mixing of stream water and groundwater. The degree of that mixing depends on static characteristics of the HZ and dynamic changes to sinks and sources of water, such as riparian ET. We explored the effects of near-stream ET in a numerical model of lateral hyporheic exchange. The primary question for this study was: will the introduction of dynamic ET to a numerical hyporheic model diminish HZ area, hyporheic exchange flux, and RTs?

Across the full range of maximum daily ET pumping rates and ambient groundwater orientations explored, ET increased HZ area and hyporheic exchange flux. RTs increased in neutral and weakly gaining and losing conditions but decreased under stronger gaining and losing conditions. Some of these changes were noticeable even at lower ET intensities: HZ areas grew substantially at ET20; areas of the intrameander zone reversed Q_y orientations in gaining scenarios at ET20; and median RT^* decreased by over 20% in strongly gaining scenarios from ET20 to ET40. The changes produced by ET in HZ area and hyporheic exchange flux were comparable to the scale of changes produced in similar modeling studies by altering aquifer characteristics, varying geomorphology, or introducing another time-varying disturbance; but only at higher ET intensities, such as ET60 and ET80. The response of RTs to ET was consistently smaller than the response produced in other studies, in some cases by orders of magnitude. This suggests that at lower ET intensities riparian ET could still be useful in productively altering HZs, but may play a secondary role in HZ modeling when compared to the effects of flood pulses [29,31].

The central mechanism for these responses was the reorientation of local minimum head values away from the stream, resulting in increased flux from the stream to the aquifer. The response of each model simulation to ET depended largely on whether ambient groundwater flow did or did not compliment these locally reoriented head gradients. Gaining scenarios demonstrated the greatest growth in HZ area; losing scenarios showed the greatest increases in F; and effects on RT were mixed, but neutral RTDs were altered the most. This study indicates ambient groundwater flow exerts a strong control on the hyporheic response to ET, but whether it enhances or diminishes those responses depends on the desired effect.

The results of this study should be taken with the understanding that some of the underlying assumptions of the model limit its applicability to real stream restoration scenarios. For the restoration of large, low-relief streams specifically, the addition of a wide belt of near-stream phreatophytes can enhance HZ area, RTs, and exchange flux, provided local $J_{y/x}$ stays below 1. Establishing an accurate model of ambient groundwater flow and calculating aquifer sensitivity are crucial to predicting how ET will affect the HZ in the long-term. For example, this model was effective at preserving the geometry of the HZ because of the low aquifer sensitivity to ET, but results from Gomez-Velez et al. [29] suggest that if sensitivity had been higher ET would have increased fluxes of stream water into the HZ, potentially by an order of magnitude. Some of the positive effects of ET were only achieved at higher ET rates, which may be impossible to achieve in climate zones with low potential ET.

Future models of sinuosity-induced hyporheic exchange should explore the consistency of effects produced by ET across a wider variety of conditions for the aquifer, stream, and ET sink. This work included the Γ sensitivity term from Gomez-Velez et al. [29] to highlight the importance of describing model sensitivity, but it was not the primary focus of the study. A global sensitivity analysis of Γ with respect to the major metrics of this study would simplify the relationships of model parameters while further investigating whether riparian ET as it is currently modeled produces reasonable results. Replacing the static stream with a dynamic, head-dependent flux boundary would allow the model to represent streams small enough to have their stage lowered by nearby ET. Stream stage fluctuations would directly affect fluxes of stream water to the aquifer, which was the primary mechanism for changes to the HZ in this study. Introducing spatial heterogeneity into aquifer characteristics and day-to-day fluctuations in the ET schedule would likely provide more realistic results that can inform future efforts to restore the ecosystem services of lateral hyporheic zones.

Supplementary Materials: The following are available online at http://www.mdpi.com/2073-4441/12/2/424/s1, a separate Supporting Information document ("SupportingInformation.docx") provides detailed descriptions of some processes undertaken in the Methods section, to make it easier to recreate the model and the results produced. Copies of the ModelMuse project file, model inputs and outputs, and novel pre- and post-processing code used in this research have been organized for public release. They will be made available by being uploaded as supplementary information to the publication.

Author Contributions: Conceptualization, all authors; methodology, all authors; software, J.K. and T.A.E.; validation, J.K. and T.A.E.; formal analysis, J.K. and T.A.E.; investigation, J.K. and T.A.E.; resources, all authors; data curation, J.K. and T.A.E.; writing—original draft preparation, all authors; writing—review and editing, all authors; visualization, all authors; supervision, T.A.E., J.G.-V. and L.K.L.; project administration, J.K. and T.A.E. All authors have read and agreed to the published version of the manuscript.

Funding: This research received no external funding. J.G.-V. is funded by the U.S. National Science Foundation (award EAR 1830172) and U.S. Department of Energy (DOE), Office of Biological and Environmental Research (BER), as part of BER's Subsurface Biogeochemistry Research Program (SBR). This contribution originates from the SBR Scientific Focus Area (SFA) at the Pacific Northwest National Laboratory (PNNL).

Acknowledgments: Special thanks to Richard B. Winston, Ph.D., for his technical support and guidance in using MODFLOW and ModelMuse via written correspondence.

Conflicts of Interest: The authors declare no conflicts of interest.

References

1. Brunke, M.; Gonser, T. The ecological significance of exchange processes between rivers and groundwater. *Freshwater Biol.* **1997**, *37*, 1–33. [CrossRef]
2. Hendricks, S.P. Microbial ecology of the hyporheic zone—A perspective integrating hydrology and biology. *J. North Am. Benthological Soc.* **1993**, *12*, 70–78. [CrossRef]
3. Harvey, J.W.; Bencala, K.E. The effect of streambed topography on surface-subsurface water exchange in mountain catchments. *Water Resour. Res.* **1993**, *29*, 89–98. [CrossRef]
4. Boano, F.; Revelli, R.; Ridolfi, L. Bedform-induced hyporheic exchange with unsteady flows. *Adv. Water Resour.* **2007**, *30*, 148–156. [CrossRef]
5. Alexander, R.B.; Böhlke, J.K.; Boyer, E.W.; David, M.B.; Harvey, J.W.; Mulholland, P.J.; Seitzinger, S.P.; Tobias, C.R.; Tonitto, C.; Wollheim, W.M. Dynamic modeling of nitrogen losses in river networks unravels the coupled effects of hydrological and biogeochemical processes. *Biogeochemistry* **2009**, *93*, 91–116. [CrossRef]
6. Fuller, C.C.; Harvey, J.W. Reactive uptake of trace metals in the hyporheic zone of a mining-contaminated stream, Pinal Creek, Arizona. *Environ. Sci. Technol.* **2000**, *34*, 1150–1155. [CrossRef]
7. National Research Council. *Riparian Areas: Functions and Strategies for Management*; The National Academies Press: Washington, DC, USA, 2002.

8. Boulton, A.J. Hyporheic Rehabilitation in Rivers: Restoring Vertical Connectivity. *Freshwater Biol.* **2007**, *52*, 632–650. [CrossRef]

9. Boulton, A.J.; Datry, T.; Kasahara, T.; Mutz, M.; Stanford, J.A. Ecology and management of the hyporheic zone: Stream-groundwater interactions of running waters and their floodplains. *J. N. Am. Benthol. Soc.* **2010**, *29*, 26–40. [CrossRef]

10. McClain, M.E.; Boyer, E.W.; Dent, C.L.; Gergel, S.E.; Grimm, N.B.; Groffman, P.M.; Hart, S.C.; Harvey, J.W.; Johnston, C.A.; Mayorga, E.; et al. Biogeochemical hot spots and hot moments at the interface of terrestrial and aquatic ecosystems. *Ecosystems* **2003**, *6*, 301–312. [CrossRef]

11. Cardenas, M.B. Stream-aquifer interactions and hyporheic exchange in gaining and losing sinuous streams. *Water Resour. Res.* **2009b**, *45*, W06429. [CrossRef]

12. Harvey, J.W.; Böhlke, J.K.; Voytek, M.A.; Scott, D.; Tobias, C.R. Hyporheic zone denitrification: Controls on effective reaction depth and contribution to whole-stream mass balance. *Water Resour. Res.* **2013**, *49*, 6298–6316. [CrossRef]

13. Boano, F.; Demaria, A.; Revelli, R.; Ridolfi, L. Biogeochemical zonation due to intrameander hyporheic flow. *Water Resour. Res.* **2010**, *46*, W02511. [CrossRef]

14. Gomez, J.D.; Wilson, J.L.; Cardenas, M.B. Residence time distributions in sinuosity-driven hyporheic zones and their biogeochemical effects. *Water Resour. Res.* **2012**, *48*, W09533. [CrossRef]

15. O'Connor, B.L.; Harvey, J.W. Scaling hyporheic exchange and its influence on biogeochemical reactions in aquatic ecosystems. *Water Resour. Res.* **2008**, *44*, W12423. [CrossRef]

16. Boano, F.; Harvey, J.W.; Marion, A.; Packman, A.I.; Revelli, R.; Ridolfi, L.; Wörman, A. Hyporheic flow and transport processes: Mechanisms, models, and biogeochemical implications. *Rev. Geophys.* **2014**, *52*, 603–679. [CrossRef]

17. Buffington, J.; Tonina, D. Hyporheic exchange in mountain rivers ii: Effects of channel morphology on mechanics, scales, and rates of exchange. *Geogr. Compass* **2009**, *3*, 1038–1062. [CrossRef]

18. Tonina, D.; Buffington, J.M. Hyporheic exchange in mountain rivers i: Mechanics and environmental effects. *Geogr. Compass* **2009**, *3*, 1063–1086. [CrossRef]

19. Cardenas, M.B. A model for lateral hyporheic flow based on valley slope and channel sinuosity. *Water Resour. Res.* **2009**, *45*, W01501. [CrossRef]

20. Cardenas, M.B. The effect of river bend morphology on flow and timescales of surface water-groundwater exchange across pointbars. *J. Hydrol.* **2008**, *362*, 134–141. [CrossRef]

21. Revelli, R.; Boano, F.; Camporeale, C.; Ridolfi, L. Intra-meander hyporheic flow in alluvial rivers. *Water Resour. Res.* **2008**, *44*, W12428. [CrossRef]

22. Boano, F.; Camporeale, C.; Revelli, R.; Ridolfi, L. Sinuosity driven hyporheic exchange in meandering rivers. *Geophys. Res. Lett.* **2006**, *33*, L18406. [CrossRef]

23. Wondzell, S.M.; Gooseff, M.N.; McGlynn, B.L. An analysis of alternative conceptual models relating hyporheic exchange flow to diel fluctuations in discharge during baseflow recession. *Hydrol. Proc.* **2010**, *24*, 686–694. [CrossRef]

24. White, W.N. *A Method of Estimating Groundwater Supplies Based on Discharge by Plants and Evaporation from Soil—Results of Investigation in Escalante Valley, Utah*; U.S. Geological Survey Water-Supply Paper; US Government Printing Office: Washington, DC, USA, 1932; Volume 659, 105p.

25. Duke, J.R.; White, J.D.; Allen, P.M.; Muttiah, R.S. Riparian influence on hyporheic-zone formation downstream of a small dam in the Blackland Prairie region of Texas. *Hydrol. Process.* **2007**, *21*, 141–150. [CrossRef]

26. Lautz, L.K. Estimating groundwater evapotranspiration rates using diurnal water-table fluctuations in a semi-arid riparian zone. *Hydrogeol. J.* **2008**, *16*, 483–497. [CrossRef]

27. Yue, W.; Wang, T.; Franz, T.E.; Chen, X. Spatiotemporal patterns of water table fluctuations and evapotranspiration induced by riparian vegetation in a semiarid area. *Water Resour. Res.* **2016**, *52*, 1948–1960. [CrossRef]

28. Butler, J.J., Jr.; Kluitenberg, G.J.; Whittemore, D.O.; Loheide II, S.P.; Jin, W.; Billinger, M.A.; Zhan, X. A field investigation of phreatophyte-induced fluctuations in the water table. *Water Resour. Res.* **2007**, *43*, W02404. [CrossRef]

29. Gomez-Velez, J.D.; Wilson, J.L.; Cardenas, M.B.; Harvey, J.W. Flow and residence times of dynamic stream bank storage and sinuosity-driven hyporheic exchange. *Water Resour. Res.* **2017**, *53*, 8572–8595. [CrossRef]

30. Gomez, J.D.; Wilson, J.L. Age distributions and dynamically changing hydrologic systems: Exploring topography-driven flow. *Water Resour. Res.* **2013**, *49*, 1503–1522. [CrossRef]

31. Larsen, L.; Harvey, J.W.; Magilo, M.M. Dynamic hyporheic exchange at intermediate timescales: Testing the relative importance of evapotranspiration and flood pulses. *Water Resour. Res.* **2014**, *50*, 318–335. [CrossRef]

32. Zarnetske, J.P.; Haggerty, R.; Wondzell, S.M.; Baker, M.A. Dynamics of nitrate production and removal as a function of residence time in the hyporheic zone. *J. Geophys. Res.* **2011**, *116*, G01025. [CrossRef]

33. Zheng, C.; Wang, P.P. *MT3DMS A Modular Three-Dimensional Multi-Species Transport Model for Simulation of Advection, Dispersion and Chemical Reactions of Contaminants in Groundwater Systems; Documentation and User's Guide*; US Army Engineer Research and Development Center Contract Report SERDP-99-1; US Army Corps of Engineers: Washington, DC, USA, 1999; 202p.

34. Pollock, D.W. *User Guide for MODPATH Version 6—A Particle-Tracking Model for MODFLOW*; US Geological Survey Techniques and Methods 6–A41; US Department of the Interior: Reston, VA, USA, 2012; 58p.

35. Welsch, D.J. *Riparian forest buffers: Function and Design for Protection and Enhancement of Water Resources*; US Department of Agriculture, Forest Service, Northern Area State & Private Forestry: Radnor, PA, USA, 1991; NA-PR-07-91.

36. Winston, R.B. *ModelMuse—A Graphical user Interface for MODFLOW-2005 and PHAST*; US Geological Survey Techniques and Methods 6-A29; US Department of the Interior: Reston, VA, USA, 2009; 52p.

37. Harbaugh, A.W. *MODFLOW-2005: The U.S. Geological Survey Modular Ground-Water Model: The Ground-Water Flow Process*; US Geological Survey Techniques and Methods 6-A16; US Department of the Interior: Reston, VA, USA, 2005; variously p.

38. Solyu, M.E.; Lenters, J.D.; Istanbulluoglu, E.; Loheide, S.P. On evapotranspiration and shallow groundwater fluctuations: A Fourier-based improvement to the White method. *Water Resour. Res.* **2012**, *48*, W06506. [CrossRef]

39. Fan, J.; Oestergaard, K.T.; Guyot, A.; Lockington, D.A. Estimating groundwater recharge and evapotranspiration from water table fluctuations under three vegetation covers in a coastal sandy aquifer of subtropical Australia. *J. Hydrol.* **2014**, *519*, 1120–1129. [CrossRef]

40. Baird, K.J.; Maddock, T. Simulating riparian evapotranspiration: A new methodology and application for groundwater models. *J. Hydrol.* **2005**, *312*, 176–190. [CrossRef]

41. Shah, N.; Nachabe, M.; Ross, M. Extinction depth and evapotranspiration from ground water under selected land covers. *Ground Water* **2007**, *45*, 329–338. [CrossRef] [PubMed]

42. Carroll, R.W.H.; Pohll, G.M.; Morton, C.G.; Huntington, J.L. Calibrating a basin-scale groundwater model to remotely sensed estimates of groundwater evapotranspiration. *J. Am. Water Resour. Assoc.* **2015**, 1–14. [CrossRef]

43. Ihaka, R.; Gentleman, R. R: A language for data analysis and graphics. *J. Comput. Graph. Stat.* **1996**, *5*, 299–314.

44. Triska, F.J.; Kennedy, V.C.; Avanzino, R.J.; Zellweger, G.W.; Bencala, K.E. Retention and transport of nutrients in a third-order stream in northwest California: Hyporheic processes. *Ecology* **1989**, *70*, 1893–1905. [CrossRef]

45. Wohl, E. *Rivers in the Landscape: Science and Management*, 1st ed.; John Wiley & Sons, Ltd.: Chichester, West Sussex, UK, 2014; pp. 133–147.

46. Kasahara, T.; Hill, A.R. Lateral hyporheic zone chemistry in an artificially constructed gravel bar and a re-meandered stream channel, southern Ontario, Canada. *J. Am. Water Resour. Assoc.* **2007**, *43*, 1257–1269. [CrossRef]

47. Kasahara, T.; Hill, A.R. Modeling the effects of lowland stream restoration projects on stream-subsurface water exchange. *Ecol. Eng.* **2008**, *32*, 310–319. [CrossRef]

48. Gomez-Velez, J.D.; Harvey, J.W. A hydrogeomorphic river network model predicts where and why hyporheic exchange is important in large basins. *Geophys. Res. Lett.* **2014**, *41*, 6403–6412. [CrossRef]

49. Heeren, D.M.; Fox, G.A.; Fox, A.K.; Storm, D.E.; Miller, R.B.; Mittelstet, A.R. Divergence and flow direction as indicators of subsurface heterogeneity and stage-dependent storage in alluvial floodplains. *Hydrol. Process.* **2014**, *28*, 1307–1317. [CrossRef]

50. Shamsuddin, M.K.N.; Sulaiman, W.N.A.; Ramli, M.F.; Kusin, F.M. Vertical hydraulic conductivity of riverbank and hyporheic zone sediment at Muda River riverbank filtration site, Malaysia. *Appl. Water. Sci.* **2019**, *9*. [CrossRef]

51. Loheide, S.P. A method for estimating subdaily evapotranspiration of shallow groundwater using diurnal water table fluctuations. *Ecohydrology* **2008**, *1*, 59–66. [CrossRef]

MDPI

St. Alban-Anlage 66

4052 Basel

Switzerland

Tel. +41 61 683 77 34

Fax +41 61 302 89 18

www.mdpi.com

Water Editorial Office

E-mail: water@mdpi.com

www.mdpi.com/journal/water